Monographs on Theoretical and Applied Genetics 13

Monographs on Theoretical and Applied Genetics

Volume 1 **Meiotic Configurations**
A Source of Information for Estimating Genetic Parameters
By J. Sybenga

Volume 2 **Pollination Mechanisms, Reproduction and Plant Breeding**
By R. Frankel and E. Galun

Volume 3 **Incompatibility in Angiosperms**
By D. de Nettancourt

Volume 4 **Gene Interactions in Development**
By L.I. Korochkin

Volume 5 **The Molecular Theory of Radiation Biology**
By K.H. Chadwick and H.P. Leenhouts

Volume 6 **Heterosis**
Reappraisal of Theory and Practice
Editor: R. Frankel

Volume 7 **Induced Mutations in Plant Breeding**
By W. Gottschalk and G. Wolff

Volume 8 **Protoplast Fusion**
Genetic Engineering in Higher Plants
By Y.Y. Gleba and K.M. Sytnik

Volume 9 **Petunia**
Editor: K.C. Sink

Volume 10 **Male Sterility in Higher Plants**
By. M.L.H. Kaul

Volume 11 **Tree Breeding: Principles and Strategies**
By G. Namkoong, H.C. Kang, and J.S. Brouard

(continued after index)

Robert F. Costantino
Robert A. Desharnais

Population Dynamics and the *Tribolium* Model: Genetics and Demography

With 79 Illustrations

Springer-Verlag
New York Berlin Heidelberg London
Paris Tokyo Hong Kong Barcelona

Robert F. Costantino
Department of Zoology
University of Rhode Island
Kingston, RI 02881
USA

Robert A. Desharnais
Department of Biology
California State University
Los Angeles, CA 90032
USA

Library of Congress Cataloging-in-Publication Data
Costantino, Robert F.
Population dynamics and the *Tribolium* model: genetics and demography / Robert F. Costantino, Robert A. Desharnais.
p. cm. – (Monographs on theoretical and applied genetics: 13)
Includes bibliographical references and indexes.
ISBN 0-387-97581-0
1. *Tribolium*. 2. Insect populations – Mathematical models. 3. Population genetics. 4. Stochastic processes – Mathematical models. I. Desharnais, Robert A. II. Title. III. Series.
QL596.T2C58 1991
595.76′47 – dc20 91-19332

Printed on acid-free paper.

Camera-ready copy provided by the authors.
Printed and bound by Edwards Brothers, Inc., Ann Arbor, MI.
Printed in the United States of America.

9 8 7 6 5 4 3 2 1

ISBN 0-387-97581-0 Springer-Verlag New York Berlin Heidelberg
ISBN 3-540-97581-0 Springer-Verlag Berlin Heidelberg New York

Preface

The study of populations is becoming increasingly focused on *dynamics*. We believe there are two reasons for this trend. The first is the impact of nonlinear dynamics with its exciting ideas and colorful language: bifurcations, domains of attraction, chaos, fractals, strange attractors. Complexity, which is so very much a part of biology, now seems to be also a part of mathematics. A second trend is the accessibility of the new concepts. The barriers to communication between theorist and experimentalist seem less impenetrable. The active participation of the experimentalist means that the theory will obtain substance. Our role is the application of the theory of dynamics to the analysis of biological populations.

We began our work early in 1979 by writing an ordinary differential equation for the rate of change in adult numbers which was based on an equilibrium model proposed a decade earlier. During the next few months we filled our notebooks with straightforward deductions from the model and its associated biological implications. Slowly, some of the biological observations were explained and papers followed on a variety of topics: genetic and demographic stability, stationary probability distributions for population size, population growth as a birth-death process, natural selection and density-dependent population growth, genetic disequilibrium, and the stationary stochastic dynamics of adult numbers.

Everywhere we used the *same* simple differential equation. Could other experimental observations be incorporated into the mathematical model? It seemed so. Our thoughts then turned to writing this monograph. We quickly realized that we needed a far richer mathematical framework than that provided by our original equation; consequently, this book opens with a general formulation in which our adult recruitment model is a special case. Nevertheless, the direct connection between observation and theory, so carefully nurtured in our initial work, still remains a central strategy of our research effort. You will find data in this book.

We wish to thank John Stanley for telling us of his experiences with R. N. Chapman and the beginning of the flour beetle work. His careful reading and sharp red pencil have improved the manuscript. Wes Taylor has spent a good deal of his time carefully reading and commenting upon drafts of this monograph. We sincerely appreciate his efforts. J. C. Gower kindly told us of his working relationship with P. H. Leslie and M. S. Bartlett. A. E. Bell, in another historical note, described how his attention was drawn to *Tribolium* by Thomas Park. Our own work forms the bulk of the book and we most sincerely acknowledge our collective collaborators: Joel Cohen, Brian Dennis, Lev Ginzburg, Alan Hastings, David Jillson, Laifu Liu, Marc Mangel, and Craig Peters. Our research has benefited immensely from the rich data-base provided by the experiments of Andrea Moffa-White. R. F. Costantino is pleased to acknowledge the support provided by Marc Mangel and the Center for Population Biology along with the Departments of Mathematics and Genetics for the appointment as Visiting Professor for the academic year 1989-90 at the University of California at Davis. R. A. Desharnais thanks Joel Cohen for five intellectually stimulating years at Rockefeller University. We both appreciate Joel's encouragement. R. A. D. would also like to acknowledge the California State University for a Creative Leave Award. We both thank the editors at Springer for their patience during an ever lengthening time for the book's completion. Finally, we pay tribute to our wives, Pat and Karen, for their constant love and support.

ROBERT F. COSTANTINO
ROBERT A. DESHARNAIS

Contents

Chapter 1 Introduction

Population dynamics is the study of the relationship between the attributes of a population and the forces which affect those attributes. The attributes include variables such as the size of the population, its age structure, and its genetic composition. The forces which affect these variables include birth and death rates, behavioral interactions among individuals, environmental fluctuations, natural selection, and interactions with other species. The complexity of these relationships makes the study of populations a difficult and fascinating challenge.

One approach to this challenge is to examine experimental populations under carefully controlled conditions. Flour beetles of the genus *Tribolium* have been used in this fashion for over sixty years. Laboratory populations are easy to culture and many species can complete their life cycle in less than a month. Population attributes such as density and age structure are easy to measure and the populations themselves can be replicated. For this reason, the flour beetle has been used in many ecological investigations, where topics such as the regulation of density and species competition have been studied.

Another approach to the study of population dynamics is mathematical modeling. Simplifying assumptions are made and, based on biological principles, equations are derived which relate one or more population attributes to forces which affect those attributes. For example, in a simple model of population growth, the rate of change in the number of individuals at any time can be expressed as a difference between recruitment and mortality rates, and these latter rates may themselves be functions of population size. Properties of the model, such as equilibria and stability, are obtained from a mathematical analysis, and, hopefully, the properties discovered provide some insight into how the forces which act within the population affect its dynamics.

Occasionally, these two approaches are combined. In the case of *Tribolium*, there has been a long tradition of collaboration between experimentalists and theoreticians, and in some instances, the same individual has fufilled both roles. In our opinion, this is an extremely productive way to proceed. The "biological models" (eg. *Tribolium*) often suggest new problems for mathematical study or provide concrete examples of hypothetical dynamic phenomena. On the other hand, mathematical models often suggest new experiments and new ways of analyzing data. This book has been written with this philosophy in mind.

Our intention in this monograph is to establish a foundation, based on both the biology of *Tribolium* and mathematical modeling, from which some of the complex questions associated with the dynamics of populations can be addressed. In particular, we deal with four areas of population theory: age structure, stochastic population dynamics, natural selection, and species interactions. Much of the biology is specific to *Tribolium*, but the forces under study – age-class interactions, population growth fluctuations, genotypic fitness differences, interspecific behavioral interactions – are ubiquitous features of populations in general. Hopefully, our results have implications which extend beyond the confines of the *Tribolium* laboratory.

The remainder of this chapter is composed of two parts. In the next section we provide a brief historical background on the use of mathematics in the study of *Tribolium*. This section is by no means a comprehensive review; our intent is to provide a general feeling for the different approaches that were taken to model the flour beetle. In §1.2 we outline in more detail the four topics that are covered in the remaining chapters.

1.1 A Brief History of the Mathematics of *Tribolium*

1.1.1 The Beginning

The flour beetle *Tribolium* was introduced as an experimental insect for the study of populations in 1928, with the publication of a paper entitled "The quantitative analysis of environmental factors" by Royal Norton Chapman, then Professor and Chief of the Division of Entomology at the University of Minnesota. As a Guggenheim Fellow in 1926-27, Chapman travelled to Europe with the stated objective, "to make an investigation of the biotic potential of insects, particularly destructive species, to changing environmental conditions" (Guggenheim Fellows 1926-27). He initially visited R. A. Fisher and A. D. Imms at the Rothamsted Experiment Station at Harpenden, England. He then moved on to work with W. R. Thompson at the European Parasite Laboratory at Hyeres, France where the experiments reported in his seminal 1928 paper were done.

Throughout the 1928 paper, Chapman emphasized the critical role of mathematical calculations. It is impossible to say whether Fisher, Imms and Thompson were responsible for Chapman's strong conviction of the need for carefully controlled laboratory experiments in ecology; it is likely that these men would have encouraged this perspective (Kingsland 1985).

Chapman was an entomologist, not a mathematician. The only formula he used was borrowed from the work of Fourier and Ohm on the transmission of electricity through various conductors. In the entomologist's version, the equilibrium size of a population is given by

$$N^* = \frac{B}{R}, \tag{1.1}$$

where the number of insects per gram of flour, N^*, is equal to the potential rate of reproduction, B, divided by the environmental resistance, R. Nicholson (1933) critized Chapman for his use of this equation. Chapman mistakenly defined the biotic potential in terms of the equilibrium density of beetles so that N^* appeared on both sides of the equation and was simply cancelled when he computed the resistance R. On the other hand, we feel that Nicholson (1933, p. 140) was too harsh when he wrote, "Chapman's hypothesis completely fails to help us to deal quantitatively with animal populations." In Chapman's defense, we note that that the popular logistic model of ecological theory,

$$\frac{dN}{dt} = N(r - aN), \tag{1.2}$$

has a stable equilibrium at

$$N^* = \frac{r}{a}, \tag{1.3}$$

where r is the *per capita* intrinsic rate of increase and a is a "density-sensitivity" parameter which measures the linear rate of decrease in the growth rate with increases in density. This is very similar to Chapman's ratio of the biotic potential to the environmental resistance.

From this first paper the studies of *Tribolium* proceeded along two lines: biological and mathematical. Reviews of the vast literature that has developed on *Tribolium* can be found in the papers of Mertz (1972), King and Dawson (1972), and Bell (1982) and in the three volume series of books by Sokoloff (1972, 1974, 1977). We begin with a brief discussion of the mathematics of *Tribolium*, comparing the methods and philosophies of the various workers. We provide a "flavor" of the early mathematical models, without the extensive analysis needed to fully appreciate this work.

It would be a mistake to assume that the mathematical research applied to *Tribolium* has proceeded chronologically from the simple to the complex. Indeed, some of the most elegant and sophisticated models appeared at the very beginning with the work of John Stanley, a student of Chapman. The spectrum of mathematical techniques applied to the beetle was immense and includes the use of deterministic differential equations (Stanley 1932a), elliptic integrals (Cox and Smith 1957), statistical genetic procedures (Bell and Moore 1958), Markov processes (Neyman et al. 1956, Leslie and Gower 1958, Ahmed 1963), second-order nonlinear difference equations (Leslie 1962), computer simulations (Landahl 1955a,b, Desharnais and Liu 1987), stochastic differential equations (Costantino and Desharnais 1981, Dennis and Costantino 1988) and nonlinear integral equations (Hastings and Costantino 1987), to name a few. "It is doubtful if Chapman had any idea of all the mathematical work which would be done in the coming decades as a succession of biomathematicians grappled with descriptions of the activity of flour beetles" (Stanley 1987, personal communication).

1.1.2 Models of Adult Numbers

We begin with an elementary scheme. Consider a seasonally breeding insect whose generations do not overlap. A population of N_A sexually mature adults, with an average rate of eggs laid per individual b, will produce an egg population of size

$$N_E = b\,N_A\,. \tag{1.4}$$

If all of these eggs, laid in generation k, survive to adulthood, then the size of the adult population in the next or $k+1$ generation is

$$N_A(k+1) = b\,N_A(k)\,, \qquad k=0,1,2,\ldots \tag{1.5}$$

The solution of the linear or *Malthusian* growth rule (1.5) is

$$N_A(k) = N_A(0)\,b^k\,, \qquad k=0,1,2,\ldots\,, \tag{1.6}$$

where $N_A(0)$ is the number of adults in generation zero. With $b<1$ the population forecast is extinction. No change in the number of animals is projected with $b=1$. With $b>1$, the number of adults is expected to increase forever.

While Malthusian growth, $b>1$, may be appropriate for a few generations, there must always be a limit to growth; consequently, understanding the mechanism regulating animal numbers is of general importance in population biology.

The growth of flour beetle populations do not obey the linear rule (1.5) – animal numbers do not increase indefinitely. Chapman (1928) demonstrated that laboratory cultures reached a fluctuating nonexpanding density. The data of Lloyd (1968), presented in Fig. 1.1, provides an example of this observation. The oscillations in the immature life stages are striking and in direct contrast to the much more constant number of adults.

What is the biology of density regulation? Two hypotheses were proposed. First, Chapman (1928) said cannibalism (Fig. 1.2) was the biological basis for the regulation of animal numbers: adult and larval stages eat eggs and pupae. Others agreed with

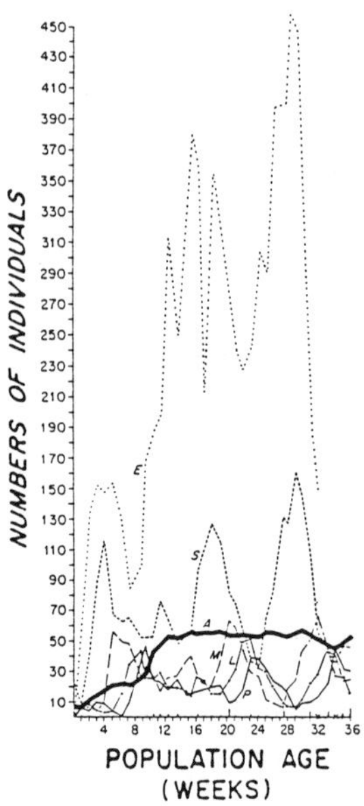

Fig. 1.1. The observed steady-state outcome of the Chicago strain of *T. castaneum*. The following abbreviations are used: E = eggs, S = small larvae, M = medium larvae, L = large larvae, P = pupae, A = adults. [From M. Lloyd (1968), Self regulation of adult numbers by cannibalism in two laboratory strains of flour beetles (*Tribolium castaneum*). *Ecology* **49**:245–259. Copyright 1968 by the Ecological Society of America.]

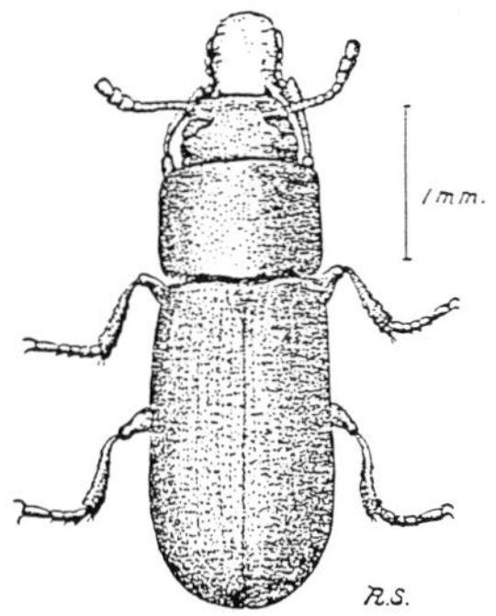

Fig. 1.2. An adult beetle eating an egg. [From T. Park (1934), Observations on the general biology of the flour beetle, *Tribolium confusum*. *Quart. Rev. Biol.* **9**:36 – 54.]

Chapman and expanded his initial interpretation: adults eat eggs (Chapman 1928, Chapman and Baird 1934, Park 1934) and females eat more eggs than males (Stanley 1942, Boyce 1946, Rich 1956), adults eat pupae and young adults (Chapman 1928, Holdaway 1932, Park, Leslie and Mertz 1964, Mertz and Davies 1968), larvae eat eggs (Chapman 1933, Chapman and Whang 1934), and the cannibalistic voracity of larvae on eggs increases with the age of the larvae (Park et al. 1965, Ho and Dawson 1966). The second idea was that fecundity was reduced with an increase in animal numbers (Allee 1931, Park 1932, MacLagan 1932, MacLagan and Dunn 1935, Crombie 1943, Boyce 1946, Birch et al. 1951, Fujita and Utida 1953, Rich 1956).

What is the mathematics of density regulation? One approach is to rewrite (1.5) as

$$N_A(k+1) = N_A(k) f(N_A(k)) , \tag{1.7}$$

where $f(N_A(k))$ is a monotone decreasing function of adult number. We must now choose a function that is consistent with the biology. In *Tribolium*, two expressions had a dominant role in describing density dependent population growth.

In the first case, cannibalism by adults was viewed as a "random collision." An adult-egg contact was interpreted to mean that the egg was eaten. The contacts were assumed to be randomly distributed among the eggs so that the probability of an egg not being eaten was computed by using the binomial distribution as

$$(1-c)^{N_A} \approx \exp(-cN_A) , \tag{1.8}$$

where c is the probability of cannibalism. The density dependent function is

$$f(N_A(k)) = b \exp[-cN_A(k)] , \tag{1.9}$$

where c is more accurately viewed as the rate of adult cannibalism on immatures. This function appeared throughout the *Tribolium* literature; examples include the work of Crombie (1946), Landahl (1955a,b), Rich (1956), Neyman et al. (1956), Sonleitner (1961), Taylor (1965, 1967, 1968, 1971), Sokal and Sonleitner (1968), Lloyd (1968),

Mertz and Davies (1968), and Park et al. (1970). Hassell (1978, p. 12), in a lucid exposition, has shown that this same argument appeared in the host-parasitoid models of Thompson (1929), Bailey (1931), Nicholson (1933) and Nicholson and Bailey (1935).

With the modification of cannibalism, the size of the adult population in generation $k+1$ is

$$N_A(k+1) = b\,N_A(k)\exp[-cN_A(k)]\,, \qquad k=0,1,2,\ldots \tag{1.10}$$

The equilibrium number of adults,

$$N_A^* = \frac{\log(b)}{c}, \tag{1.11}$$

is stable if

$$1 < b < 7.389\,. \tag{1.12}$$

This discrete-time model was first applied to *Tribolium* by Neyman et al. (1956). A continuous-time delay-differential equation analogous to this discrete-time version was introduced by Taylor (1965).

For years, Leslie had used the Verhulst assumption of density dependence (Leslie 1948, 1957, 1958, 1959; Leslie and Gower 1958, 1960). He brought this argument forward when he addressed flour beetle growth. Leslie (1962) wrote the density dependent function as

$$f(N_A(k)) = \frac{b}{1+\alpha N_A(k)}, \tag{1.13}$$

where α represents the density effect. The mathematics reflected the proposed biological mechanism of density regulation: the birthrate b is reduced with an increase in adult density. This was the view held by Park, and in his collaborations with the biomathematicians Leslie and Bartlett, this interpretation was part of the mathematical formulation. Substituting (1.13) into (1.7) yields

$$N_A(k+1) = \frac{b\,N_A(k)}{1+\alpha N_A(k)}, \qquad k=0,1,2,\ldots \tag{1.14}$$

as the second model for adult number. The Verhulst form of density dependence appeared in many *Tribolium* papers, including those of Gause (1931), Crombie (1946, 1947), Bartlett (1957, 1960a,b), Barnett (1962), Park et al. (1964), Leslie et al. (1968), and Mangel and Ludwig (1977).

Leslie's (1948) derivation of this model was based on Verhulst's continuous-time logistic equation (1.2). In a similar manner, Fujita and Utida (1953) derived this discrete-time form of the logistic, suggested its application to *Tribolium*, and gave the following analytical closed-form solution of this nonlinear difference equation:

$$N_A(k) = \frac{b-1}{\alpha - b^{-k}[\alpha - (b-1)/N_A(0)]}. \tag{1.15}$$

For $b > 1$, population number is expected to approach the equilibrium

$$N_A^* = \frac{b-1}{\alpha} \tag{1.16}$$

geometrically without oscillations.

Building on the discrete time logistic equation (1.14), Leslie (1962) also used a second-order nonlinear difference equation to compute adult number:

$$N_A(k+1) = u\,N_A(k) + [N_A(k)+u\,N_A(k-1)]\left(\frac{v}{2}\right)\left(\frac{b}{1+\alpha N_A(k-1)} - 1\right). \tag{1.17}$$

Here, u and v are the survival factors of the adult and immature age classes, respectively. The parameters b and α have the same interpretation as in (1.14). This can also be viewed as a density-dependent age structured model with two age classes – adults and immatures.

We cannot defend either the exponential model (1.10) or the logistic model (1.14) as adequate characterizations of all aspects of *Tribolium* population growth – that would be expecting far too much of these elementary schemes. However, the models do capture the density dependent feature of the adults and are capable of describing the observed growth pattern in adult number (Fig. 1.1). More importantly, they represent an attempt to combine simple and general models of population growth with experimental data from the *Tribolium* laboratory system.

1.1.3 Models of Age Structure

Age structure in beetle populations was first studied by John Stanley (1932a). Twenty-three years after Stanley's pioneering paper, Herbert Landahl (1955a,b) recognized the general nature of the theory proposed by Stanley and within this framework introduced numerical solutions as a means of studying the temporal pattern of population numbers. Very similar models were studied by Watt (1955), Taylor (1965, 1967, 1968, 1971), Niven (1967, 1969), Sonleitner (1977) and Fujii (1978). These computer models were mathematically complex and, while they did describe the principal features of the data, were analytically intractable.

We shall not attempt a detailed description of the many age-structured models that were applied to *Tribolium*. Sokoloff (1974, pp. 536–562) provides a chonological survey of most of these mathematical efforts. However, as an example of the dynamic complexity that can arise from this class of models (Ebenman 1988), we will consider some simple examples.

A comment by Bartlett (1960a, p. 50), merely an aside in a book that stressed stochastic processes, suggested that the observed fluctuations of animal numbers could be due to age-structure. To demonstrate that this was, at least, a possibility, Bartlett (1960a) wrote

$$\begin{pmatrix} N_I(k+1) \\ \\ N_A(k+1) \end{pmatrix} = \begin{pmatrix} 0 & f(N_A(k)) \\ \\ 1 & 1-\mu \end{pmatrix} \begin{pmatrix} N_I(k) \\ \\ N_A(k) \end{pmatrix}, \tag{1.18}$$

where $N_I(k)$ is the number of immature beetles in generation k, $N_A(k)$ is the number of adults, $f(N_A(k))$ is the birth-rate of the adults, and μ is the adult mortality rate. Note that the immature age-class does not reproduce. For the density-dependent function, Bartlett chose

$$f(N_A(k)) = b - \alpha N_A(k) \,. \tag{1.19}$$

The fecundity of the adult age class was expressed as a linearly decreasing function of the number of adults, i.e., the Verhulst assumption. An alternative representation of this effect would be to write the *per capita* birth rates using a negative exponential form. The dynamics of this latter model were studied by Guckenheimer et al. (1977).

For the Bartlett model (1.18) the equilibrium

$$\begin{aligned} N_I^* &= \frac{\mu\,(b-\mu)}{\alpha} \\ N_A^* &= \frac{b-\mu}{\alpha} \end{aligned} \tag{1.20}$$

is locally stable if

$$b - 2\mu < 1 \,. \tag{1.21}$$

Despite its simple form, the dynamics of Bartlett's model can be quite complex. In Fig. 1.3 a series of nine numerical examples of (1.18) are plotted on the "phase plane" (N_I, N_A). Starting with $b = 2.5$ the population quickly settles into a stable fixed point (Fig. 1.3a). As b is increased to 2.8, the stable equilibrium bifurcates. At $b = 2.85$ we observe sustained oscillations in immature and adult numbers (Fig. 1.3b). The continuous cycle starts out being rather elliptical and then at $b = 3.00$ deforms into a quadrilateral (Fig. 1.3c). As b increases, the points "condense" into the corners of the quadrilateral yielding a four-point cycle (Fig. 1.3d). This cycle bifurcates into eight points (Fig. 1.3e), then sixteen points, thirty-two points, etc., until tiny "islands" of attractors are formed. These islands grow and "fuse" until, at $b = 3.35$, four large islands exist (Fig. 1.3f). The islands lose their clear resolution (Fig. 1.3g,h) and grow in complexity. At $b = 3.45$ an

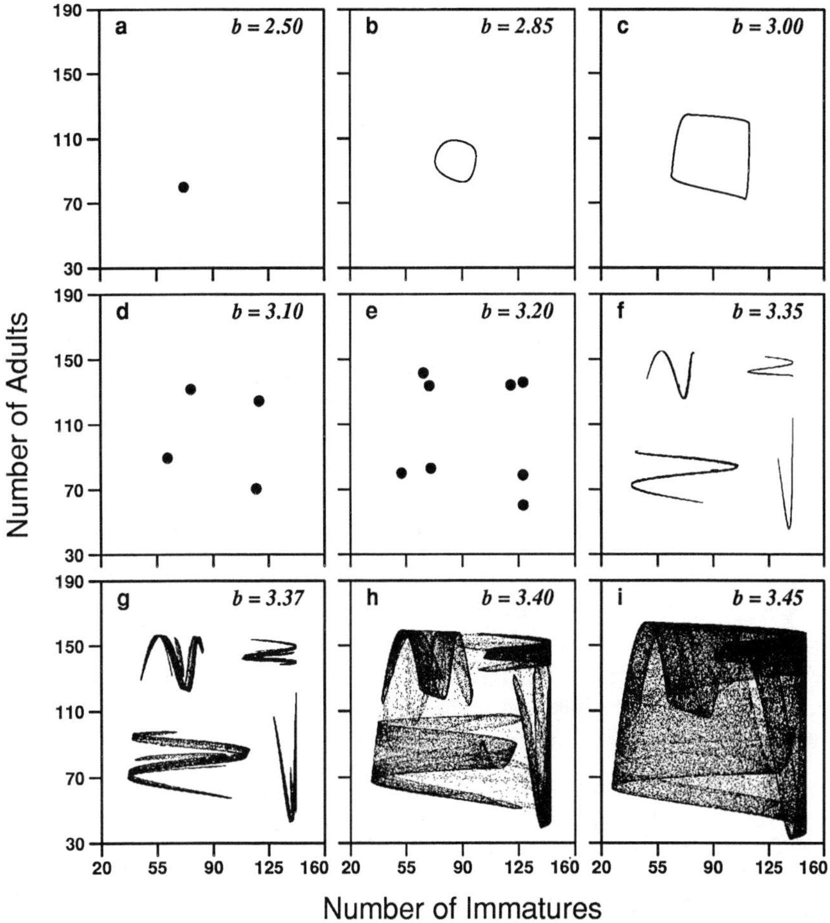

Fig. 1.3. "Bartlett's attractors." Phase plane trajectories for Bartlett's (1960a) two age class model (1.18) with $\mu = 0.9$, $\alpha = 0.02$, and nine different values of b.

intricately detailed attractor exists (Fig. 1.3i). As the dynamic complexity unfolds with higher and higher b-values, there is an intriguing similarity between the shape of the cycle in Fig. 1.3c and the outline of the attractor in Fig. 1.3i. Although, in his analysis of the model, Bartlett did not go beyond noting the continuous cycle Fig. 1.3b, it was clear that fluctuations in animal numbers could be attributed to the interactions among various age classes.

In the Bartlett model (1.18), the number of eggs laid decreased as the population increased. This was consistent with the experimental results of both Park (1932) and Maclagan (1932). Chapman and Whang (1934) disagreed. In their experiments they were able to maintain a constant rate of oviposition and concluded that the decrease in the number of eggs was due to larvae and adults eating eggs. Using the exponential form (1.9), we can write a nonlinear two-age class matrix model compatible with the latter observations:

$$\begin{pmatrix} N_I(k+1) \\ \\ N_A(k+1) \end{pmatrix} = \begin{pmatrix} 0 & b \\ \\ \exp[-c_I N_I(k) - c_A N_A(k)] & 1-\mu \end{pmatrix} \begin{pmatrix} N_I(k) \\ \\ N_A(k) \end{pmatrix}. \tag{1.22}$$

Here, the number of eggs laid per adult per unit time, b, is not a function of population size. Instead, density dependence occurs through immature survivorship, where c_I and c_A are the cannibalism and/or inhibition rates for immatures and adults, respectively. Note that crowding of immatures also leads to a decrease in the number of recruits (Jillson and Costantino 1980). The equilibrium number of adults is given by

$$N_A^* = \frac{\log(b/\mu)}{c_A + b c_I}, \tag{1.23}$$

and $N_I^* = bN_A^*$ for the equilibrium number of immatures. Like other difference equations, this model can exhibit a multiplicity of behaviors which include continuous closed-loop cycles, discrete point cycles, and strange attractors. With its negative exponential assumption, (1.22) is more typical of the models of Landahl (1955a,b) and others. A numerical example of this model is presented in Chapter 9.

1.1.4 Beetle Movement, Food Preference and Tunnels

The models reflected in the work of J. Neyman, P. H. Leslie, and M. Bartlett are characteristic of the vast majority that are associated with *Tribolium*. The nine papers in John Stanley's series on the mathematics of *Tribolium* are qualitatively different. Stanley recognized, and attempted to deal with, the immense biological complexity of the beetle cultures. He commented on his research perspective (Stanley 1985, personal communication) as follows:

> My own approach has always been what one might call "voyeuristic" in that, when I begin to study some complicated system to see "who does what and to whom," I seem to see the system laid out on a flat surface as it were, so that I look into it like a person looking down from a high view point. The variables used are then, as it were, "impersonations" of the bugs, people or whatnot operating in the system, together with the forces, actions and whatnot which make up their interactions. I have always felt that if one is to understand the "sociology" of a population, one must, so to speak, "get down among the bugs and see what they do!" This is very well shown in my paper: "A further study of the 're-tunneling' problem." I must admit that using this approach usually leads to an intractable set of differential and other equations, but at least one can vary the values of variables, and then trace through to see what then happens.

Concerned about the lack of biological reality in mathematical models, Stanley focused on such things as beetle movement, food preference and tunnels, i.e., a kind of micro-geographical description of the organization and behavior of the beetles in their habitat. For example, the random collision between adults and eggs was invoked by Stanley twenty-three years before the Neyman, Park and Scott (1956) paper was published. But a collision, in Stanley's view, was not interpreted to mean that the egg was necessarily eaten. The beetle had a choice of two foods: eggs and flour. Stanley's "food choice" argument defined, for both eggs and flour, rates of encounter, rates of consumption, maintenance ration (of nutrition) and the notion of food preference. The first six papers of Stanley (1932a,b, 1934, 1941, 1942, 1943) were centered on various aspects of his food choice hypothesis, although he also dealt with questions concerning animal numbers.

The exponential model (1.10) captures the cannibalistic feature of the adults but does not suggest how temperature, humidity, and/or size of the habitat may influence the rate of egg cannibalism. We can obtain some insight on these associations by referring to Stanley's (1932a) food choice conjecture. If, for both eggs and flour, the rates of encounter are set equal to the rates of consumption, then Stanley's food choice is equivalent to Neyman's random collision and the average rate of egg eating per adult is given by

$$c = \left(\frac{s\,\pi}{V_F}\right)\left(\frac{\sigma_E + \sigma_A}{2}\right)^2, \tag{1.24}$$

where, s is the mean speed of a beetle, σ_E and σ_A are the mean widths of an egg and an adult, respectively, and V_F is the total volume of flour. For *T. confusum*, $s = 4$ mm/sec, $\sigma_E = 0.4$ mm and $\sigma_A = 1.02$ mm (Stanley 1932a, 1964). In a volume of $V_F = 1000$ cubic millimeters, we find $c = 0.0063$. The rate of egg cannibalism is set by the size of an egg, the size of an adult, the volume of flour to be searched and the speed of adult movement. The latter may be altered by such environmental factors as temperature and relative humidity.

The movement of adults and larvae, within the total volume of flour available to the animals, creates a network of tunnels which in turn has a profound affect on the geographical distribution of eggs, larvae, pupae, and adults. Tunnelling was the topic of Stanley's (1949, 1964, 1966) three final papers in his pioneering series.

To get an idea of Stanley's approach to the mathematics of *Tribolium*, we ask you to consider a tunnel (Fig. 1.4) generated by the movement of an adult beetle in a mass of flour of volume V_F. To compute the volume of tunnel established by the beetle, let $A(x)$ denote the area of a cross-section of the tunnel by the plane determined by an arbitrary value of x and let dx be a thin slice of the tunnel. An approximation to the volume of the slice is $A(x)dx$. Hence, the volume of tunnel is

$$V_T = \int_0^l A(x)\,\mathrm{d}x\,. \tag{1.25}$$

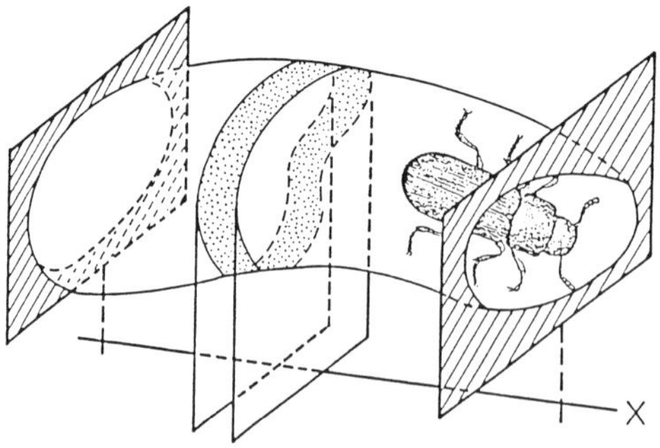

Fig. 1.4. Adult movement and tunnel formation. The variable x represents the length of a tunnel. A thin slice of tunnel of width dx is illustrated in the center of the diagram.

The length of the tunnel, l, is also the distance travelled during time t, so we can write $l = st$ where s is the mean speed of a beetle (including rest periods) in millimeters/second while boring a new tunnel (Stanley 1949). If, as a first approximation, we assume an adult bores out a circular tunnel, then

$$V_T(t) = \int_0^{st} \pi R^2 \mathrm{d}x = \pi R^2 st , \tag{1.26}$$

where R is the mean radius of a tunnel. The total production of tunnel volume is equal to the mean speed of a beetle, s, multiplied by the duration of travel, t, times the cross-sectional area of a tunnel, πR^2.

While watching the beetle forming the initial tunnel this argument seems satisfactory. However, after forming a small tunnel network the animal will not continue to construct new tunnels, but will begin to move back through already established tunnels which, of course, does not increase the total volume of tunnel $V_T(t)$. Assuming adults move randomly, the rate of change of $V_T(t)$ is proportional to the volume of untunnelled flour:

$$\frac{\mathrm{d}V_T}{\mathrm{d}t} = (\pi R^2 s)\left(\frac{V_F - V_T}{V_F}\right). \tag{1.27}$$

Integrating (1.27) we have

$$V_T(t) = V_F\left[1 - \exp\left(-\frac{\pi R^2 st}{V_F}\right)\right]. \tag{1.28}$$

The total volume of tunnel is a monotonically increasing function and eventually the entire volume of flour will in theory be converted into tunnels. As emphasized by Stanley (1949), this is a rather crude first analysis of tunnel formation. The theory was both modified and extended in Stanley (1964, 1966).

While Stanley was the first to study habitat dynamics he was not alone. Cox and Smith (1957) envisioned an experiment to study the spatial distribution of beetles in a circular container with unit radius r. They imagined that the beetle followed a straight path and when it hit the side of the container the insect had two options: with probability p it could return along its original path, or with probability $1-p$ it could move a distance τ, possibly zero, along the boundary before choosing a new direction, independent of its previous direction. The distance τ has a distribution function $H(\tau)$. The angle θ that the beetle's movement makes to the normal to the boundary at the point of departure is assumed to be uniformly distributed. Thus, when the beetle hits the boundary, the orientation of its new path is equally likely among all possible choices. The resulting spatial distribution of beetles was expected to be almost uniform near the center ($r < 1/2$) with an increasing concentration of animals toward the boundary of the habitat.

Other questions, similar in perspective to those of John Stanley, were also addressed. Stanley (1932a) argued that under specified laboratory culture conditions the expected number of eggs will increase and asymptotically approach an equilibrium density. Chiang (1957) formulated this problem in terms of a stochastic birth-death process and described the transient and steady-state probability distribution for the number of eggs. The retunnelling problem described by Stanley (1949) was interpreted by Ahmed (1963) as a Markov chain with three states: S_1 is tunnelling, S_2 is stationary, and S_3 is retunnelling. Ahmed placed a beetle in fresh flour and then recorded the location of the animal and the times and the types of transitions that occurred. Given that there was initially a small network of tunnels available, this analysis suggested that an adult beetle will be moving 80% of the time (retunnelling a bit more than tunnelling), and will be stationary 20% of the time.

Unlike the models of adult numbers or age structure, these models focused on behaviors which were specific to laboratory populations of *Tribolium*. They traded a loss of generality for an increase in biological reality.

1.1.5 Species Interactions

Charles Darwin, in Chapter III, of *The Origin of Species*, wrote, "As the species of the same genus usually have, though by no means invariability, much similarity in habitats and constitution, and always in structure, the struggle will generally be more severe between them, if they come into competition with each other, than between the species of distinct genera." With that quotation, Thomas Park (1948, 1954, 1957) began his first series of reports on the experimental studies of competition between populations of the flour beetles *Tribolium confusum* Duval abd *Tribolium castaneum* Herbst.

A new area of population research with the beetle was opened. The question was no longer restricted to how these insects regulate their numbers but was extended to the identification of the factors that determine the long term outcome of mixed species cultures. What is the basis of the competition: food, space, temperature, cannibalism, fertility, adult mortality? Will these species coexist? More generally, is it possible to predict the outcome when two species are grown together? The pioneering papers of Park drew the attention of several biomathematicians (Landahl 1955b, Bartlett 1957,

Taylor 1967, Niven 1969). In this brief introduction we will consider Park's collaborative effort with J. Neyman and E. Scott in the United States and his association with P. H. Leslie, M. S. Bartlett and J. C. Gower in England.

Jerzy Neyman, Thomas Park and Elizabeth Scott (1956) proposed a straightforward extension of equation (1.10):

$$\begin{aligned} N_1(k+1) &= b_1 N_1(k) \exp[-c_{11}N_1(k) - c_{12}N_2(k)] , \\ N_2(k+1) &= b_2 N_2(k) \exp[-c_{21}N_1(k) - c_{22}N_2(k)] , \end{aligned} \tag{1.29}$$

where, for species i, N_i is the number of adults, b_i is the rate at which potential recruits are produced per adult and c_{ij} is the cannibalism rate at which an adult of species j consumes eggs, pupae or callow adults of species i. Notice that if the adults of one species do not interact in any way with the immature life stages of the other species, i.e., $c_{12} = c_{21} = 0$, then we recover the single species model (1.10). This is a discrete time model with nonoverlapping generations. These equations, while not applied to *Tribolium* beyond the Neyman paper, have been studied by Murphy (1968), Maynard-Smith (1974), Hassell and Commins (1976), Commins and Hassell (1976) and Hassell (1978). Recently Hofbauer et al. (1987) have examined this form for an arbitrary number of species.

A key biological assumption in the analysis of Neyman et al. (1956) is that cannibalism is indiscriminate, i.e., $c_{11} = c_{21} = c_1$ and $c_{22} = c_{12} = c_2$. In the light of more recent data, as we note in Chapter 8, this statement is untenable.

Another important theoretical paper associated with *Tribolium* competition is Leslie's (1962) publication, "A stochastic model for two competing species of *Tribolium* and its application to some experimental data." Five papers supported the latter work: Leslie (1958, 1960), Leslie and Gower (1958, 1960) and Bartlett et al. (1960). These six papers, which all appeared in *Biometrika* within a four year interval, form an important step in our general understanding of the properties of stochastic models for both single and two interacting species. We asked J. C. Gower, presently head of the biomathematics division and statistics department at Rothamsted Experimental Station, to describe the "British group." Gower (1988, personal communication) wrote,

> Leslie was the key man in this work. Rothamsted acquired a computer in 1954 and this was the first in Britain (and must have been one of the first in the world) that was used for biological and agricultural research. I came here in 1955, having completed a post-graduate Diploma in Statistics under Bartlett in Manchester. At that time my main job was to see how the computer could be exploited for statistical work and we encouraged collaboration with people with interesting research projects that had a substantial computing component. Leslie at the Bureau of Animal Population in Oxford was one such person, so he and I got together to do the work that was eventually published in *Biometrika*. Oxford is less than 50 miles from here and Leslie lived some 15 miles from Oxford in the right direction, so he would frequently drive over the Chilterns to discuss things with me. The stage was reached when neither of us knew enough about stochastic processes to make the progress needed. Leslie then approached Bartlett who was (and still is) a leading expert in the field, well-known to both of us. Bartlett sorted out the mathematics and left the rest to Leslie

> and I. Basically, Bartlett did the harder bits of mathematics, Leslie then provided the biological insights and I did the computing, though we all provided some input in the various areas.

Leslie (1962) proposed using the second-order nonlinear difference equation (1.17) to compute adult number. It is instructive to note that Leslie (1962, p. 16) abandoned this model for the discrete-time Lotka-Volterra scheme (1.14) extended to two species:

$$\begin{aligned} N_1(k+1) &= \frac{b_1 N_1(k)}{1+\alpha_{11}N_1(k)+\alpha_{12}N_2(k)}, \\ N_2(k+1) &= \frac{b_2 N_2(k)}{1+\alpha_{21}N_1(k)+\alpha_{22}N_2(k)}, \end{aligned} \tag{1.30}$$

where b_1 and α_{11} are the logistic parameters for species 1 living alone, and b_2 and α_{22} are those for species 2 while α_{12} and α_{21} are the interaction coefficients (Leslie 1958). Leslie (1962) did not follow the Neyman et al. (1956) assumption of indiscriminate cannibalism. Using the data, he found that cannibalism was discriminate, i.e., $\alpha_{11} \neq \alpha_{21}$ and $\alpha_{22} \neq \alpha_{12}$; furthermore, the coexistence equilibrium was unstable.

These are the basic deterministic features of the stochastic model used in the classic *Tribolium* papers of Park et al. (1964) and Leslie et al. (1968). Models of this type are still studied (Elmes 1973; Mertz et al. 1976; Smith and Mead 1979, 1980). More importantly, questions concerning the quasi-stationary state, stability, stochastic fluctuations and time to extinction, which were recognized and associated with the flour beetle over thirty years ago, remain topics of active research today (Gurney and Nisbet 1985, Roozen 1987).

1.1.6 Population Genetics

The tradition of mathematical work associated with *Tribolium*, as established by Stanley, Neyman, Leslie, Landahl, Bartlett and others, addressed a broad spectrum of population problems including the temporal pattern of growth, mechanisms regulating density, beetle movement, food choice, habitat alteration by tunnel formation, the spatial distribution of the insects, and species competition. In all of these topics, population size was the main focus of attention. After all, the flour beetle was promoted as an experimental insect for the analysis of the causes of fluctuations in population size.

Throughout this early work one feature was missing: genetics. It is not clear why this was so. William Provine (1986), in his treatise on Sewall Wright's role in evolutionary biology, makes it clear that population size was important to the geneticists. "Fisher and Wright, whatever their differences at this time, agreed perfectly that the effective population size, or population number as Fisher called it, was fundamentally important in the statistical analysis of evolution in nature. Much of the history of population genetics in the ensuing decades would indeed hinge upon theoretical and practical field investigations of effective population size" (Provine, 1986 p. 255).

In fact, the population geneticist J.B.S. Haldane (1953) made a theoretical point that is a part of modern ecology's research effort: for an equilibrium to be stable, the density regulating mechanism must not be too sensitive. Haldane's (1953) population model is

$$N(k+1) = bN(k)^{1-\alpha}, \tag{1.31}$$

which has the explicit solution

$$N(k) = N(0)^{(1-\alpha)^k} b^{\alpha^{-1}[1-(1-\alpha)^k]}. \tag{1.32}$$

The stability of the equilibrium

$$N^* = b^{1/\alpha} \tag{1.33}$$

is sensitive to the magnitude of the density regulation α (Fig. 1.5). If the dependence is weak, $\alpha = 1/2$, the equilibrium $N^* = b^2$ is a global attractor with no oscillations. As α is increased, damped oscillations appear but N^* is still stable until, at $\alpha = 2$, the equilibrium

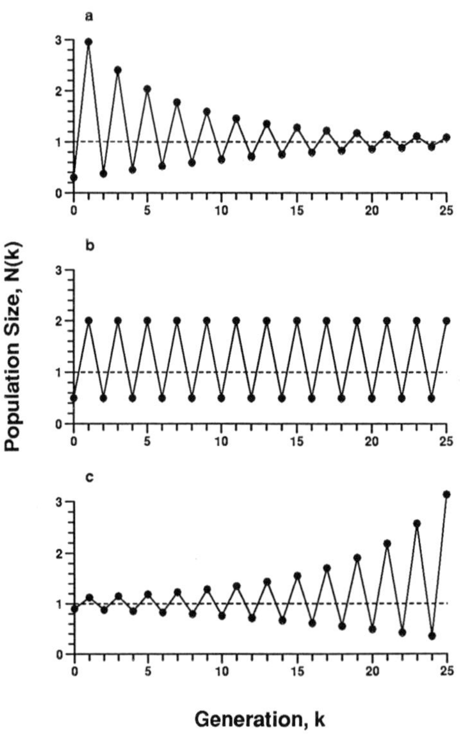

Fig. 1.5. Examples of Haldane's model (1.31). The numerical examples use $b=1$ and (a) $\alpha=1.9, N(0)=0.3$; (b) $\alpha=2, N(0)=0.5$; and (c) $\alpha=2.1, N(0)=0.9$. The dashed line represents the equilibrium $N^*=1$.

$N^* = b^{1/2}$ becomes unstable and population numbers oscillate in a two-point cycle. Beyond $\alpha = 2$, for example at $\alpha = 3$, we find $N^* = b^{1/3}$ is not only unstable, but the smallest deviation from N^* results in violent, continually expanding, undamped oscillations. This model is not appropriate for the flour beetle; nevertheless, Haldane's research raised an important question. Can population oscillations, which are so pronounced in the *Tribolium* data (Fig. 1.1), be a consequence of the magnitude of the density regulating mechanism?

While the association between population genetics and effective population size was recognized, research with the flour beetle was not extended in that direction for many years. Part of the reason for the delay was that by 1928, when *Tribolium* was introduced as a laboratory organism, Mendelian genetics was well established and *Drosophila* had already assumed a central role in genetical research. Furthermore, the theory of population genetics was based on very general concepts and many specific biological details (such as cannibalism) were not explicitly part of the argument. As a result, there was no development of a "population genetics unique to *Tribolium*" or other organism.

One of the first people to use the flour beetle in genetical research was A. E. Bell. He described his introduction to *Tribolium* (Bell, 1985 personal communication) as follows:

> Since the major objective of our laboratory was to obtain results which were transferable to economic species, we sought to broaden our studies to include another laboratory animal.... *Tribolium* was not seriously considered until Thomas Park gave the 1953 Sigma Xi Lecture at Purdue and described his ecological studies on the competition of *T. castaneum* and *T. confusum* in different environments. After the lecture I visited with him for about an hour about *Tribolium* biology, generation cycles, culturing techniques, quantitative variability, and our research interests. With his encouragement and a two week loan of his *Tribolium* technician for training in the basic culturing procedures, we enthusiastically launched our preliminary *Tribolium* work in the Spring of 1954.

Many of the early genetic experiments with the flour beetle involved the evaluation of selection theory for single quantitative genetic traits (Bell 1982). Equations such as that for the expected change in the mean phenotypic value, ΔG, from selection in one generation,

$$\Delta G = h^2 S , \tag{1.34}$$

were evaluated by comparing predictions based on parameter estimates for the heritability, h^2, and selection differential, S, made on individuals of the parent generation to the observed mean value of the trait in the offspring. In general, short term responses were in agreement with the theory, most likely, because the value of the heritability (the ratio of the additive genetic variance to the phenotypic variance of the trait) changes quite slowly as compared with the mean phenotypic value. In this context, the flour beetle experiments served to connect genetic theory and programs of applied animal breeding (Bell 1983).

Experiments on the selection of mutant and wild type alleles were also conducted. These included the pearl eye mutant of *T. castaneum* (Kollros 1944), the striped, black and split mutants of *T. confusum* (McDonald and Peer 1960, 1961), the body color mutant sooty of *T. castaneum* (Schlager 1963, Sokal and Huber 1963, Sokal and Karten 1964) and the black body color mutant of *T. castaneum* (Sokal and Sonleitner 1968). In each of these studies allele frequency was monitored in a manner similar to the *Drosophila* population cage experiments (L'Hertier et al. 1937). However, perhaps due to the influence of the ecologists Chapman and Park, in each of these *Tribolium* studies population size was also recorded. Selection was evaluated in the context of the complex growth dynamics of the beetle. But a theory, with at least some biological detail specific to the beetle, was not available.

The classical discrete generation model of natural selection at one diallelic locus in large randomly-mating population is given by the equation

$$p(k+1) = \left(\frac{p(k)}{\overline{W}(k)}\right)[p(k)W_{11} + q(k)W_{12}] \,, \tag{1.35}$$

where $p(k)$ is the frequency of one of the alleles, $q(k) = 1 - p(k)$ is the frequency of the second allele, W_{11}, W_{12} and W_{22} are the fitnesses of the three genotypes, and

$$\overline{W}(k) \quad = \quad p(k)^2 W_{11} + 2p(k)q(k)W_{12} + q(k)^2 W_{22} \tag{1.36}$$

is the mean population fitness. For each genotype, fitness is defined as the genetic contribution of the individual to the next generation. With respect to the dynamics of allele frequencies, it is easy to see from (1.36) that only the relative magnitudes of the fitnesses are important. Unfortunately, this fact hampered the integration of population genetics and ecology. If one views fitness in absolute terms, then the average population fitness equals the *per capita* population growth rate:

$$N(k+1) = N(k)\,\overline{W}(k) \,. \tag{1.37}$$

Thus, the dynamics of natural selection and population growth are coupled.

If we allow the fitness values to be constants, then we can identify a serious flaw in the application of this analysis to *Tribolium*: the population growth equation is Malthusian. Fortunately, the constant fitness assumption can be modified. Unfortunately, it was over 30 years after Chapman's work before a density-dependent measure of fitness was first proposed (MacArthur 1962). In fact, it was not until the 1970's that a concerted effort was made to integrate the models of population genetics and ecology (Anderson 1971, Charlesworth 1971, Roughgarden 1971).

In keeping with this synthesis of population growth and genetics, Desharnais and Costantino (1980) were the first to propose a *Tribolium* model of natural selection in a density-regulated population. Their choice for the genetic fitness expression was based on the cannibalism model (1.10):

$$W_{ij}(N(k)) = 1 + b_{ij}\exp[-c\,N(k)] - \mu . \tag{1.38}$$

In this expression, b_{ij} is the expected number of pupae with genotype ij, c is the rate at which a single adult will eat a single pupa during one time interval, and μ is the death rate for adults. Substituting (1.38) into equations (1.35) – (1.37) yields the following equations for allele frequency and population size:

$$\begin{aligned} p(k+1) &= \frac{p(k)\,(1+[p(k)b_{11}+q(k)b_{12}]\exp[-cN(k)] - \mu)}{1+\bar{b}(k)\exp[-cN(k)] - \mu}, \\ N(k+1) &= N(k)\,(1+\bar{b}(k)\exp[-cN(k)] - \mu), \end{aligned} \tag{1.39}$$

where the average "birth" rate is given by

$$\bar{b}(k) = p(k)^2 b_{11} + 2p(k)q(k)b_{12} + q(k)^2 b_{22} . \tag{1.40}$$

A stable, non-trivial equilibrium for both allele frequency and population size (p^*, N^*) exists provided

$$b_{11} < b_{12} > b_{22} \qquad \text{and} \qquad \mu < \bar{b}^* < \mu\exp\left(\frac{2}{\mu}\right). \tag{1.41}$$

The significance of these predictions is that they are made in the context of a known biological mechanism of density dependence – cannibalism.

In closing this review, we return to the mathematical model with which we started. Notice that if $\mu = 1$, then the generations do not overlap and the equations (1.39) reduce to

$$\begin{aligned} p(k+1) &= \left(\frac{p(k)}{\bar{b}(k)}\right)[p(k)b_{11}+q(k)b_{12}], \\ N(k+1) &= N(k)\,\bar{b}(k)\exp[-cN(k)] . \end{aligned} \tag{1.42}$$

The equilibrium number of adults is given by

$$N^* = \frac{\log(\bar{b}^*)}{c}, \tag{1.43}$$

and the stability condition for population size, from (1.41), is

$$1 < \bar{b}^* < 7.3890 . \tag{1.44}$$

This model is identical to (1.10) which was first proposed by Neyman et al. (1956) – with the exception of genetic variability in the recruitment rate b. In this special case, the genetic equation reduces to the classical model (1.35). It is somewhat ironic, that with the new synthesis of ecology and genetics, we recover the simplicity of the original models in each discipline.

1.2 Topics in the Monograph

The experimentalists, led by Royal Chapman and Thomas Park, have clearly demonstrated the general applicability of the flour beetle for the study of populations (Sokoloff 1972, 1974, 1977). The versatility of this insect makes it quite appropriate to address many population problems: growth and the regulation of density, natural selection, species interactions, age-structured populations, concepts of equilibrium, the impact of environmental diversity on population growth, and the application of quantitative genetic theory to both animal breeding and evolutionary biology.

The biomathematicians have also made their contribution to the study of populations. Based on the known cannibalistic behavior of larvae and adults, they introduced expressions of the form $\exp(-cN)$ to characterize density regulation. More importantly, in collaboration with the experimentalists, they introduced the idea of using *nonlinear* population models in the design and interpretation of *Tribolium* laboratory experiments.

Nonlinearity has become an important theme in modern population research. The central role of nonlinearity is reflected in the works of Goel et al. (1971), May (1974, 1976), Gurtin and MacCamy (1974, 1979), Griffel (1976), Rorres (1976), Guckenheimer et al. (1976), Swick (1976, 1977, 1980), Oster (1978), Gurney and Nisbet (1980), Schaffer et al. (1986), Webb (1985), Lasota and Mackey (1985), West (1985), Devaney (1986), Metz and Diekmann (1986), Tuljapurkar (1987) and Jensen (1987) to mention just a few. Much of this work has been devoted to the development of new techniques for determining the consequences of nonlinearity and novel applications of these techniques.

Recognizing the importance of nonlinearity in population dynamics, we have a new challenge. It is a challenge which was not explicitly stated in the early years of *Tribolium* research: to interpret the flour beetle data in the context of a nonlinear population theory. We now describe the four research topics which are covered in the remaining chapters of this book.

1.2.1 Life Stage Interactions

One manifestation of nonlinearity in *Tribolium*, as well as in other insect species, is the behavioral interaction among life stages. The analysis and implications of these interactions is the subject of Chapters 2 and 3.

Chapter 2 begins with a general nonlinear age-structured model. This model, with its assumptions of density-dependent fecundity and mortality rates, is specialized to account for the cannibalism and crowding effects in laboratory populations of *Tribolium*. Parameter estimates are obtained for the model and population projections are compared to experimental data. A numerical analysis of the model provides some insight into the dynamical behavior of *Tribolium* populations. A main conclusion of this work is that flour beetles can exhibit stable demographic limit cycles.

In Chapter 3 we reduce the biological detail and develop a model of egg-larval cannibalism which is both analytically and numerically tractable. The model consists of a system of nonlinear Volterra integral equations which are amenable to a bifurcation analysis. This model introduces a new phenomenon into population theory: multiple demographic attractors. The analysis predicts the coexistence of a stable equilibrium point with a stable limit cycle. The implications of this result are investigated numerically and with parameter estimates from eight genetic strains of *Tribolium*.

1.2.2 Stochastic Population Growth

Stochastic models provide a framework for making probabilistic statements about the dynamics of a population and allow an analysis of population abundance variability. In stochastic models with density-dependence, the concept of stable equilibrium is replaced by a steady-state probability distribution. In Chapters 4 and 5 we view the dynamics of adult numbers in *Tribolium* populations as a nonlinear stochastic process.

One of the insights obtained from Chapters 2 and 3 is that the demographic dynamics of *Tribolium* populations can be viewed on two time scales: a "fast" time scale for the oscillations of the immature age classes and a "slow" time scale for adult numbers. We begin Chapter 4 with the development of a simple "adults only" model. This model, which is a continuous-time analog of (1.10), is converted to a stochastic differential equation for the dynamics of adult numbers. An analysis of the stochastic model yields a steady-state probability distribution which can be approximated with a gamma probability density function. Statistical procedures are developed to estimate parameters and make comparisons among populations. The gamma hypothesis is tested with several experimental data sets and applications of the statistical procedures are demonstrated. Stochastic approximations for the analysis of stationary time series data and time-dependent distributions are also discussed.

A more realistic alternative to the stochastic differential equation is the analysis of population growth as a stochastic birth-death process. In this framework, population size is defined as a nonnegative integer and births and deaths are viewed as discrete transitions. In Chapter 5 we model adult numbers as a birth-death process where the transition probabilities are density-dependent. We obtain "quasi-stationary" probability distributions, extinction probabilities, and approximations for the approach to the stationary state. Data are examined in the context of these predictions. Chapter 5 concludes with a general discussion of stochastic models.

1.2.3 Natural Selection

An exciting development in population theory has been the synthesis of genetic models of natural selection with population growth. With this synthesis, new insights into the dynamics of genetically heterogeneous populations have been obtained (Slatkin and Maynard Smith 1979, Ginzburg 1983). In Chapters 6 and 7 we explore some of the interactions between natural selection and population growth.

In Chapter 6 we focus on density-dependent natural selection in the context of the biology of *Tribolium.* We investigate the implications of the location of genetic variability in the parameters of the population growth model. Experimental results on the corn oil sensitive mutant of *T. castaneum* are presented. We end the chapter with a discussion of a fundamental theoretical prediction: the maximization by natural selection of adult numbers in *Tribolium.*

Genetic disequilibrium is the subject of Chapter 7. We make use of Ginzburg's (1977) entropy function to evaluate the effects of genetic initial conditions on the dynamics of population growth. Haldane's (1957) "cost of natural selection" is revisited and both denisty-independent and density-dependent models are examined. Experimental data on *Tribolium* populations are used to test several theoretical predictions. Recent extensions to the theory are also discussed. The chapter closes with a critique of the theory.

1.2.4 Species Interactions

In their extensive studies of competition in *Tribolium,* Thomas Park and his colleagues made some intriguing observations. When *T. castaneum* and *T. confusum* are cultured together, one or the other species persists; coexistence is never realized. Furthermore, the outcome of competition was sometimes indeterminate, with *T. castaneum* surviving in some of the replicates and *T. confusum* surviving in others (Park 1962). In Chapter 8 we consider these observations in the context of a general theory of species interaction appropriate for *Tribolium.* We show that when the biological basis of the interaction is cannibalism, coexistence is an unlikely outcome. Possible explanations for indeterminism are discussed.

We also consider in Chapter 8 an entirely different biological mechanism of species interaction in *Tribolium*: the inhibition of metamorphosis. In the species *T. brevicornis,* the larval stage is extended when adult densities are high (Jillson and Costantino 1980). With both our model and experimental data, we show that in cultures composed of *T. castaneum* and *T. brevicornis,* competitive exclusion is also the rule, but inhibition in *T. brevicornis* makes it the certain loser. We present data for two inhibiting species, *T. madens* and *T. brevicornis,* which suggest, for the first time, the possibility of coexistence.

Many questions are raised in this monograph and open problems remain in every area of inquiry. Possible avenues of future research are explored in Chapter 9.

Have we come full circle so that in the late 1980's we are looking at many of the same questions posed in the early 1930's? That may be so. We hope to show, throughout the monograph, that the combination of mathematical modeling and *Tribolium* experimentation is an effective strategy for addressing many general problems of modern population biology.

Chapter 2 Biology and Dynamics of Age Structure

It would be hard to imagine a much simpler experimental system. Place some flour and yeast in a glass container with a few graminivorous beetles. Keep the container in a dark incubator at a constant temperature and humidity. At regular intervals, count the beetles and change the flour. An incubator, a scale, and a few plates and sifters are all that is needed. Yet despite its simplicity, this experimental system has the potential for producing a variety of interesting population phenomena including stable equilibria, Hopf bifurcations, multiple attractors, and stable limit cycles. How the physiology and behavior of flour beetles of the genus *Tribolium* can lead to these topics is the subject of this chapter and the next.

In this chapter, we begin with a presentation of some general models for the dynamics of an age-structured population. We proceed to specialize the recruitment and mortality functions based on the biology of *Tribolium*. Using data from the literature, we simplify the model and estimate its parameters. We examine the dynamics of age structure by comparing computer simulations to experimental data. This leads us to the conclusion that laboratory populations of *Tribolium* exhibit stable demographic limit cycles for a wide range of experimental conditions. The results of this chapter are based largely on the work of Desharnais and Liu (1987).

2.1 Models for Age-Structured Populations

2.1.1 McKendrick-von Foerster Equation

We begin our analysis of age-structured dynamics with the McKendrick-von Foerster equation. This model, which was made popular by von Foerster (1959), is based on the pioneering work of Sharpe and Lotka (1911), Lotka (1925), and McKendrick (1926). Although it was originally formulated as a linear model, it is easily extended to the case of density-dependence. For recent mathematical developments of this model, readers are referred to the book by Webb (1985).

Let the function $n(x,t)$ represent the density of individuals of age x at time t. The McKendrick-von Foerster model describes the aging and death of the population as

$$\lim_{h \to 0} \frac{n(x+h,t+h)-n(x,t)}{h} = -\mu(x,t)n(x,t)\,. \tag{2.1}$$

The nonnegative function $\mu(x,t)$ is the per capita mortality rate for individuals of age x at time t. If $n(x,t)$ is differentiable, then

$$\frac{\partial n(x,t)}{\partial t}+\frac{\partial n(x,t)}{\partial x} = -\mu(x,t)n(x,t)\,. \tag{2.2}$$

This first-order partial differential equation is sometimes called the "transport model" for population demography, since it is related to diffusion models in chemistry and physics.

In the McKendrick-von Foerster model, reproduction is described as a boundary condition at age zero:

$$n(0,t) = B(t)\,, \tag{2.3}$$

where $B(t)$ is the total birth rate of the population. Letting $b(x,t)$ represent the number of offspring born to an individual of age x at time t, the total birth rate is given by

$$B(t) = \int_0^{\omega} n(x,t)b(x,t)\,\mathrm{d}x\,, \tag{2.4}$$

where ω is the maximum attainable age.

In populations composed of males and females, some special assumptions must be made. If one sex, usually males, is in abundant supply, then this sex is ignored, and the equations are used to model the numbers of the limiting sex. Alternatively, if the mortality rates are similar for the two sexes, then a constant sex ratio can be assumed, and the fertility function $b(x,t)$ is adjusted to account for total numbers of individuals. The former procedure is commonly used by human demographers. We shall use the latter procedure in our applications to *Tribolium.*

Finally, to complete the model, one must specify an initial age distribution

$$n(x,0) = \phi(x)\,, \tag{2.5}$$

where $\phi(x)$ is a nonnegative function of age.

The McKendrick-von Foerster equations can be represented in integral form using the method of characteristics. For details of the derivation, readers are referred to Webb (1985, pp. 3 – 14). The integral equations are given by

$$n(x,t) = \begin{cases} n(x-t,0)\exp\left[-\int_{x-t}^{x} \mu(s,s+t-x)\,\mathrm{d}s\right], & x \geq t, \\ n(0,t-x)\exp\left[-\int_{0}^{x} \mu(s,s+t-x)\,\mathrm{d}s\right], & x < t. \end{cases} \tag{2.6}$$

where $n(x-t,0)$ and $n(0,t-x)$ are defined in (2.3)–(2.5). Equation (2.6) emphasizes the fact that the population of age x at time t is determined by the survivorship of colonists of age $x-t$ at time zero ($x \geq t$) or individuals born $t-x$ time units earlier ($x<t$).

2.1.2 The Leslie Matrix Model

Another approach to modeling age-structured populations is to consider time and age at evenly-spaced discrete points. The population is represented by a $\omega \times 1$ vector $\mathbf{n}(t)$ whose elements are the number of individuals in each age class. A matrix is used to project the age vector one unit forward in time. This scheme, which was introduced into population biology by Bernardelli (1941), Lewis (1942), and Leslie (1945), can be represented by the matrix equation

$$\mathbf{n}(t+1) = \mathbf{M}(t)\mathbf{n}(t), \tag{2.7}$$

where $\mathbf{M}(t)$ is the $\omega \times \omega$ Leslie matrix

$$\mathbf{M}(t) = \begin{pmatrix} b(1,t) & b(2,t) & b(3,t) & \cdots & b(\omega-1,t) & b(\omega,t) \\ s(1,t) & 0 & 0 & \cdots & 0 & 0 \\ 0 & s(2,t) & 0 & \cdots & 0 & 0 \\ 0 & 0 & s(3,t) & \cdots & 0 & 0 \\ \cdot & \cdot & \cdot & \cdot & \cdot & \cdot \\ \cdot & \cdot & \cdot & \cdot & \cdot & \cdot \\ \cdot & \cdot & \cdot & \cdot & \cdot & \cdot \\ 0 & 0 & 0 & \cdots & s(\omega-1,t) & 0 \end{pmatrix}. \tag{2.8}$$

The element $b(x,t)$ is the average number of offspring produced per indvidual of age x during the interval from t to $t+1$, and $s(x,t)$ is probability that an individual of age x survives through the same interval. Given an initial age vector $\mathbf{n}(0)$, expression (2.7) can be iterated to yield the age vector at some arbitrary future time t.

The Leslie matrix model is essentially a discrete-time analogue of the McKendrick-von Foerster equations, with mortality rates $\mu(x,t) = -\log[s(x,t)]$. In most cases, the continuous model is more mathematically tractable since the methods of calculus can be used in its analysis. However, the discrete model is much easier to use in numerical work. If the number of age classes is fairly large, the Leslie model is usually an adequate approximation to the McKendrick-von Foerster equations. We shall take advantage of both approaches in our applications to *Tribolium*.

2.2 Life Stage Interactions

2.2.1 Life Cycle

Like most beetles, *Tribolium* undergo a holometabolous development, from egg through larval and pupal stages, to the imago or adult (Fig. 2.1). The duration of each stage and the number of larval instars depend upon many environmental factors including temperature, humidity, and food quality and availability. There are also differences between species and genetic strains. Even under constant environmental conditions, there are differences among individuals. Sokoloff (1974) provides an excellent summary of much of the information on variation in developmental rates.

For the purposes of modeling, assume that within a population each life stage is of constant duration. Let $\omega > 0$ represent the maximum age of a beetle. Let E be a connected subset of the interval $[0,\omega]$ such that a beetle of age x is an egg if and only if $x \in E$. Let the subintervals L, P, and A be defined similarly for the larval, pupal, and adult stages, respectively. The duration of each life stage S will be denoted as D_S, where $S \in \{E, L, P, A\}$. Formally,

$$D_S = \int_S 1 \, \mathrm{d}x \,, \qquad S \in \{E, L, P, A\} \,. \tag{2.9}$$

In the analogous discrete time model, E, L, P, and A are sets of consecutive integers and D_S is the number of elements in set S.

For the purpose of making contact with experimental data, it is convenient to have variables which describe the total density of each life stage. We will use $N_S(t)$ to denote the density of life stage S, where

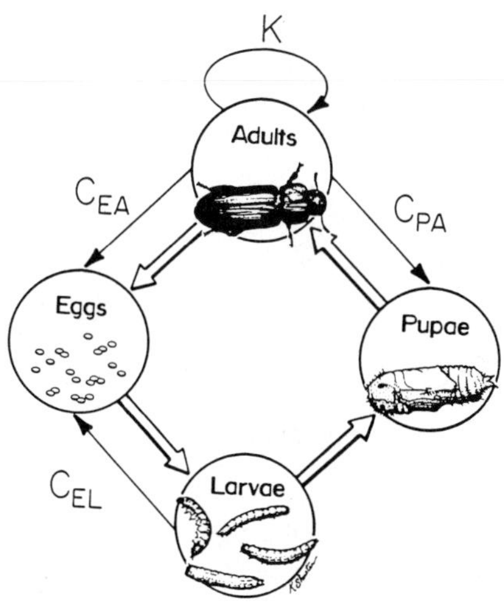

Fig. 2.1. The *Tribolium* life cycle. The wide arrows indicate transitions from one life stage to the next and the narrow arrows indicate the density-dependent interactions among the life stages. K is the rate at which adult crowding reduces fecundity and C_{IJ} represents the rate at which life stage J cannibalizes life stage I. [Modified from R. A. Desharnais and R. F. Costantino (1980), Genetic analysis of a population of *Tribolium*. VII. Stablility. Response to genetic and demographic perturbations. *Can. J. Genet. Cytol.* **22**: 577–589.]

$$N_S(t) = \int_S n(x,t)\,\mathrm{d}x, \qquad S \in \{E, L, P, A\}\ . \tag{2.10}$$

In the Leslie matrix model, the integrals are replaced by summations.

2.2.2 Mortality

In the general continuous and discrete formulations, mortality rates are considered arbitrary time-dependent functions. We will assume these rates can be partitioned into time-independent and time-dependent components. We will refer to the time-independent component as *natural mortality*. Under constant environmental conditions, time-dependent mortality occurs mainly through cannibalism. Thus mortality rates are *density-dependent* – they vary with the density and age distribution of the population.

As we discussed in the previous chapter, cannibalism is well-documented in *Tribolium* populations. Park *et al.* (1965) have shown that under crowded conditions adults and larvae cannibalize eggs and adults cannibalize pupae. In some genetic strains, larvae also cannibalize pupae, but the magnitude of this interaction seems less important. In light of these behaviors, we will make the following simplifying assumptions: (i) larvae and adults cannibalize eggs at rates independent of the ages of the eggs; (ii) adults cannibalize pupae at rates independent of the ages of the pupae; and (iii) larval and adult mortality is density-independent. Although other life stage interactions may affect mortality, under most experimental conditions, assumptions (i)–(iii) capture the most significant factors determining individual survival.

Let $\mu_0(x)$ represent the rate of natural mortality at age x. The rate at which a beetle of age y in life stage J eats individuals in life stage I will be denoted as $c_{IJ}(y)$. From assumptions (i)–(iii) above, we have

$$\mu(x,t) = \begin{cases} \mu_0(x) + \int_L c_{EL}(y) n(y,t)\,\mathrm{d}y + \int_A c_{EA}(y) n(y,t)\,\mathrm{d}y\ , & x \in E\ , \\ \mu_0(x)\ , & x \in L\ , \\ \mu_0(x) + \int_A c_{PA}(y) n(y,t)\,\mathrm{d}y\ , & x \in P\ , \\ \mu_0(x)\ , & x \in A\ . \end{cases} \tag{2.11}$$

As usual, the discrete case is handled by replacing the integrals with summations. These cannibalistic sources of mortality are diagrammed in Fig. 2.1.

2.2.3 Fecundity

As with mortality, fecundity rates are also dependent on beetle age and population density. After making adjustments for cannibalism, Rich (1956) has shown that the fecundity of females decreases with increases in the density of adults. Assuming once again that these effects can be partitioned into density-independent and density-dependent components, we let $b_0(x)$ represent the density-independent fecundity rate of a beetle of age x and $k(x,y)$ represent the rate at which a beetle of age y reduces the fecundity of a beetle of age x. If this "crowding" effect is proportional to density, then the time-dependent birth rate is given by

$$b(x,t) = b_0(x)\exp\left[-\int_0^{\omega} k(x,y)n(y,t)\,\mathrm{d}y\right]. \tag{2.12}$$

With this equation, all life stages can contribute to the crowding effect, although only adult crowding has been studied in detail. The effect of adults on fecundity is represented by the "self-loop" for adults in Fig. 2.1. The discrete case uses a summation in (2.12) in place of the integral.

This equation completes the specialization of the McKendrick-von Foerster model to *Tribolium*. Other life stage interactions, such as the cannibalism of pupae by larvae or larval-larval cannibalism, could be included by a modification of these equations.

2.2.4 Equilibrium

For the *Tribolium* model described above, we define a *non-trivial equilibrium age structure* as any combination of a nonzero density and age distribution that remains constant over time. Let $n^*(x)$ represent an equilibrium age structure and let the functions $b^*(x)$ and $\mu^*(x)$ denote the age-dependent birth and mortality rates at this equilibrium. From (2.3), (2.4) and (2.6), an equilibrium is any nonnegative function which satisfies

$$\begin{aligned} n^*(x) &= n^*(0)\exp\left[-\int_0^x \mu^*(s)\,\mathrm{d}s\right], \\ n^*(0) &= \int_A n^*(x)b^*(x)\,\mathrm{d}x\,. \end{aligned} \tag{2.13}$$

Under what conditions does the model possess a non-trivial equilibrium? Can the model possess multiple non-trivial equilibria? Answers to these questions are given by the following result.

Let $b_0(x)$ and $\mu_0(x)$ denote the density-independent birth and mortality rates, respectively. Define the density-independent net reproductive rate as

$$R_0 = \int_A b_0(x) l_0(x)\, dx\ , \tag{2.14}$$

where

$$l_0(x) = \exp\left[-\int_0^x \mu_0(y)\, dy\right] . \tag{2.15}$$

A nontrivial equilibrium age structure $n^*(x)$ exists if and only if $R_0 > 1$. If a nontrivial equilibrium exists, it is unique.

If all density-dependence were removed from the model, one would obtain the classic linear McKendrick-von Foerster equations with birth rates given by $b_0(x)$ and mortality rates given by $\mu_0(x)$. In this case, the net reproductive rate R_0 gives the total expected number of offspring per individual under density-independent conditions. The function $l_0(x)$ gives the probability of surviving from age zero to age x. Without density dependence, the population can either grow exponentially, remain constant, or decay to extinction. If the birth and mortality rates allow growth ($R_0 > 1$), then the above result guarantees the existence of a unique non-trivial equilibrium.

Desharnais and Liu (1987) derived an equivalent result for the Leslie matrix version of the model. The proof for continuous time is presented in an appendix to this chapter. In this proof we show that the problem of computing $n^*(x)$ can be reduced to the problem of solving for the unique real root of a simple nonlinear integral equation.

2.3 Parameterization

In order to explore the dynamics of the model and to make comparisons with experimental data, the functions describing fecundity, mortality, cannibalism, and crowding must be specified. A commonly used method is to *parameterize* the functions so that only a small number of quantities need to be estimated. In this section, our strategy has been to simplify the model by keeping the number of parameters as small as possible while retaining the significant aspects of *Tribolium* physiology and behavior.

Later in this chapter, computer simulations will be made to compare the model to experimental data. These data are from the corn oil sensitive (*cos*) strain of *Tribolium castaneum* (Desharnais and Costantino 1980, Desharnais and Liu 1987). Details of the biology of the *cos* mutant are given by Costantino *et al.* (1967, 1968), Costantino *et al.* (1970), Costantino and Rowe (1972), Scully and Costantino (1975), and Moffa and Costantino (1977). Wherever possible, parameter estimates were obtained which reflect the biology of this particular genetic strain of beetles under the culture conditions of the experimental data. However, in some instances, it was necessary to use data from different genetic strains and species of *Tribolium*. Therefore, the estimated parameters must be considered crude approximations.

2.3.1 Life Cycle

The duration of each life stage will be fixed at values which are consistent with the biology of the *cos* strain of *Tribolium castaneum.* Using a basic time unit of one day we have $E = [0,3)$, $L = [3,23)$, $P = [23,26)$, and $A = [26,\omega=300)$. For the durations of each life stage, we have $D_E = 3$, $D_L = 20$, $D_P = 3$, $D_A = 274$. We emphasize that these values are particular to the *cos* mutant under culture conditions of approximately 33°C and 56% relative humidity.

2.3.2 Fecundity

The fecundity of *Tribolium* females tends to peak early and decrease with age. The simplest assumption is a linear rate of decrease, so we use

$$b_0(x) = \begin{cases} \alpha - \beta(x - D_I), & x \in A, \quad x \leq D_I + \alpha / \beta, \\ \\ 0, & \text{elsewhere,} \end{cases} \tag{2.16}$$

where α is the peak fecundity rate, β is the rate of decrease in fecundity with age, and $D_I = D_E + D_L + D_P$ is total duration of the immature life stages. In this parameterization, we assume that newly emerged adults of age D_I have a maximum fecundity of α which then decreases linearly with slope β until it reaches a value of zero at age $D_I + \alpha/\beta$. Thereafter, fecundity remains at zero. With (2.16), we use two parameters to describe the entire reproductive schedule of adults.

In Fig. 2.2a we have plotted the data of Moffa (1976, p. 51) for the *cos* strain of *Tribolium castaneum.* Since *Tribolium* populations generally maintain a 1:1 sex ratio, we express fecundity rates as one-half the number of eggs laid per female per day. Parameters α and β were estimated using linear least-squares regression. The regression line of Fig. 2.2a is highly significant ($P < 0.001$). The estimates of α and β and their standard errors are given in Table 2.1. For these data, the linear assumption is a good approximation.

The effects of adult crowding on fecundity are described using the function $k(x,y)$ in (2.12). The simplest approximation is to assume that all adults contribute equally to this effect and that all females are equally sensitive to crowding. In this case,

$$k(x,y) = \begin{cases} K, & x, y \in A, \\ \\ 0, & \text{elsewhere,} \end{cases} \tag{2.17}$$

where K is a nonnegative constant. This assumes that only adults contribute to the effects of crowding on fecundity. Substituion of (2.17) into (2.12) yields the prediction that the logarithm of fecundity rates will decrease linearly with increases in total adult density.

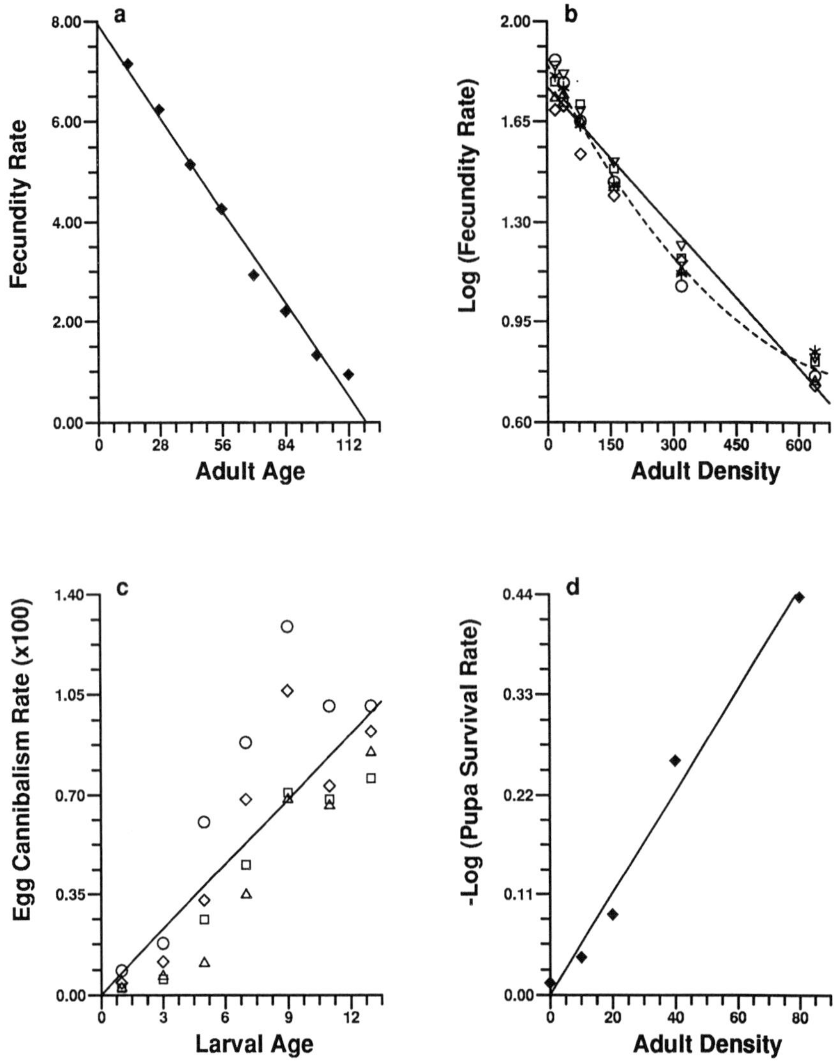

Fig. 2.2. Data used to estimate several of the parameters of the *Tribolium* age-structured model. See text for details. [From R. A. Desharnais and L. Liu (1987), Stable demographic limit cycles in laboratory populations of *Tribolium castaneum*. *J. An. Ecol.* **56**: 885–906.]

The data of Rich (1956) on *T. confusum* can be used to obtain a rough estimate of the magnitude of the crowding effect. Rich confined adults and color-marked eggs at various adult densities and for various durations of exposure. The destruction of marked eggs allowed estimates of egg-eating rates and counts of unmarked eggs allowed estimates of female fecundity. With this design, fecundity rates could be corrected for

Table 2.1. Parameter estimates for the *Tribolium* model

Parameter	Description	Estimated Value (± s.e.)	Source of Data
μ_L	Larval mortality	0.0251 ± 0.0020	Present study
μ_A	Adult mortality	0.0130 ± 0.0009	Desharnais and Costantino (1980)
α	Maximum fecundity	7.96 ± 0.20	Moffa (1976)
β	Decrease in fecundity with age	0.0664 ± 0.0028	Moffa (1976)
K	Decrease in fecundity from crowding	0.00164 ± 0.00006	Rich (1956)
C_{EL}'	Larval cannibalism of eggs (slope)	0.000760 ± 0.000047	Park *et al.* (1965)
C_{EA}	Adult cannibalism of eggs	0.00252 ± 0.00016	Rich (1956)
C_{PA}	Adult cannibalism of pupa	0.00558 ± 0.00026	Jillson and Costantino (1980)

cannibalism. Using the fecundity data of Rich (1956, Table IV), we have plotted the natural logarithm of the daily fecundity rate as a function of adult density in Fig. 2.2b. So that all parameter estimates are commensurate with the population experiments to be described later, we express density in numbers per 20g of media. In Fig. 2.2b, the different symbols represent different durations of exposure. Pooling these data, we fit linear (solid line) and quadratic (broken line) functions by the method of least squares. Although both functions fit well ($P<0.001$), a test for lack of fit (Searle 1971, p. 103) suggests that the linear model is inadequate (lack of fit significant at $P<0.001$). Nevertheless, to preserve the simplicity of expression (2.17), we used the linear regression. Parameter K was estimated as the absolute value of the slope of the solid line in Fig. 2.2b. The estimated value and its standard error appear in Table 2.1.

2.3.3 Natural Mortality

We now consider survivorship in the absence of cannibalism. As a first approximation, we assume a constant mortality rate within each age class. Let μ_S denote the density-independent mortality rate for life stage S, where $S \in \{E, L, P, A\}$. As a general rule, under laboratory culture conditions, natural mortality rates are low among the immature life stages. For eggs and pupae, we set $\mu_E = \mu_P = 0$. However, for the *cos* strain of *T. castaneum*, larval viability is decreased when the media contains unsaturated fatty acids (Scully and Costantino 1975). To estimate larval survivorship, we used unpublished data from a previous study (Desharnais and Costantino 1980). We isolated adults from each of four of our control populations and placed them on standard media (95% flour and 5% brewer's yeast by weight) for 24 hours. A total of 250 eggs were collected from each group of adults and placed on corn oil media (90% flour, 5% yeast, and 5% corn oil by weight). Larvae were recovered 16 days later. For each batch of eggs, we used the natural logarithm of the survivorship divided by the duration of time spent as a larva

(13 days = 16 days – 3 days as eggs) as a crude estimate of the larval mortality rate. The average of the four mortality rates is our estimate of μ_L. This average and its standard error appear in Table 2.1. For adult mortality, we used the census data from the four control populations of Desharnais and Costantino (1980). Since the populations were censused every two weeks, we divided the number of dead adults at week $n+2$ by the average number of live adults at weeks n and $n+2$ for $n=0,2,\ldots,36$. We subtracted these ratios from one to get survivorships, took their natural logarithms, and divided by the census period (14 days) to obtain mortality rates. The average of these 76 observations is our estimate μ_A (Table 2.1). To check for density dependence, we plotted these mortality rates against the corresponding number of live adults. No functional realtionship was noticeable ($r=-0.055$, $P>0.6$).

In summary, we use in (2.11)

$$\mu_0(x) = \begin{cases} 0\,, & x \in E\,, \\ \mu_L\,, & x \in L\,, \\ 0\,, & x \in P\,, \\ \mu_A\,, & x \in A\,, \end{cases} \tag{2.18}$$

where the parameters μ_L and μ_A are constant natural mortality rates for larvae and adults, respectively.

2.3.4 Cannibalism

Accurate estimates of the cannibalism rates in *Tribolium* are difficult to obtain. Our estimates come from *assay* experiments in which fixed numbers of cannibals and victims are confined together in a small volume of media over a short period of time, and the survivorship of the vicitms is measured at the end of the experiment. While this type of experiment ignores the potentially complex interactions that might occur when all life stages are living together, it still provides a quantitative estimate of cannibalistic voracity which can be used to investigate the consequences of this behavior on population dynamics.

We estimated the rates of egg cannibalism by larvae. From hatching to pupation, larvae increase in size and in the voracity at which they eat eggs. Park et al. (1965) examined the age specific rates at which larvae of several genetic strains eat eggs. In their Table 10, they report the percentage of 100 eggs of *T. castaneum* eaten by groups of fifty larvae of *T. castaneum* after forty-eight hours in eight grams of media. To convert to cannibalism rates per larva per day for twenty grams of media, we subtracted the percentage eaten from one, took the natural logarithm, and divided by 250 ([50 larvae / 8 grams] × [2 days] × [20 grams]). These rates are plotted as a function of larval age in Fig. 2.2c. The different symbols represent the four genetic strains. As an approximation, we assumed that egg cannibalism rates by larvae increase linearly from zero as larvae get older. Let C_{EL}' denote the slope of this relationship. Our estimate of C_{EL}' is the slope of the regression line in Fig. 2.2c, which is statistically significant ($P<0.001$). This estimate and its standard error can be found in Table 2.1. Although a test for lack of fit suggests that a nonlinear curve would not reduce the sums of squares significantly ($P > 0.09$), a visual inspection of the data in Fig. 2.2c suggests that a sigmoidal curve may be more appropriate as a general description of the relationship between larval age and cannibalism rates. Our assumption of a linear rate of increase gives

$$c_{EL}(y) = C_{EL}'(y - D_E), \qquad y \in L, \tag{2.19}$$

for the age-specific cannibalism rate $c_{EL}(y)$ in (2.11). Recall that D_E is the duration of the egg stage.

For adults eating eggs, we assumed a common cannibalism rate for all adult ages: $c_{EA}(y) = C_{EA}$ for $y \in A$. In his study of egg cannibalism by adults of *T. confusum*, Rich (1956, Table IV) computed 24 hour per capita rates for eight grams of media. Although his rates vary with adult density and duration of exposure to cannibalism, we scaled his estimates to correspond to twenty grams of media, and pooled these values to obtain an overall average. This estimate of C_{EA} and its standard error appear in Table 2.1.

Finally, we turn to cannibalism of pupae by adults. Again we have assumed no age dependence among adults. For (2.11), we let $c_{EP}(y) = C_{EP}$ for $y \in A$. Using the *cos* strain of *T. castaneum*, Jillson and Costantino (1980) exposed groups of 26 larvae to varying densities of adults on 26 grams of standard media. In Fig. 2.2d we have plotted the negative logarithm of the survival rate through the duration of the pupal stage (−log[proportion surviving]/3 days as pupae) against the adult density per 20 grams of media. A least-squares regression confirms a significant linear relationship ($P < 0.001$) which goes through the origin. Our estimate of C_{PA} and its standard error appear in Table 2.1.

To summarize, we have the following parameterization the age-specific birth and mortality rates $b(x,t)$ and $\mu(x,t)$:

$$b(x,t) = \begin{cases} [\alpha - \beta(x - D_I)] \exp(-KN_A(t)), & x \in A, \quad x \le D_I + \alpha/\beta, \\ 0, & \text{elsewhere}, \end{cases} \tag{2.20}$$

$$\mu(x,t) = \begin{cases} C_{EL}' \int_L (y - D_E) n(y,t)\, dy + C_{EA} N_A(t), & x \in E, \\ \mu_L, & x \in L, \\ C_{PA} N_A(t), & x \in P, \\ \mu_A, & x \in A. \end{cases} \tag{2.21}$$

The parameter estimates are given in Table 2.1. Figure 2.1 shows the association between the parameters of density dependence and the life stage interactions.

2.4 Experimental Data and Computer Simulations

In the analysis that follows, we compare the behavior of our model to the experimental data of Desharnais and Costantino (1980). They assessed demographic stability in laboratory populations of the *cos* strain of *T. castaneum* by deliberately perturbing populations which were near equilibrium. We begin with a brief outline of the experimental design.

2.4.1 Experimental Design

Thirteen cultures were initiated with 64 young adults, 16 pupae, 20 large larvae, and 70 small larvae. All beetles were homozygous for the *cos/cos* allele. Each population was contained in a one-half pint milk bottle with 20 grams of corn oil media (90% wheat flour, 5% brewer's yeast, and 5% liquid corn oil) and kept in an unlighted incubator at 33 ± 1°C and 56 ± 11% relative humidity. Every two weeks all age classes, except eggs, were censused and all age classes, including eggs, were placed in fresh media. This procedure was followed for 38 weeks.

After ten weeks of culture, each population was randomly assigned to one of four treatments. Three of the treatments, each with three replicates, involved demographic perturbations and one treatment, with four replicates, served as a control. For the demographic perturbations we (i) added 100 adults, (ii) removed all adults, and (iii) removed all immatures. The control cultures were not disturbed.

The census data appear below. We have combined the data into three age groups: adults, large larvae plus pupae, and small larvae. Since the populations exhibited oscillations which are sometimes out of phase, we present the data for individual replicates. When viewing these data, it should be kept in mind that the census period is coarse in relation to the period of the oscillations predicted by the model. We discuss this problem in more detail below.

2.4.2 Discretization

With the parameter estimates for our nonlinear age-structured model of the dynamics of *Tribolium* populations, we are now in a position to examine the dynamics of the model and compare these results to demographic data. Unfortunately, the model is sufficiently complex to make a complete analytical treatment unfeasible at this time. Instead, we shall rely on computer simulations. For numerical computations, the discrete Leslie matrix model is much simpler to use, so we begin with a discretization of the continuous model.

The discrete Leslie model is given by (2.7) and (2.8). To complete the model, we need only specify the "birth rates" $b(x,t)$, $x=1,2,\ldots,\omega$, and survivorships $s(x,t)$, $x=1,2,\ldots,\omega-1$. Recall that in the discrete case, the ages for the life stages are defined by sets of consecutive integers, E, L, P, and A; the duration D_S is the number of elements in set S; and the total density of life stage S is given by

$$N_S(t) = \sum_{x \in S} n(x,t), \qquad S \in \{E, L, P, A\}. \tag{2.22}$$

The birth rates are given by the discrete-time equivalent of equation (2.20),

$$b(x,t) = \begin{cases} [\alpha - \beta(x - D_I)] \exp(-KN_A(t)), & x \in A, \quad x \le D_I + \text{int}(\alpha/\beta), \\ 0, & \text{elsewhere}, \end{cases} \tag{2.23}$$

where $D_I = D_E + D_L + D_P$ and "int" is the integer function. For survivorship, we use the approximation $s(x,t) = \exp[-\mu(x,t)]$. From (2.21) we have

$$s(x,t) = \begin{cases} \exp\left(-C_{EL}' \sum_{y \in L} (y - D_E) n(y,t) - C_{EA} N_A(t)\right), & x \in E, \\ \exp(-\mu_L), & x \in L, \\ \exp(-C_{PA} N_A(t)), & x \in P, \\ \exp(-\mu_A), & x \in A, \quad x \ne \omega. \end{cases} \tag{2.24}$$

Using (2.23) and (2.24) in $\mathbf{M}(t)$, the age vector is obtained by iterating $\mathbf{n}(t+1) = \mathbf{M}(t)\mathbf{n}(t)$.

2.4.3 Equilibrium and Stability

The existence of a unique nontrivial equilibrium age structure can be verified by using the parameter estimates of Table 2.1 in (2.16) and (2.18) and computing the density-independent net reproductive rate R_0 of (2.14). For the continuous model, we obtained $R_0 = 183.1$, which is half the expected number of eggs that an adult female beetle will produce over its lifetime. The density-independent Mathusian growth rate r is the solution to

$$\int_A e^{-rx} b_0(x) l_0(x)\, dx = 1, \tag{2.25}$$

where $l_0(x)$ is given by (2.15). We obtained $r = 0.132$ per day, which compares favorably with other esitmates for *T. castaneum* under similar conditions of temperature and humidity (Sokoloff 1974, Table 11.22). Using the same parameter estimates in the discrete matrix model gives $R_0 = 185.5$ and $r = 0.131$ (Desharnais and Liu 1987). Since $R_0 > 1$, or equivalently $r > 0$, the existence and uniqueness of $n^*(x)$ is guaranteed.

We computed the equilibrium $n^*(x)$ using the parameter estimates and equations (2.32)–(2.41) of the appendix. To compare this result with the experimental data, we divided the larval interval ($3 \le x < 23$) into two intervals of equal size: small larvae ($3 \le x < 13$) and large larvae ($13 \le x < 23$). By life stage, the model predicts an equilibrium of 89.36 adults, 68.94 large larvae plus pupae, and 77.87 small larvae. The equivalent results for the Leslie matrix model are 87.87 adults, 68.53 large larvae plus pupae, and 75.15 small larvae (Desharnais and Liu 1987). These values seem reasonable, given the demographic data presented below.

To evaluate stability, we linearized the discrete Leslie model (2.7) in the neighborhood of the equilibrium vector $\mathbf{n}^*$. If we let $\Delta(t) = \mathbf{n}(t) - \mathbf{n}^*$ represent the deviations from equilibrium, substitution into (2.7) gives

$$\Delta(t+1) = \mathbf{Z}\Delta(t) + O(|\Delta|^2), \qquad (2.26)$$

where

$$\mathbf{Z} = \{z(x,y)\} = \left\{ \frac{\partial n(x,t+1)}{\partial n(y,t)} \right\}_{\mathbf{n}(t)=\mathbf{n}^*}, \qquad x,y = 1,2,\ldots,\omega. \qquad (2.27)$$

The $\omega \times \omega$ matrix $\mathbf{Z}$ gives the coefficients of the linearized model. The first row of $\mathbf{Z}$ is given by

$$z(1,y) = b^*(y) - \sum_{v=1}^{\omega} k(v,y)n^*(v)b^*(v), \qquad (2.28)$$

where

$$b^*(y) = b_0(y)\exp\left[-\sum_{v=1}^{\omega} k(y,v)n^*(v) \right],$$

and the remaining elements are given by

$$z(x+1,y) = \begin{cases} n^*(x+1)/n^*(x), & x = y,\ x \neq \omega, \\ -c_{EL}(y)n^*(x+1), & x \in E,\ y \in L, \\ -c_{EA}(y)n^*(x+1), & x \in E,\ y \in A, \\ -c_{PA}(y)n^*(x+1), & x \in P,\ y \in A, \\ 0, & \text{elsewhere}. \end{cases} \qquad (2.29)$$

The nontrivial equilibrium $\mathbf{n}^*$ is asymptotically locally stable if and only if the modulus of the dominant eigenvalue of $\mathbf{Z}$ is less than one.

To evaluate stability, we used $\mathbf{n}^*$ in (2.28) and (2.29) to compute the elements of the matrix $\mathbf{Z}$. The eigenvalues of this 300×300 matrix were kindly computed by Jane Cullum at the IBM Research Center in Yorktown Heights, New York, using a new algorithm for large sparse nonsymmetric matrices (Cullum and Willoughby 1986). Using $i = \sqrt{-1}$, we report the following observations: (i) all the eigenvalues are distinct; (ii) all the eigenvalues are complex conjugates; (iii) the dominant pair of eigenvalues are $\lambda_1, \lambda_2 = 1.0001 \pm 0.2277i$; (iv) the largest subdominant pair of eigenvalues are $\lambda_3, \lambda_4 = 0.9872 \pm 0.0197i$; (v) $\text{mod}(\lambda_1) > 1 > \text{mod}(\lambda_3)$, that is, only the dominant pair of eigenvalues has a modulus that exceeds unity. Since $\text{mod}(\lambda_1) = 1.0257$, we predict that $\mathbf{n}^*$ for the *Tribolium* populations is an *unstable* equilibrium. Finally, since λ_1 and λ_2 dominate the long-term behavior of the linearized model (2.26), we can use these eigenvalues to obtain an estimate of the period of the cycles near $\mathbf{n}^*$ (Gurney and Nisbet 1985). Letting $\log(\lambda_1) = a + bi$, the period is approximately $p = \pi/b$. In this case, $\log(\lambda_1) = 0.0254 + 0.2234i$ and $p = 28.06$ days. This is close to the generation time of 27 days and compares favorably with the period of 26.76 days obtained from a spectral analysis of the simulation results. Details of the spectral analysis are presented in §2.4.6.

2.4.4 Demographic Data and Simulations

We now examine the census data of Desharnais and Costantino (1980). The adult densities for the four control populations are plotted in Figs. 2.3a–d. The horizontal dashed line in each figure is the expected equilibrium number of adults. We iterated the model (2.7) using the same initial conditions as the experimental populations, with a uniform distribution of individuals within the small larval, large larval, and pupal age groups, and with all adults at 27 days of age. The adult densities from the simulation are presented in Fig. 2.3e. Although the experimental data show some fluctuations in adult density, there are no regular small amplitude oscillations as in the simulation results. From the observed differences between replicates, it seems possible that stochastic sources of variation overwhelm these oscillations. It is also possible that the census period is too coarse to detect small amplitude oscillations. In any case, the overall densities of the experimental populations agree quite well with the simulation results. In particular, it is interesting that the adult densities fluctuate at levels which *exceed* the predicted equilibrium value (dashed line). This suggests that a time average of adult density would be poor predictor of the equilibrium in an unstable system.

The densities of the immature age classes for the control populations are plotted in Figs. 2.4a–d. The solid curves represent the small larvae and the broken curves are for the large larvae plus pupae. There is a definite regularity to the fluctuations of the immature age classes in the experimental populations. These oscillations are very large in magnitude, sometimes going from near zero in one census to over 400 individuals in the next. As with the adult data, irregularities in the observed oscillations could be due in part to stochastic effects. For example, the replicates in Figs. 2.4a and 2.4b seem to be in phase in the first half of the experiment, but out of phase at the end of the experiment,

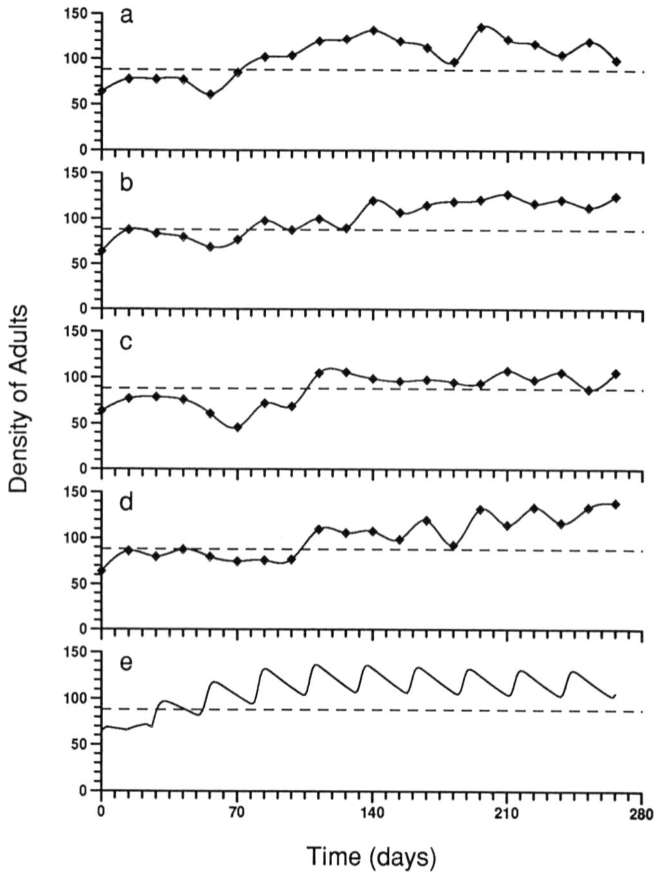

Fig. 2.3. Density of adults versus time for the four control populations (a–d) and the matrix version of the *Tribolium* model (e). The census points (diamonds) were interpolated using cubic splines. The horizontal dashed lines represent the predicted equilibrium density of adults. [From R. A. Desharnais and L. Liu (1987), Stable demographic limit cycles in laboratory populations of *Tribolium castaneum*. *J. An. Ecol.* **56**: 885–906.]

suggesting that the periodicity may drift. However, with oscillations of this magnitude and period, another potential source of irregularities is the discrete nature of the censusing procedure. The observations are spaced at roughly half the period. Under these conditions, it is possible for two or more adjacent observations to capture densities between the peaks and troughs, giving the illusion of little or no fluctuation.

The simulation results appear in Fig. 2.4e. Qualitatively, there is a good agreement between the model's behavior and the dynamics of the *Tribolium* populations. The amplitudes of the oscillations and their period are roughly equal. The oscillations of the small larvae are also out of phase with the oscillations of large larvae plus pupae. It is not possible to directly compare the shapes of the oscillations, since the curves for the

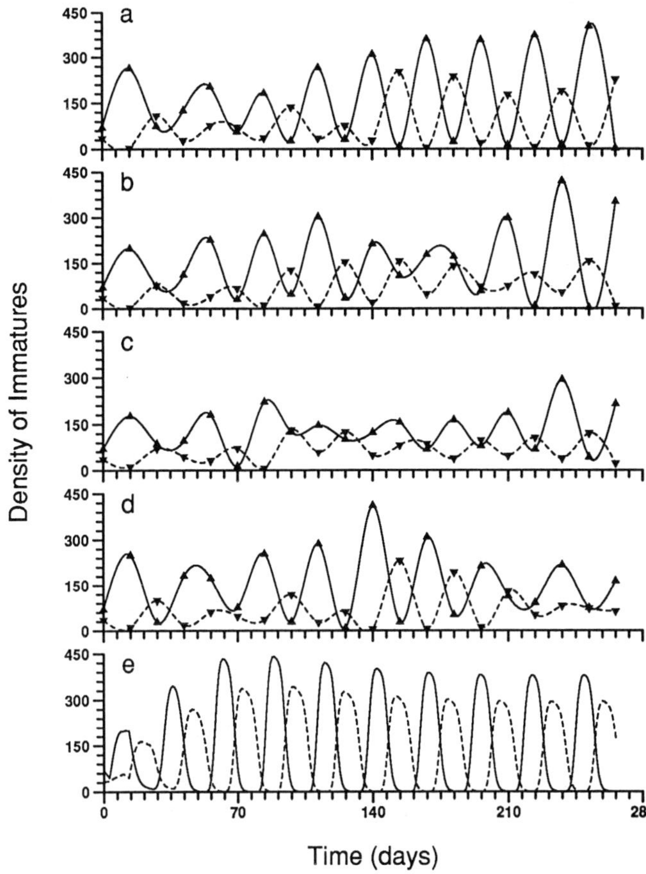

Fig. 2.4. Density of small larvae (solid curves) and large larvae plus pupae (broken curves) versus time for the four control populations (a–d) and the matrix version of the *Tribolium* model (e). The census points (triangles) were interpolated using cubic splines. [From R. A. Desharnais and L. Liu (1987), Stable demographic limit cycles in laboratory populations of *Tribolium castaneum*. *J. An. Ecol.* **56**: 885–906.]

experimental data are determined by cubic splines. However, when the simulation data is periodically sampled every fourteen days and these data are interpolated with splines, there is a drift in amplitude of the peaks and troughs, indicating that the periodicity of the simulation data is not an integral multiple of the census period. A precise estimate of the periodicity of the simulation results is presented in §2.4.6.

2.4.5 Perturbations

We now consider the demographic perturbations. In three relicates composing one treatment, 100 adults were added at day 70. The adult densities for these populations

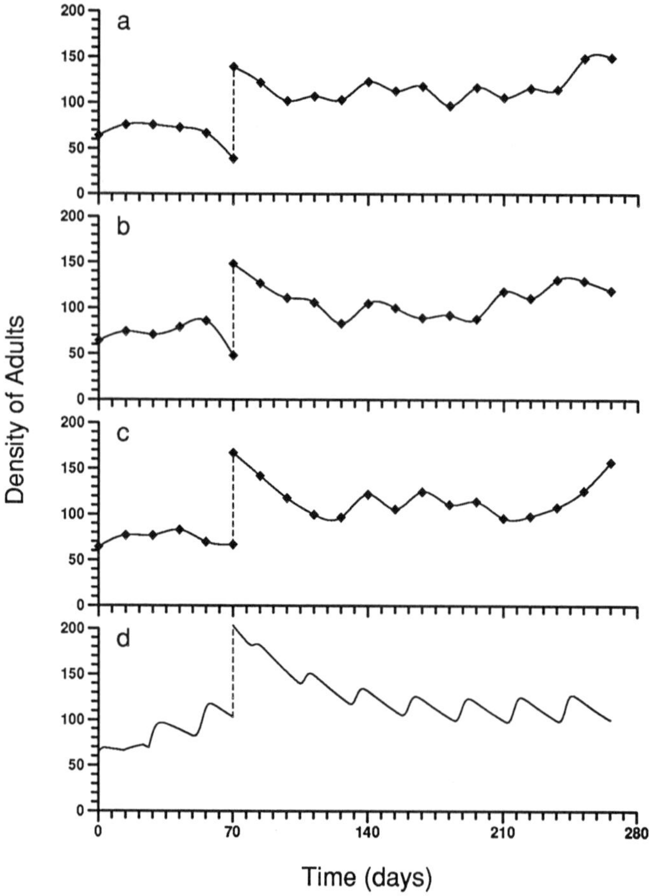

Fig. 2.5. Density of adults versus time for the three experimental populations (a–c) and the matrix version of the *Tribolium* model (d) where 100 adults were added at day 70. The census points (diamonds) were interpolated using cubic splines. The vertical dashed lines represent the perturbation. [From R. A. Desharnais and L. Liu (1987), Stable demographic limit cycles in laboratory populations of *Tribolium castaneum*. *J. An. Ecol.* **56**: 885–906.]

are presented in Figs. 2.5a–c. The vertical dashed line represents the perturbation. After the perturbation,adult numbers decreased to a density level which is comparable to the controls. In Fig. 2.5d we show the results for the model (2.7), which was subjected to the same perturbation. The rate of return of the adult densities to the unperturbed levels is approximately the same for the matrix model and the experimental populations. In the census periods following the perturbation the densities of the immature stages continue to oscillate in a manner similar to the control populations (Desharnais and Liu 1987, Table 2).

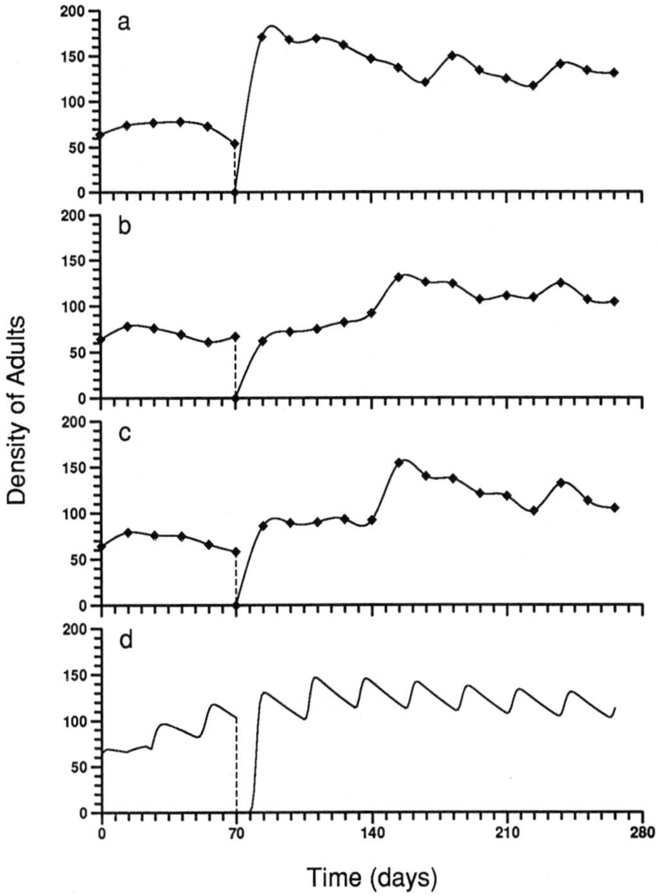

Fig. 2.6. Density of adults versus time for the three experimental populations (a–c) and the matrix version of the *Tribolium* model (d) where all adults were removed at day 70. The census points (diamonds) were interpolated using cubic splines. The vertical dashed lines represent the perturbation. [From R. A. Desharnais and L. Liu (1987), Stable demographic limit cycles in laboratory populations of *Tribolium castaneum*. *J. An. Ecol.* **56**: 885–906.]

In Figs. 2.6a–c, we present the results for the three populations in which all adults were removed at day seventy. In two of the populations (Figs. 2.6b,c), there was a rapid return of adult density to the unperturbed level. This same pattern is predicted by the model (Fig. 2.6d). The remaining experimental culture overshoots the unperturbed density level and then gradually decreases to the same level as the controls (Fig. 2.6a). An examination of the census data reveals that there was a large cohort of potential recruits (182 large larvae plus pupae) in this population at the time that the adults were removed. In all three experimental populations, the oscillations of the immature life stages persist after the perturbations (Desharnais and Liu 1987, Table 2).

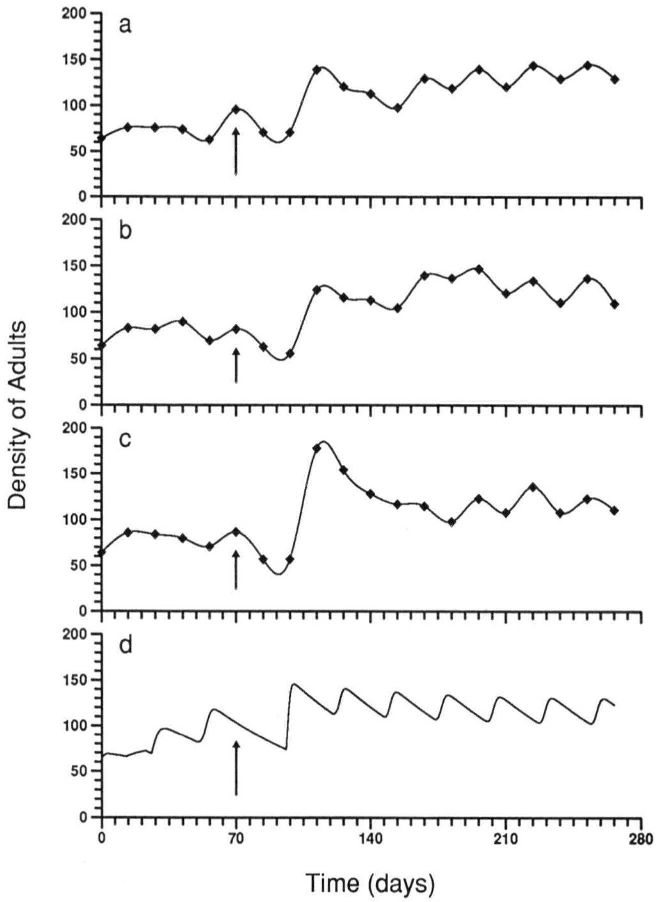

Fig. 2.7. Density of adults versus time for the three experimental populations (a–c) and the matrix version of the *Tribolium* model (d) where all immatures were removed at day seventy. The census points (diamonds) were interpolated using cubic splines. The vertical arrows represent the perturbation. [From R. A. Desharnais and L. Liu (1987), Stable demographic limit cycles in laboratory populations of *Tribolium castaneum*. *J. An. Ecol.* **56**: 885–906.]

In the last treatment, all immatures were removed at day 70. The adult densities for these populations are presented in Fig. 2.7. As expected, there was a decline in adult density following the perturbation. In two of the cultures (Figs. 2.7a,b), this decline was followed by a return to the adult densities of the controls. The model displayed similar behavior (Fig. 2.7d). In the population of Fig. 2.7c, a small overshoot followed the decline in adult density before a return to the unperturbed level. In this treatment, the removal of immatures may have tended to temporarily synchronize the immature cycles with the census periods. This could explain the increased regularity of the fluctuations in adult density.

2.4.6 Stable Limit Cycle

The local stability analysis and perturbation results suggest that these *Tribolium* populations possess a stable oscillatory attractor. To visualize this idea, we iterated the model (2.7) using an initial condition which was very close to equilibrium and plotted the proportions of adults, large larvae plus pupae and small larvae in Fig. 2.8. The circle in the upper right indicates the counterclockwise population trajectory as it spirals away from the equilibrium. We tried several other initial conditions, and the trajectories always approach the same loop as the simulation in Fig. 2.8. It appears that the asymptotic behavior of the model can be characterized as a stable demographic limit cycle.

We conducted a spectral analysis to ascertain the period of the limit cycle. The model was iterated 50,000 times to guarantee asymptotic behavior. The results from the subsequent $T = 1{,}001$ iterations were used to examine the periodicity. Let $N_A(1), N_A(2), \ldots, N_A(T)$ represent the sequence of adult densities. Define

$$d(t) = N_A(t) - T^{-1} \sum_{t=1}^{T} N_A \qquad \text{and} \qquad v = \sum_{t=1}^{T} d(t)^2 .$$

The quantities $a(\tau)$ and $b(\tau)$ can be thought of as "correlations" of $d(t)$ with $\cos(\tau t)$ and $\sin(\tau t)$, respectively, where

$$\begin{aligned} a(\tau) &= (vT)^{-1/2} \sum_{t=1}^{T} d(t) \cos(\tau t) , \\ b(\tau) &= (vT)^{-1/2} \sum_{t=1}^{T} d(t) \sin(\tau t) . \end{aligned} \tag{2.30}$$

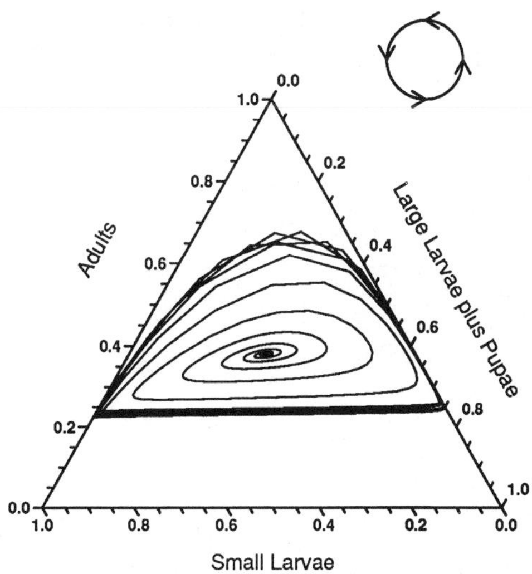

Fig. 2.8. Triangular coordinate plot of the proportion of adults, large larvae plus pupae, and small larvae predicted by the matrix version of the *Tribolium* model when the initial condition is near the unstable equilibrium. The circle in the upper right indicates the direction of change over time. [From R. A. Desharnais and L. Liu (1987), Stable demographic limit cycles in laboratory populations of *Tribolium castaneum*. *J. An. Ecol.* **56**: 885–906.]

The spectral density at angular frequency τ is given by

$$\Omega(\tau) = a(\tau)^2 + b(\tau)^2 \tag{2.31}$$

(Kendall and Stuart 1968, pp. 410–412). The angular frequency τ is related to the period p by $p=(2\pi)/\tau$. For the adult densities, we obtained a large sharp peak in $\Omega(\tau)$ at $\tau=0.234810$ which corresponds to a period of $p=26.7586$ days. An identical result was obtained using the number of small larvae. To corroborate this estimate, we computed the number of days between adjacent maxima of adult numbers. Adult peaks occurred every 26–27 days with an average of 26.75 days. As might be expected, the periodicity of the limit cycle is close to the duration of the life cycle (27 days).

2.4.7 Bifurcation Points

How far away are these *Tribolium* populations from the stable region of parameter space? How do the life history and behavioral traits affect stability? These questions were addressed by observing the behavior of our model for various parameter values. Since the simplified version of the model still possesses eight parameters (Table 2.1), we made no attempt to map out the entire region of parameter space. Instead, we defined the estimates of Table 2.1 as the default values, and varied each parameter individually. The method of bisection was used to estimate the thresholds at which the model bifurcates from a stable point equilibrium into a stable limit cycle.

The bifurcation points for each parameter appear in Table 2.2. Oscillations persist for all possible values of β and μ_A when the remaining parameters maintain their default values. For the other parameters, we have reported the equilibrium densities at the bifurcation point and indicated whether or not the model is stabilized or destabilized when the parameter is increased above its bifurcation value.

Table 2.2. Bifurcation values for each parameter

Parameter†	Bifurcation value	Ratio to estimated value	Equilibrium densities at bifurcation‡	Effect of increasing the parameter
μ_L	0.0788	3.14	(66.0, 47.6, 92.1)	stabilize
α	0.718	0.0902	(24.1, 6.8, 7.1)	destabilize
K	0.0211	12.9	(73.0, 44.8, 48.7)	stabilize
C_{EL}'	0.000146	0.192	(132.2, 212.1, 238.1)	destabilize
C_{EA}	0.00896	3.56	(73.1, 44.9, 48.8)	stabilize
C_{PA}	0.187	33.5	(6.3, 36.5, 41.9)	stabilize

† Oscillations persist for all values of β and μ_A when the remaining parameters have their estimated values.
‡ Number of adults, large larvae plus pupae, and small larvae, respectively.

The parameters α and C_{EL}' are destabilizing and the remaining parameters are stabilizing. This suggests that increases in the reproductive rate of flour beetles has a destabilizing effect while increases in mortality rates are stabilizing. The same phenomenon is true in predator-prey systems (Rosenzweig 1971) and in discrete generation single species models (May 1974). As Chapman and Whang (1934) conjectured from some of the earliest experiments with *Tribolium*, an increase in the cannibalism rate of larvae on eggs can lead to sustained oscillations. It is interesting that the other forms of cannibalism actually have a stabilizing effect. This suggests that the relationship between cannibalism and stability depends upon the relative rates at which the active life stages (larvae and adults) consume the inactive ones (eggs and pupae). However, we can not rule out the possibility that cannibalism of eggs or pupae by adults are destabilizing for other combinations of parameter values.

In some cases, the bifucation points are not too far from the estimated values. A 3.14 factor increase in the larval mortality rate, μ_L, would be sufficient for stability. A 3.56 factor increase in the rate at which adults eat eggs, C_{EA}, would have a similar effect. Decreasing the cannibalism rate of eggs by larvae, C_{EL}', to 19.2% of its estimated value would also result in a stable equilibrium. Finally, we can reduce the total fecundity and maintain the last age of reproduction by scaling both α and β by the same factor. Using the method of bisection, we found that stability is obtained when the total fecundity is 18.4% of its default value.

2.5 Summary and Conclusions

In this chapter we presented a nonlinear age-structured model which describes the physiological and behaviorial interactions among the various life stages in laboratory populations of *Tribolium*. We showed the existence of a unique nontrivial equilibrium. The parameters of this model were estimated, and model simulations were compared to experimental data. Both the simulations and the experimental populations exhibit stable oscillations. Simulations of the model for a variety of parameter values yielded points of bifurcation from a stable point equilibrium to a stable limit cycle. A comparison of the bifurcation points to the estimated parameter values suggests that the actual populations may not be too far from the stable region of parameter space (Table 2.2).

There exists a large amount of variability in rates of reproduction, mortality, and cannibalism among the various genetic strains and species of *Tribolium* (Park, Mertz and Petrusewicz 1961; Sokoloff 1974). Environmental conditions also affect these biological attributes. This raises the interesting possibility of doing "comparative population dynamics," where one experimentally probes for transitions from stable equilibria to stable limit cycles. The numerical work of this chapter suggests that these transition points occur at parameter values which are of the same order of magnitude as those currently existing in laboratory populations. Studies of this sort would help place the ideas of Hopf bifurcations and stable limit cycles in population biology on a firmer experimental footing.

In any attempt to model a biological system, one must confront a tradeoff between analytical tractability and realism. In this chapter, we tried to retain what we feel are the most important aspects of the biology of *Tribolium* in relation to their population dynamics while keeping the model as simple as possible. Considering that our formulation of the model is based on several crude assumptions and that several of our parameter estimates come from published data for different genetic strains and different species, the agreement between the model simulations and the experimental data is quite good. The relatively small number of parameters in this model allowed us to retain some generality and to probe parameter space for insight into the dynamical properties of *Tribolium* populations. Nevertheless, our work falls far short of a complete analytical treatment. In the next chapter, we continue the discussion of *Tribolium* demographic dynamics on a more analytical level by focusing on the interaction between eggs and larvae.

2.6 Appendix

In this appendix we derive necessary and sufficient conditions for the existence and uniqueness of a nontrivial equilibrium for the continuous-time model (2.2)–(2.3) with mortality rates (2.11) and fecundity rates (2.12). The derivation follows Desharnais and Liu (1987) for the equivalent discrete-time equations.

A nontrivial equilibrium is defined as any real function $n^*(x)>0$ which satisfies (2.13). Here and throughout, a superscript * means that the function is evaluated at equilibrium.

To simplify the algebra, let $\rho(x,y)$ be the probability of surviving from age x to age y $(x \leq y)$ under density-independent sources of mortality; that is,

$$\rho(x,y) = \exp\left[-\int_x^y \mu_0(s)\,\mathrm{d}s \right], \qquad x \leq y . \tag{2.32}$$

For adults and larvae

$$n^*(x) = \begin{cases} n^*(D_E)\rho(D_E,x)\,, & x \in L\,, \\ n^*(D_I)\rho(D_I,x)\,, & x \in A\,, \end{cases} \tag{2.33}$$

where $D_I = D_E + D_L + D_P$. Using (2.32) and (2.33) in (2.13) gives

$$n^*(0) = n^*(D_I) \int_A b_0(x)\rho(D_I,x)\exp\left[-n^*(D_I)\bar{k}^*(x)\right]\mathrm{d}x\,, \tag{2.34}$$

$$n^*(x) = \begin{cases} n^*(0)\rho(0,x)\exp\{-x[n^*(D_E)\overline{c}^*_{EL} + n^*(D_I)\overline{c}^*_{EA}]\}, & x \in E, \\ n^*(D_E)\rho(D_E,x)\exp\{-(x-D_E-D_L)[n^*(D_I)\overline{c}^*_{PA}]\}, & x \in P, \end{cases} \tag{2.35}$$

for eggs and pupae, where

$$\overline{k}^*(x) = \int_A k(x,y)\rho(D_I,y)\,\mathrm{d}y, \tag{2.36}$$

$$\overline{c}^*_{EL} = \int_L c_{EL}(y)\rho(D_E,y)\,\mathrm{d}y, \tag{2.37}$$

$$\overline{c}^*_{EA} = \int_A c_{EA}(y)\rho(D_I,y)\,\mathrm{d}y, \tag{2.38}$$

$$\overline{c}^*_{PA} = \int_A c_{PA}(y)\rho(D_I,y)\,\mathrm{d}y. \tag{2.39}$$

Equations (2.36)–(2.39) give mean crowding and cannibalism rates, weighted by survivorship. From (2.35), with $x = D_I$,

$$n^*(D_E) = n^*(D_I)\rho(D_E,D_I)^{-1}\exp[D_P n^*(D_I)\overline{c}^*_{PA}]. \tag{2.40}$$

Substituting (2.34) and (2.40) into (2.35) with $x = D_E$ gives

$$\int_A b_0(x)\rho(0,x)\exp\left\{-\eta^*\left[\overline{k}^*(x) + \xi_1 + \xi_2\exp(\xi_3\eta^*)\right]\right\}\mathrm{d}x = 1, \tag{2.41}$$

where

$$\begin{aligned} \xi_1 &= D_P\overline{c}^*_{PA} + D_E\overline{c}^*_{EA}, \\ \xi_2 &= D_E\rho(D_E,D_I,)^{-1}\overline{c}^*_{EL}, \\ \xi_3 &= D_P\overline{c}^*_{PA}, \\ \eta^* &= n^*(D_I). \end{aligned}$$

Replace η^* with η consider the left side of (2.41) as a function $R(\eta)$. From the nonnegativity of $b_0(x)$, $\rho(0,x)$, $k(x,y)$, ξ_1, ξ_2, and ξ_3, it is easy to see that

$$\frac{\mathrm{d}R(\eta)}{\mathrm{d}\eta} < 0 \quad \text{for} \quad \eta > 0, \tag{2.42}$$

$$\lim_{\eta\to\infty} R(\eta) = 0, \tag{2.43}$$

$$R_0 = R(0) = \int_A b_0(x)\rho(0,x)\,dx\ , \tag{2.44}$$

where R_0 is the expected number of offspring per adult under density independence. (R_0 is called the *net reproductive rate* in standard demographic theory.) Clearly, if $R_0 < 1$, the population will go extinct with or without density dependence.

If a nontrivial equilibrium exists, $n^*(D_l) = \eta^* > 0$ and (2.41) implies $R(\eta^*) = 1$. From (2.42), $R(0) = R_0 > 1$. Conversely, if $R_0 > 1$, then (2.42)–(2.44) imply the existence of a unique real root η^* for $R(\eta) = 1$. If we let $n^*(D_l) = \eta^*$, we can use (2.40), (2.35), (2.34), and (2.33) to determine $n^*(x)$ uniquely.

Chapter 3 Demographic Oscillations

In Chapter 2 we showed how the behavioral interactions among the life stages in *Tribolium* populations can lead to stable demographic oscillations. In particular, our numerical results implicated egg cannibalism by larvae as the driving force for the oscillations.

In this Chapter we examine in more detail the dynamics of egg cannibalism by larvae. We review the recent analytical results of Hastings (1987) and Hastings and Costantino (1987) on an egg-larval "submodel" for *Tribolium*. In contrast with the work of the previous chapter, this model yields explicit results for local stability and the properties of the bifurcation to a stable limit cycle. Numerical results are presented which characterize the analytical results and extend them to more biologically complex situations. We estimate the location in parameter space of several genetic strains of *T. confusum* and *T. castaneum* and discuss the unresolved problems and the implications for further experimental work.

3.1 Egg-Larval Submodel

3.1.1 Fast and Slow Time Scales

One of the striking observations from Chapter 2 is the difference between the relative magnitudes of the oscillations of the immature and adult age classes. For example, in both the experimental data and the computer simulations, larval numbers show huge peaks and troughs while adult numbers show small amplitude oscillations around a relatively smooth curve (Figs. 2.3–2.7). In fact, in the experimental data for adult numbers it is often difficult to distinguish between cyclic oscillations and random fluctuations.

To illustrate this phenomenon, let $N_S(t)$ be the density of individuals in life stage S and consider a "moving average" of duration τ defined as

$$\overline{N}_S(t) = T^{-1} \int_{-T/2}^{T/2} N_S(t+\zeta)\mathrm{d}\zeta\,, \qquad T = \min(\tau, 2t)\,. \tag{3.1}$$

[In the discrete case, $T=\min(\tau,2t+1)$ and the integral is replaced by a summation from $\zeta=-(T-1)/2$ to $(T-1)/2$.] Using this equation, we iterated the matrix model from Chapter 2 with the parameter estimates from Table 2.1 and compared the time trajectories of larvae and adults with their moving averages of duration $\tau=27$ days. These results

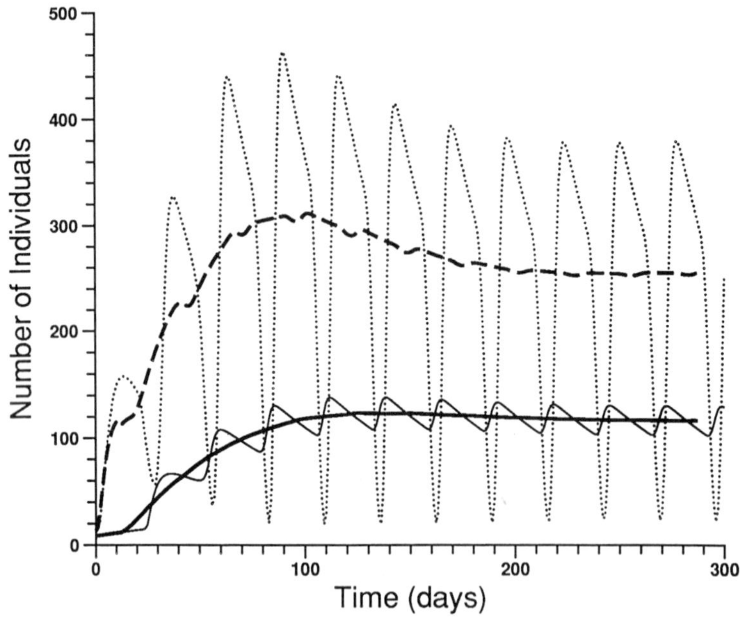

Fig. 3.1. A comparison of the numbers of adults and larvae with their moving averages for the simulation model of Chapter 2. Plotted are adult numbers (thin solid curve), the moving average for adults (heavy solid curve), larval numbers (dotted curve), and the moving average for larvae (heavy broken curve). The moving averages were computed using (3.1) with $\tau = 27$ days.

are plotted in Fig. 3.1. Clearly, it would be very misleading to characterize larval numbers as a smooth curve with a constant asymptotic value. However, this assumption is not as bad for adults, since the oscillations never deviate far from the average. Indeed, it appears that the dynamics of adult numbers are dominated by changes which occur over a *slow time scale*, while the dynamics of the immature age classes are dominated by changes which occur over a *fast time scale*.

Why does this difference in time scales exist? The main reason is that adults are long-lived relative to D_I, the developmental period. For example, the estimate of adult mortality, μ_A, from Table 2.1 places the average age of an adult cohort at $1/\mu_A = 76.9$ days, whereas in the same model the average age of an uniform distribution of immatures is $D_I/2 = 13.5$ days, with eggs and pupae each lasting only 3 days. In the most extreme case where new recruits into the adult population arrive as instantaneous "pulses" every D_I time units, the maximum distance from peak to trough in the adult oscillations is governed by the adult mortality rate. For example, if a pulse of recuitment occurs at time t_1, adult numbers prior to the next pulse are given by

$$N_A(t) = N(t_1)\exp[-\mu_A(t - t_1)] \,, \qquad t_1 \le t < t_1 + D_I \,. \tag{3.2}$$

If the adult oscillations are cyclic with period D_I, then, using (3.2) in (3.1), we have

$$\frac{\mu_A D_I}{1-\exp(-\mu_A D_I)} \le \frac{N_A(t)}{\overline{N}_A(t)} \le \frac{\mu_A D_I \exp(-\mu_A D_I)}{1-\exp(-\mu_A D_I)}. \tag{3.3}$$

If the product $\mu_A D_I \ll 1$, then a first order approximation to (3.3) is

$$1 - \frac{\mu_A D_I}{2} \le \frac{N_A(t)}{\overline{N}_A(t)} \le 1 + \frac{\mu_A D_I}{2}. \tag{3.4}$$

Hence, the quantity $\mu_A D_I/2$ can serve as an index of the relative importance of the fast to slow time scales for the dynamics of adult numbers. For the parameter estimates of Chapter 2, $\mu_A D_I/2 = 0.176$ and the steady-state simulation results of Fig. 3.1 fall well within the bounds given by both (3.3) and (3.4).

Life history information can be be used to predict the relative importance of the fast to slow time scales for several genetic strains of *Tribolium*. In Table 3.1 we present estimates of the developmental periods, adult mortality rates, indices $\mu_A D_I/2$ of fast to slow time scales, and relative bounds for adult oscillations as given by (3.3). The estimates of the developmental periods are taken from the data of Park et al. (1961, Table 6). In this same paper, the authors present the median duration of life for adult beetles of each strain (Park et al. 1961, Table 8). To convert these estimates to mortality rates, we note that if adult survivorship follows an exponential distribution, then the mortality rate is related to the median by $\mu_A = -\log(0.5)/\text{median}$. Standard errors were not provided for either of these estimates.

The results in Table 3.1 suggest that adult oscillations will be small relative to the time-average of adult density. The largest index is for strain cIV, and this value suggests oscillations with a magnitude of less than 7% of the average adult density. Hence, for modeling purposes, a reasonable assumption might be to focus on the fast and slow time scales independently, with the dynamics of the immature age classes dominating on the

Table 3.1. The relative importance of fast to slow time scales for adult population dynamics in eight genetic strains of *Tribolium*

Genetic strain	Developmental period (days)	Adult mortality rate (per day)	Index of fast to slow time scales	Relative bounds for adult oscillations
cI	28.12	0.003057	0.04298	(1.0436, 0.9576)
cII	34.87	0.002584	0.04505	(1.0457, 0.9556)
cIII	29.76	0.003093	0.04602	(1.0467, 0.9547)
cIV	31.09	0.004188	0.06511	(1.0665, 0.9363)
bI	31.31	0.002236	0.03500	(1.0354, 0.9363)
bII	32.02	0.002241	0.03587	(1.0363, 0.9646)
bIII	31.30	0.002051	0.03209	(1.0324, 0.9682)
bIV	32.24	0.001908	0.03076	(1.0311, 0.9696)

fast time scale and adult population dynamics dominating on the slow time scale. In fact, this is the approach we take for the remainder of this monograph, with the present Chapter dealing with the fast time scale and subsequent chapters analyzing various aspects of adult population dynamics.

3.1.2 Some Simplifying Assumptions

In order to obtain a mathematically tractable model of the immature age classes, we must make some simplifying assumptions. If one accepts the idea of a fast and slow time scales for *Tribolium* population dynamics, then a reasonable assumption for an analysis on the fast time scale is that adult numbers remain constant. Given this assumption, we can write the rate of total reproduction as

$$n(0,t) = B \,, \tag{3.5}$$

where, as in Chapter 2, $n(0,t)$ is number of individuals of age 0 at time t and B is the total birth rate. Here B is a constant representing the net rate at which new eggs enter the population. Expression (3.5) will serve as boundary condition for the McKendrick-von Foerster equation.

Following the analyses of Hastings (1987) and Hastings and Costantino (1987), we shall also assume that the rate at which a larva eats eggs is independent of the age of the larva. This assumption is unrealistic because the cannibalistic voracity of larvae is related to larval size. In Chapter 2 we described larval cannibalism as a linear function of age. Unfortunately, the analysis of egg-larval dynamics becomes extremely complicated with any sort age-dependent cannibalism. In order to make some progress, we shall accept a constant cannibalism rate as a crude approximation. We address the problem of age-dependent cannibalism again in §3.4.1.

Given these assumptions, we take as our model the McKendrick-von Foerster equation (2.2) with the boundary condition (3.5) and mortality rates

$$\mu(x,t) = \begin{cases} \mu_E + cN_L(t)\,, & x \in E\,, \\ \mu_L\,, & x \in L\,. \end{cases} \tag{3.6}$$

Here c is the rate at which larvae eat eggs and $N_L(t)$ is the number of larvae at time t. Since pupae do not actively affect the survivorship of eggs or larvae, we need not consider their dynamics in our analysis. We will refer to this formulation as the *egg-larval submodel.*

3.1.3 Integral Equations

The egg-larval submodel can be expressed in integral form. Substituting the mortality rates (3.6) into (2.6) and using the boundary condition (3.5) gives

$$n(x,t) = \begin{cases} B \exp\left[-\mu_E x - c \int_0^x N_L(t-s)\,ds \right], & x \in E, \\ n(D_E, t-x+D_E) \exp[-\mu_L (x - D_E)], & x \in L. \end{cases} \tag{3.7}$$

For $x \in E$, the number of eggs is equal to the total number of eggs laid, B, times the survivorship rates for natural mortality and cannibalism by larvae. For $x \in L$, the number of larvae is determined by the number of eggs which hatch, $n(D_E, t-x+D_E)$, times the survival rate for natural mortality. The rate at which eggs hatch is given by

$$n(D_E, t) = B \exp\left[-\mu_E D_E - c \int_0^{D_E} N_L(t-s)\,ds \right]. \tag{3.8}$$

The total number of larvae can be obtained by integrating (3.7) for $x \in L$. Let the variable $y = x - D_E$ represent the age of larvae from the time of hatching. Integrating (3.7) for $y = 0$ to D_L gives

$$N_L(t) = B \exp(-\mu_E D_E) \int_0^{D_L} \exp\left[-\mu_L y - c \int_0^{D_E} N_L(t-s-y)\,ds \right] dy \tag{3.9}$$

for the total number of larvae. Arrival at this single integral equation is made possible by the assumption of an age-independent cannibalism rate by larvae.

3.2 Analytical Results

In this section we present the analytical results for the egg-larval submodel which are based on the work of Hastings (1987). We present a general sketch of the analysis. Hastings (1987) should be consulted for detailed proofs.

3.2.1 Equilibrium

If $B > 0$, the model (3.9) has a constant equilibrium solution. If we let N_L^* represent the equilibrium number of larvae, from (3.9) we have

$$N_L^* = B \exp(-\mu_E D_E - c D_E N_L^*) \, [1 - \exp(-\mu_L D_L)] \, / \, \mu_L . \tag{3.10}$$

Since the right side of (3.10) is a positive monotonic decreasing function of N_L^* for $N_L^* > 0$, there exists a single equilibrium. Given N_L^*, the equilibrium number of eggs can be computed using

$$N_E^* = B\,[1 - \exp(-\mu_E D_E - cD_E N_L^*)] \,/\, (\mu_E + cN_L^*)\,. \tag{3.11}$$

3.2.2 Local Stability

An analysis of the local stability of the equilibrium (3.10) requires an additional assumption. In what follows, we assume

$$\mu_L = 0\,. \tag{3.12}$$

For many species and genetic strains of *Tribolium*, the death rate of larvae is small, and (3.12) can be considered a reasonable approximation. Moreover, Hastings (1987) has shown that the results reported in this section hold for small values of μ_L in the neighborhood of (3.12). In this case, one can view the following results as the lowest order term in an expansion around $\mu_L = 0$.

The stability analysis is aided by the introduction of an additional variable. Let

$$v(t) = \int_0^{D_E} N_L(t-s)\,ds \tag{3.13}$$

represent the amount of "larval life" which affects the viability of eggs which are currently hatching. Using (3.13) in (3.9) gives

$$N_L(t) = B\exp(-\mu_E D_E)\int_0^{D_L} \exp[-c\,v(t-y)]\,dy \tag{3.14}$$

for the total number of larvae. With the assumption (3.12), the equilibrium is given by

$$\begin{aligned} N_L^* &= D_L B \exp(-\mu_E D_E - cD_E N_L^*)\,, \\ v^* &= D_E N_L^*\,. \end{aligned} \tag{3.15}$$

Consider small deviations from the equilibrium (3.15). Let these deviations be defined by

$$\begin{aligned} \Delta_1(t) &= N_L(t) - N_L^*\,, \\ \Delta_2(t) &= v(t) - v^*\,. \end{aligned} \tag{3.16}$$

Substituting (3.16) into (3.13) and (3.14), expanding the exponential function as a series, and ignoring higher order terms one obtains

$$\Delta_1(t) = -\gamma \int_0^{D_L} \Delta_2(t-s)\mathrm{d}s\,,$$

$$\Delta_2(t) = \int_0^{D_E} \Delta_1(t-s)\mathrm{d}s \tag{3.17}$$

for the linearized model, where $\gamma = cN_L^*/D_L$. The composite parameter γ will be treated as a bifurcation parameter in the stability analysis.

Assume the deviations can be represented as a linear combination of exponential functions:

$$\Delta_h(t) = \sum_{j=1}^{\infty} \xi_{hj} \exp(\lambda_j t)\,, \qquad h = 1, 2\,, \tag{3.18}$$

where the λ_j's are the eigenvalues of the linearized system and the ξ_{hj}'s are constants, possibly complex, which depend on initial conditions. The equilibrium is asymptotically locally stable if and only if all the eigenvalues have negative real parts. Substituting (3.18) into (3.17) and integrating gives

$$\sum_{j=1}^{\infty} \xi_{1j} \exp(\lambda_j t) = -\gamma \sum_{j=1}^{\infty} \xi_{2j} \exp(\lambda_j t)\,[1 - \exp(\lambda_j D_L)] \,/\, \lambda_j\,,$$

$$\sum_{j=1}^{\infty} \xi_{2j} \exp(\lambda_j t) = \sum_{j=1}^{\infty} \xi_{1j} \exp(\lambda_j t)\,[1 - \exp(\lambda_j D_E)] \,/\, \lambda_j\,. \tag{3.19}$$

This can be rearranged as

$$\sum_{j=1}^{\infty} \exp(\lambda_j t)\,\{\xi_{1j} + \xi_{2j}\gamma[1 - \exp(\lambda_j D_L)] \,/\, \lambda_j\} = 0\,,$$

$$\sum_{j=1}^{\infty} \exp(\lambda_j t)\,\{\xi_{2j} - \xi_{1j}[1 - \exp(\lambda_j D_E)] \,/\, \lambda_j\} = 0\,. \tag{3.20}$$

These expressions must hold for all $t \geq 0$. Therefore, the terms in curly brackets must equal zero for $j = 1, 2, \ldots, \infty$. If λ is any eigenvalue of the linearized equations, then it must be true that

$$\xi_{1j} / \xi_{2j} = -\gamma [1 - \exp(\lambda D_L)] / \lambda ,$$
$$\xi_{2j} / \xi_{1j} = [1 - \exp(\lambda D_E)] / \lambda . \quad (3.21)$$

The product of these two equations gives us the characteristic equation

$$1 + \gamma [1 - \exp(\lambda D_E)] [1 - \exp(\lambda D_L)] / \lambda^2 = 0 \quad (3.22)$$

for the eigenvalues.

Expression (3.22) can be used to obtain local stability conditions for the system (3.13)–(3.14). Hastings (1987) has shown that for sufficiently small values of γ, all the solutions to (3.22) have negative real parts. Furthermore, direct substitution shows that $\lambda = 0$ is never a solution to (3.22). Thus, if the system becomes unstable as γ increases, then there must be a point at which a pair of eigenvalues crosses the imaginary axis. To illustrate, let $\lambda = a \pm bi$, where $i = \sqrt{-1}$, and consider Fig. 3.2. The eigenvalues are points in the complex plane. As gamma increases, by continuity, a pair of eigenvalues cross the imaginary axis at $b = \pm\omega$. Denote this value of γ as γ^*. If the eigenvalue crosses the imaginary axis with a positive velocity, i.e., $d\lambda/d\gamma > 0$ at $\lambda = \omega i$, then γ^* is the *bifurcation point* for the equilibrium.

Substitution of $\lambda = \omega i$ into (3.22) leads to a pair of equations corresponding to the real and imaginary parts:

$$0 = -\omega^2 + \gamma [1 - \cos(D_L\omega) - \cos(D_E\omega) + \cos((D_L + D_E)\omega)] ,$$
$$0 = \sin(D_L\omega) + \sin(D_E\omega) - \sin((D_E + D_L)\omega) . \quad (3.23)$$

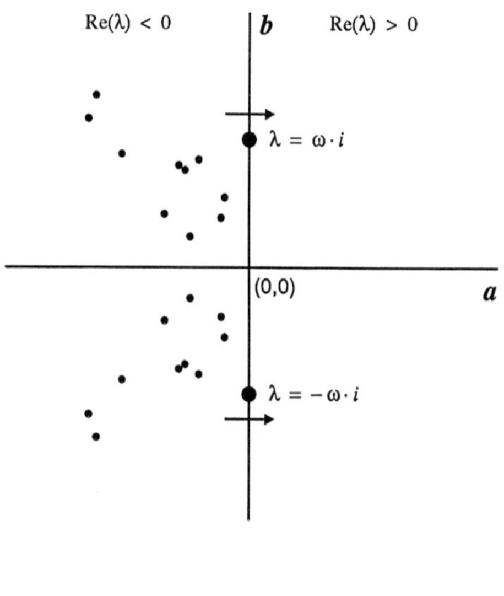

Fig. 3.2. A diagrammatic representation of the eigenvalues of the linearized egg larval sub-model. At the bifurcation point, the leading eigenvalues (large circles) cross the imaginary axis in the direction given by the arrows.

Hastings (1987) derived the solutions

$$\omega(j) = \frac{2j\pi}{D_E + D_L},$$

$$\gamma(j) = \frac{\omega(j)^2}{2 - \cos[D_E\omega(j)] - \cos[D_L\omega(j)]}, \tag{3.24}$$

where j is any positive integer. Furthermore, Hastings showed that $\gamma(1) < \gamma(j)$ for $j > 1$ and $d\lambda/d\gamma > 0$ at $\gamma = \gamma(1)$ and $\lambda = i\omega(1)$. Hence $\gamma^* = \gamma(1)$ is the point at which a Hopf bifurcation occurs.

Since γ^* is the smallest value of γ at which stability is lost, this quantity gives the stability boundary of the model. Recall that $\gamma = cN_L^*/D_L$. Using (3.15) for N_L^*, one obtains

$$B < (\gamma^*/c)\exp(\mu_E D_E + \gamma^* D_L D_E) \tag{3.25}$$

as a necessary and sufficient condition for local asymptotic stability. Note that γ^* is computed from (3.24) with $j = 1$. Hence (3.25) gives a solution of closed form for the stability boundary.

The stability boundary is plotted in Fig. 3.3 for $\mu_E = 0$. Each curve represents a different duration of the egg stage with the circled numbers giving the duration in days. Points falling below the curve represent stable parameter space; points falling above the curve are unstable. For very short durations of the larval stage, large rates of egg production are needed to destabilize the system. However, the bifurcations values for egg production rates decrease rapidly for $D_L < 4$ days, eventually reaching a minimum before increasing again. Thus intermediate values of larval durations are the most destabilizing. In general, increasing the duration of the egg stage has a destabilizing effect, except when the larval stage is of very short duration. In all cases, an increase in the egg production rate is destabilizing. Finally, since the parameters B and c can be interchanged in (3.25), the stability boundaries for cannibalism rate as function of egg and larval duration will be qualitatively the same as the curves in Fig. 3.3 for egg production rate.

3.2.3 Nonlinear Analysis

The loss of stability through a Hopf bifurcation implies the existence periodic cycles of small amplitude when γ is close to γ^*. However, these cycles can arise in two different ways. If the cycles appear when γ exceeds the critical value γ^*, then the bifurcation is *supercritical* and the cycles are stable. If the cycles appear when $\gamma < \gamma^*$, then the bifurcation is *subcritical* and the cycles are unstable. Determining whether the bifurcation is supercritical or subcritical involves computing the *direction of the bifurcation.*

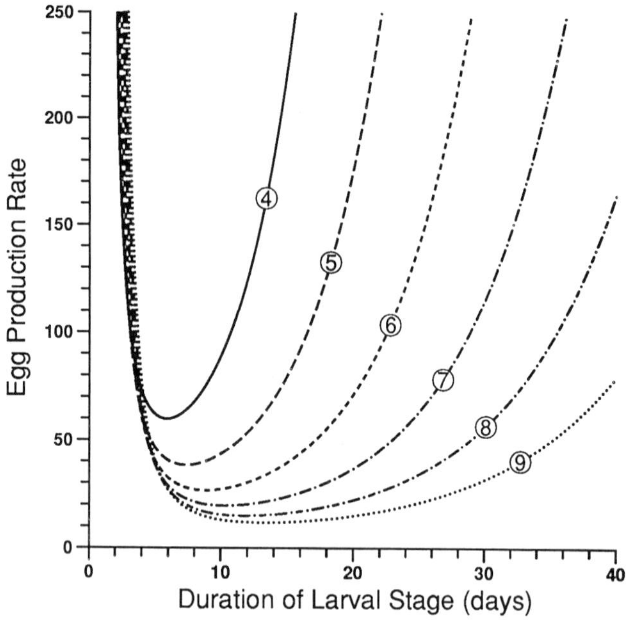

Fig. 3.3. Stability boundaries for the egg-larval submodel as given by (3.24) and (3.25). The different curves are for different durations of the egg stage (circled number in days). Parameter values above the curve lead to unstable equilibria; values below lead to locally stable equilibria. [Modified from A. Hastings and R. F. Costantino (1987), Cannibalistic egg-larval interaction in *Tribolium*: an explanation for the oscillations in population numbers. *Am. Nat.* **130**: 36–52.]

The directions of the bifurcation are illustrated in Fig. 3.4. In the case of a supercritical bifurcation, $\gamma > \gamma^*$ implies an unstable equilibrium point. Trajectories which move away from this equilibrium will approach the small amplitude periodic orbit which surrounds this equilibrium. This orbit is a *stable attractor* of trajectories in its neighborhood. In the case of a subcritical Hopf bifurcation, $\gamma < \gamma^*$ implies a stable equilibrium point. The periodic orbit which surrounds the equilibrium must be unstable. In fact, this unstable orbit defines the *domain of attraction* for the stable equilibrium. Since all trajectories of the system are bounded, there must one or more other attractors which exist outside the unstable orbit. In the case of the egg-larval submodel, the system cannot approach the origin; if two or more stable attractors exist, they are nontrivial.

Hastings (1987) used a formula derived by Diekmann and van Gils (1984) to compute the direction of the bifurcation in the egg-larval submodel. The formulas are rather complicated; readers are referred to the original publications for details. Remarkably, when $\mu_L = 0$, *the egg-larval submodel will always undergo a subcritical Hopf bifuration.* This implies the existence of *multiple stable attractors* in the vicinity of the bifurcation point. Furthermore, Hastings (1987) showed that this result also holds when μ_L is small.

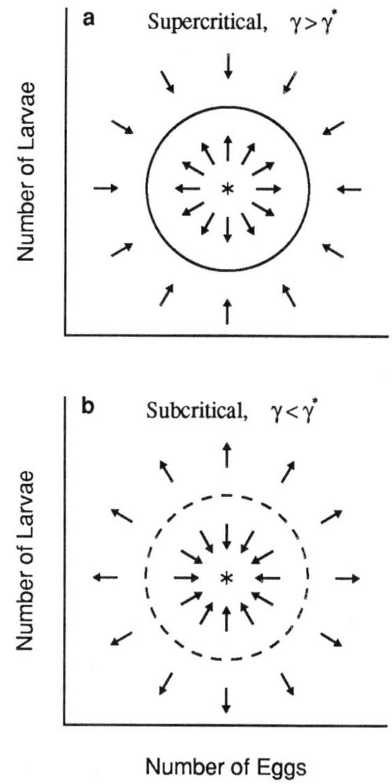

Fig. 3.4. A diagrammatic representation of the dynamics in the neighborhood of the equilibrium when parameter values are near the bifurcation point. When the bifurcation is (a) supercritical, a stable orbit (solid circle) surrounds an unstable equilibrium (asterisk) for parameter values above the bifurcation point ($\gamma > \gamma^*$). When the bifurcation is (b) subcritical, an unstable orbit (dashed circle) surrounds a stable equilibrium (asterisk) for parameter values below the bifurcation point ($\gamma < \gamma^*$). For the subcritical case, trajectories outside the unstable orbit are bounded and so they must approach a second attractor. Arrows indicate the movement of trajectories towards or away from the point or cycle.

A qualitative picture of the Hopf bifurcation can be obtained by considering the magnitude of the oscillations as a function of the bifurcation parameter. We call a plot of this type a *bifurcation portrait.* In Fig. 3.5 we illustrate hypothetical bifurcation portraits for the supercritical and subcritical cases. It should be kept in mind that these qualitative pictures are only valid for the neighborhood of the equilibrium point, although the numerical results of the next section suggest that they are appropriate descriptions of the global behavior of the model.

The supercritical bifurcation is represented in Fig. 3.5a. The bifucation point separates parameter space into two regions: (i) the interval labelled "I" in which a stable equilibrium exists and the magnitude of the oscillations is zero; and (ii) the interval labelled "III" in which the equilibrium is unstable and the magnitude of the oscillations grow as the bifurcation parameter is increased. In each region a single attractor exists – all initial conditions approach this attractor.

The subcritical bifurcation is represented in Fig. 3.5b. In this case three regions exist: (i) the interval labelled "I" in which the only attractor is a stable point; (ii) the interval "II" in which there exists both a stable equilibrium point and a stable oscillatory attractor; and (iii) the interval "III" beyond the bifurcation point where only stable oscillations persist. In region II an unstable orbit (dashed curve) separates the two stable attractors – the fate of a trajectory depends upon initial conditions.

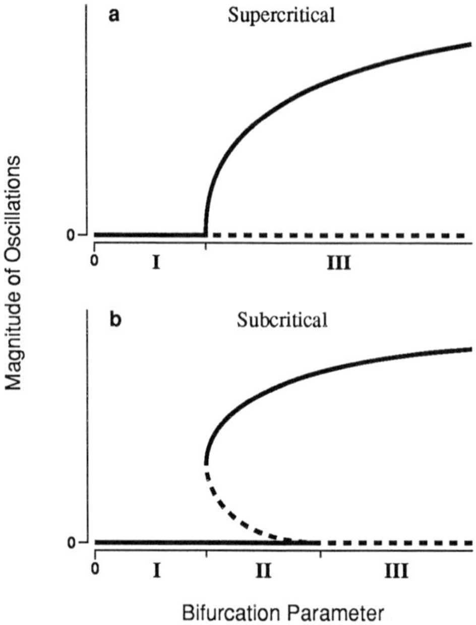

Fig. 3.5. Qualitative representations of the bifurcation portraits in the case of a (a) supercritical and (b) subcritical bifurcation. The solid lines are stable solutions and the dashed lines are unstable solutions. In region **I**, a stable equilibrium point is the only solution. In region **III**, a stable orbit and an unstable equilbrium point are the possible solutions. In the subcritical case, region **II** is an interval of parameter space where three solutions exist: a stable equilibrium point, an unstable orbit, and a stable orbit. [Modified from A. Hastings and R. F. Costantino (1987), Cannibalistic egg-larval interaction in *Tribolium*: an explanation for the oscillations in population numbers. *Am. Nat.* **130**: 36–52.]

The subcritical bifurcation can lead to some interesting dynamics. Consider an bifurcation parameter just to the left of the boundary between regions I and II (Fig. 3.5b). An inital condition which leads to a stable equilibrium point can also lead to large oscillations if the bifurcation parameter is increased slightly so that it falls in region II. Similarly, within region II, initial conditions near the boundary of the domains of attraction for the point and stable cycle may differ only slightly and lead to quite different asymptotic behavior. Finally, trajectories near the unstable orbit may persist on this unstable orbit for a long period of time before spiraling out to the stable orbit or spiraling in to the equilibrium point. These possibilites are illustrated in the next section.

3.3 Numerical Results

The prediction of a subcritical Hopf bifurcation in the egg-larval submodel raises many interesting questions. Do multiple attractors exist for a biologically significant range of parameter values? What is the range of initial conditions for each attractor? In region II of the subcritical bifurcation portrait (Fig. 3.5b), are the magnitudes of the stable oscillations significant relative to the size of the equilibrium densities? Can the bifurcation portrait be mapped out? In this section we address these questions by investigating numerically a discrete-time version of the egg-larval submodel.

3.3.1 Difference Equations

Numerical work is greatly simplified using discrete-time difference equations to represent the dynamics of eggs and larvae. Let the basic time unit of one day be divided evenly into an integral number of subintervals and let δ denote the width of a subinterval. Let E represent the set of points $\{\delta, 2\delta, 3\delta, \ldots, D_E\}$ and let $L=\{D_E+\delta, D_E+2\delta, D_E+3\delta, \ldots, D_E+D_L\}$. The discrete-time equations are given by

$$\begin{aligned} n(\delta, t+\delta) &= B\delta\,, \\ n(x+\delta, t+\delta) &= n(x,t)\exp[-\delta(\mu_E + cN_L(t))]\,, \qquad x\in E\,, \\ n(x+\delta, t+\delta) &= n(x,t)\exp[-\delta\mu_L]\,, \qquad x\in L\,,\ x\neq D_E+D_L\,. \end{aligned} \tag{3.26}$$

Smaller values of δ give better approximations to the continous-time model.

The equilibrium for the discrete-time model is found in the usual way by setting $n(x,t)=n^*(x)$ for all values of x and t. The result is

$$n^*(x) = \begin{cases} B\delta\exp[-(x-\delta)(\mu_E + cN_L^*)]\,, & x\in E\,, \\ B\delta\exp[-D_E(\mu_E+cN_L^*) - (x-D_E-\delta)\mu_L]\,, & x\in L\,. \end{cases} \tag{3.27}$$

Summing the equations for $x\in E$ and $x\in L$ gives

$$\begin{aligned} N_E^* &= B\delta\left(\frac{1-\exp[-D_E(\mu_E+cN_L^*)]}{1-\exp[-\delta(\mu_E+cN_L^*)]}\right), \qquad x\in E\,, \\ N_L^* &= B\delta\exp[-D_E(\mu_E+cN_L^*)]\left(\frac{1-\exp[-D_L\mu_L]}{1-\exp[-\delta\mu_L]}\right), \qquad x\in L\,, \end{aligned} \tag{3.28}$$

for the equilibrium densities of eggs and larvae. The second equation can be solved for N_L^*. In the limit as $\delta\rightarrow 0$, N_L^* and N_E^* in (3.28) approach the corresponding continuous-time values given in (3.10) and (3.11). For the special case where $\mu_L=0$, N_L^* is identical in the discrete and continuous models irrespective of δ.

Stability of the discrete-time equations can be examined by linearizing the model in the neighborhood of the equilibrium. Let $\mathbf{n}(t)=\{n(\delta,t),\ n(2\delta,t),\ n(3\delta,t),\ \ldots,\ n(D_E+D_L,t)\}^{\mathrm{T}}$ be a column vector of length $(D_E+D_L)/\delta$. Assuming $\mu_L=0$, local stability of the equlibrium $\mathbf{n}^*$ is determined by the eigenvalues of the following matrix:

$$\mathbf{Z} = \{z(x,y)\} = \left\{\frac{\partial n(x,t+\delta)}{\partial n(y,t)}\right\}_{\mathbf{n}(t)=\mathbf{n}^*}, \qquad x,y=\delta, 2\delta, 3\delta, \ldots, D_E+D_L\,. \tag{3.29}$$

For the first row of this matrix,

$$z(\delta, y) = 0\,, \qquad y\in E, L\,; \tag{3.30}$$

the remaining elements are given by

$$z(x+\delta, y) = \begin{cases} \rho\,, & x=y,\ x\in E, \\ 1\,, & x=y,\ x\in L,\ x\neq D_E+D_L, \\ -\delta^2 cB\rho^{x/\delta}\,, & x\in E,\ y\in L, \\ 0\,, & \text{elsewhere,} \end{cases} \tag{3.31}$$

where $\rho = \exp[-\delta(\mu_E + cN_L^*)]$. If λ is an eigenvalue of $\mathbf{Z}$ and $\mathbf{w}$ is the corresponding eigenvector, then

$$\mathbf{Z}\,\mathbf{w} = \lambda\,\mathbf{w}\,. \tag{3.32}$$

In the appendix, we show that using (3.30) and (3.31) in (3.32) gives

$$\lambda^{(D_E+D_L)/\delta-1} + \delta^2\gamma\left(\frac{1-\lambda^{D_E/\delta}}{1-\lambda}\right)\left(\frac{1-\lambda^{D_L/\delta}}{1-\lambda}\right) = 0\,, \tag{3.33}$$

for the characteristic equation of $\mathbf{Z}$, where $\gamma = cN_L^*/D_L$. This expression is the discrete-time analogue of (3.22) for the continuous model.

In the discrete model, the equilibrium is locally asymptotically stable if and only if all the eigenvalues are less than one in modulus. Hence the stability boundary is defined by $\mathrm{mod}(\lambda) = 1$. This implies $\lambda = \exp(\omega i)$, where $i = \sqrt{-1}$. As Hastings (1987) did with the continuous model, we can substitute this boundary value for λ into the characteristic equation and solve for ω and γ. The result is

$$\begin{aligned} \omega(j) &= \frac{2j\pi\delta}{D_E+D_L}\,, \\ \gamma(j) &= \frac{2\{1-\cos[\omega(j)]\}}{\delta^2\{2-\cos[D_E\omega(j)/\delta]-\cos[D_L\omega(j)/\delta]\}}\,, \end{aligned} \tag{3.34}$$

where j is a positive integer. Details of this derivation can be found in the appendix.

Note the similarity of (3.34) with the continuous version (3.24). Expanding the term $\cos[\omega(j)]$ in (3.34) as $1-\omega^2/2+O(\delta^4)$ and taking the limit as $\delta\to 0$, one obtains an expression for $\gamma(j)$ which is identical to (3.24). Hence, for small values of δ, the bifurcation points $\gamma^* = \gamma(1)$ of the discrete and continuous models are very nearly equal. If δ is large, one must verify that $\gamma(1) < \gamma(j)$ for $j > 1$ before using $\gamma^* = \gamma(1)$ in the discrete case. Since (3.15) for N_L^* holds for both the continuous and discrete models when $\mu_L = 0$, the inequality (3.25) gives a closed form solution for the stability boundary in the discrete case as well.

We have no results for the discrete model which are equivalent to Hastings' (1987) nonlinear analysis of the continuous model. However, in our numerical work with the discrete model, we have always found a *subcritical* Hopf bifurcation. These results strongly suggest that the discrete egg-larval submodel has the same qualitative properties as its continuous-time counterpart, especially when δ is small.

3.3.2 Multiple Attractors

In the case of a subcritical Hopf bifurcation, the nonlinear analysis only tells us that, in the neighborhood of γ^*, an unstable cycle surrounds the stable equilibrium point when $\gamma<\gamma^*$. The fact that the trajectories of the model are bounded implies the existence of one or more attractors on the outside of the unstable cycle. However, numerical studies must be used to characterize the model's behavior in this region. In this section we provide the details of a typical numerical example which illustrates the properties of the egg-larval submodel.

Parameter values were chosen which approximate the estimates for the cI genetic strain of *T. castaneum* described by Park et al. (1961). Since the natural mortality rates of the immature age classes is generally low, we let $\mu_L=\mu_E=0$. For the durations of the egg and larval stages we used $D_E=4$ days and $D_L=18$ days, respectively. The rate of egg cannibalism by larvae was set to be $c=0.025$ per larvae per day, a value which is typical of medium to large sized larvae (Park et al. 1965). For the total rate of egg production we let $B=150$ eggs per day, which was determined by trial and error to equal a value which yields oscillations but falls below the bifurcation value given by (3.25). Finally, for the purposes of iteration, we broke each one day interval into eight equal parts, so that $\delta=0.125$ days.

The parameter values can be used to make predictions for the model. Since $\mu_L=0$, we used a simple root finding algorithm to solve (3.15) for N_L^*. This value was substituted into (3.11) to compute N_E^*. In this way, we obtained $N_E^*=141.649$ and $N_L^*=41.704$ for the equilibrium number of eggs and larvae in the continuous model. Using (3.28), the analogous results for the discrete model are $N_E^*=151.079$ and $N_L^*=41.704$. The predicted bifurcation point is given by (3.24) in the continuous case and (3.34) in the discrete case for $j=1$. Values of $\gamma^*=0.069765$ and $\gamma^*=0.069757$ were obtained for the continuous and discrete cases, respectively. Letting B^* denote the bifurcation value for egg production, we used (3.25) to obtain $B^*=423.827$ for the continuous case and $B^*=423.556$ for the discrete case. Clearly, the differences between the continuous and discrete models are trivial with respect to the stability boundaries.

Every simulation must begin with an initial distribution of eggs and larvae. To simplify matters, we only varied the initial totals for eggs and larvae and used the following rule to determine the age distributions within these two life stages. The initial distribution of eggs was proportional to (3.27) with the initial number of larvae used in place of N_L^*. For the larval age classes, the initial distribution was uniform. With this rule, the model is in equilibrium when the initial total numbers of eggs and larvae equal their respective equilibrium values.

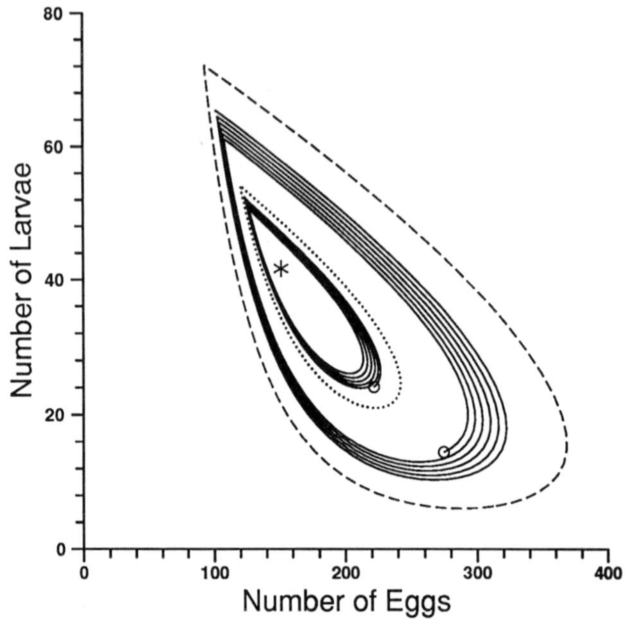

Fig. 3.6. Simulations of the discrete egg-larval submodel. The solid curves are portions of two trajectories with different starting points (open circles). The inner trajectory spirals toward the stable equilibrium point (asterisk). The outer trajectory spirals towards the stable cycle (dashed curve). These two stable attractors are separated by an unstable cycle (dotted curve). See text for parameter values.

Two simulations of the discrete model are presented in Fig. 3.6. The solid curves represent portions of two trajectories with different initial conditions. The open circles give the start of each trajectory. The asterisk in this figure represents the equilibrium point. The dashed curve represents a stable cycle and the dotted curve an unstable cycle. The stable cycle was determined by starting with various initial conditions far from the equilibrium point and iterating the difference equations many thousands of times. In each case, the trajectories were attracted to the same cycle. The unstable cycle was obtained by varying the initial conditions along a line between the origin, which approaches the stable cycle, and the equilibrium point. Using the method of bisection, we searched for an intial condition along this line which remained "balanced" on the unstable cycle. This unstable cycle separates the "phase space" of the model into two domains of attraction (Fig 3.6). Trajectories within the dotted curve slowly spiral inward to the equilibrium point. Trajectories outside the dotted curve but within the dashed curve spiral outward to the stable orbit. Trajectories outside the dashed curve spiral inward to the stable orbit. No other attractors were found.

The numerical example of Fig. 3.6 addresses some of the questions posed earlier. From the size of the unstable cycle (dotted curve), we know that there can exist significantly large domains of attraction for *both* the stable point and the stable cycle. When multiple attractors do exist, the size of the stable cycle can indeed be quite large relative to the equilibrium densities. Finally, we note that the rate of egg production in this example, $B = 150$ eggs per day, is far below the bifurcation point of $B^* = 424$ eggs per day. This demonstrates that multiple attractors can be found in a very large volume of parameter space. In the next section, we give a precise estimate of the range of the parameter B that leads to multiple attractors.

3.3.3 Bifurcation Portrait

In §3.2.3 we discussed the general properties of a subcritical Hopf bifurcation and showed how the behavior of the model can be characterized using a bifurcation portrait (Fig. 3.5). We now show how numerical methods can be used to produce a bifurcation portrait for a given set of parameter values.

The egg production rate, B, was used as the bifurcation parameter and the other parameters were assigned the same values as in the previous section ($\mu_L = \mu_E = 0, D_E = 4, D_L = 18, c = 0.025$). As discussed above, the bifurcation occurs at $B^* = 424$. Thus the equilibrium point remains stable for $B \in (0, 423]$. To find the stable oscillator, we used an initial condition of no eggs or larvae and followed the time course of the simulations until a stable cycle was achieved. The bisection method described above was used to locate an initial condition between the origin and the equilibrium point which produced a trajectory that was balanced on the unstable orbit. The magnitudes of the oscillations were measured using the square root of the area of the smallest rectangle with sides parallel to the axes which bounds the cycle divided by the square root of the product of the equlibrium densities; that is,

$$\text{magnitude} = \frac{\left[\overline{\lim_{t \to \infty}} N_E(t) - \underline{\lim_{t \to \infty}} N_E(t)\right]^{1/2} \left[\overline{\lim_{t \to \infty}} N_L(t) - \underline{\lim_{t \to \infty}} N_L(t)\right]^{1/2}}{[N_E^* N_L^*]^{1/2}}, \tag{3.35}$$

where $\overline{\lim}$ and $\underline{\lim}$ are the limits superior and inferior, respectively. This definition of the magnitude is a dimensionless quantity which is scaled to equilibrium densities.

The actual bifurcation portrait appears in Fig. 3.7. The solid lines represent the stable attractors and the dashed lines represent the unstable solutions. A stable orbit was first detected at $B = 134$, where it possesses a fairly large amplitude. As B increases beyond this point, the magnitude of these stable oscillations also increase slowly. The unstable orbit first appears at $B = 134$, where its magnitude is close to that of the stable orbit. As B increases towards the bifurcation point, the size of the unstable orbit rapidly decreases to zero. This represents a rapid contraction of the domain of attraction of the equilibrium point as B increases. The interval of multiple attraction corresponds to the region of overlap between the stable cycle and the stable equilibrium point; it spans the range of egg production rates from $B = 134$ to $B = 424$. This represents a significant portion of biologically feasible parameter space.

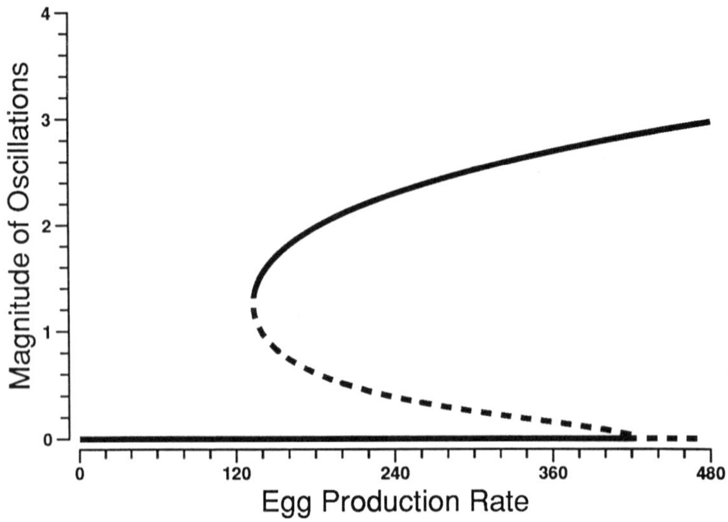

Fig. 3.7. An actual bifurcation portrait for the egg-larval submodel. The magnitude of the oscillations was measured using (3.35). The solid lines are stable attractors and the dashed lines are unstable solutions. The stable oscillations begin at $B = 134$ eggs per day and the bifurcation occurs at $B^* = 424$ eggs per day. See text for parameter values.

3.4 Biological Relevance

Do the dynamics of the egg-larval submodel have any biological relevance? As yet, we have no simple answer to this question. The analyses of this Chapter represent relatively recent work, and as in most fields of science, the theory must await experimental evidence. However, we can at least begin to address this question in several different ways. First, we can focus on the major assumptions of the model and ask whether or not they really make a difference. In particular, we will relax the assumptions of age-independent cannibalism and a constant adult population number and see if the behavior of the model is drastically changed. Second, we can try to ascertain the location of real populations in the parameter space of the egg-larval submodel and see what type of behavior is predicted. If it turns out that no real population can ever be found in the region of multiple attractors, then there may be little reason to continue an investigation of the egg-larval submodel. Finally, we can begin to list avenues for future research, including experimental possibilities and theoretical questions and extensions.

3.4.1 Age-Dependent Cannibalism

In light of what is known about the biology of *Tribolium*, the most unrealistic assumption of the egg-larval submodel concerns the age-independence of larval cannibalism rates. In Park et al. (1965), it is shown that larval cannibalism rates increase with age and eventually reach an upper asymptote (Fig. 2.2). In the previous chapter, we suggested that these cannibalism rates could be described using a sigmoidal curve. We pursue that idea further in the present section.

A convenient sigmoidal model is the logistic function. Let $c(y)$ represent the rate at which a larva of age y eats eggs. The function

$$c(y) = \frac{c_\infty}{1 + [c_\infty / c_0 - 1] \exp[-c_r(y - D_E)]} \tag{3.36}$$

can be used to describe the rate of cannibalism. The parameter c_0 represents the cannibalism rate of newly hatched larvae; c_r represents the rate of increase of cannibalistic voracity with age; and c_∞ represents the asymptotic rate of cannibalism. This three parameter function will replace the assumption of a constant rate of cannibalism in the simulations described below.

The difference equations must be modified slightly to account for age-specific cannibalism rates. The new equations are

$$\begin{aligned} n(\delta, t+\delta) &= B\delta, \\ n(x+\delta, t+\delta) &= n(x,t)\exp\Big[-\delta\Big(\mu_E + \sum_{y\in L} c(y)n(y,t)\Big)\Big], \qquad x \in E, \\ n(x+\delta, t+\delta) &= n(x,t)\exp[-\delta\mu_L], \qquad x \in L,\ x \neq D_E + D_L, \end{aligned} \tag{3.37}$$

where $c(y)$ is defined by (3.36). All other parameters have the same meanings as in (3.26). The equilibrium is given by

$$n^*(x) = \begin{cases} B\delta\exp[-(x-\delta)(\mu_E + C_L^*)], & x \in E, \\ B\delta\exp[-D_E(\mu_E + C_L^*) - (x - D_E - \delta)\mu_L], & x \in L, \end{cases} \tag{3.38}$$

where C_L^* is defined using

$$C_L^* = \sum_{y\in L} c(y)n^*(y). \tag{3.39}$$

Multiplying $n^*(x)$ in (3.38) by $c(x)$ for $x \in L$ and summing gives

$$C_L^* = B\delta\exp[-D_E(\mu_E + C_L^*)] \sum_{x\in L} c(x)\exp[-(x - D_E - \delta)\mu_L]. \tag{3.40}$$

This nonlinear equation can be solved numerically to obtain a unique value for C_L^*. This value can then be substituted into (3.38) to yield the equilibrium.

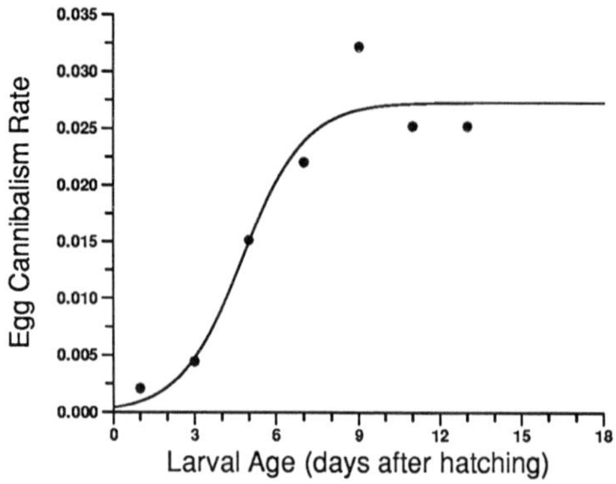

Fig. 3.8. Egg cannibalism rates as a function of larval age for the cI genetic strain of *T. castaneum*. The solid line represents the logistic function (3.36) which was fitted using the method of nonlinear least squares.

In order to obtain parameter values which can be employed in simulations, we used the method of nonlinear least squares to fit (3.36) to the data for strain cI of *T. castaneum* from Park et al. (1965, Table 10). These data were converted to cannibalism rates per larva per day in 8 grams of flour and appear in Fig. 3.8 as a function of larval age. The curve in Fig. 3.8 represents the function which was fitted using nonlinear least squares. We obtained parameter estimates of $c_0 = 0.000418 \pm 0.000680$, $c_r = 0.871125 \pm 0.341464$, and $c_\infty = 0.027339 \pm 0.002078$. For the remaining parameters we used $\mu_E = \mu_L = 0$, $D_E = 4$, $D_L = 18$, and $\delta = 0.125$. These latter estimates also approximate the values for strain cI and are compatible with the simulations of §3.3.2 and §3.3.3.

Bifurcation portraits were produced using the methods described in §3.3.3. In Fig. 3.9a, the bifurcation portrait is based on the parameter estimates obtained from Fig. 3.8. For this case, the bifurcation is supercritical. The bifurcation point has decreased dramatically from $B^* = 424$ in Fig. 3.7 to $B^* = 6.5$. This means a stable equilibrium can only exist for a total egg production rate of under 6.5 eggs per day! Furthermore, no multiple attractors are predicted. Next we increased the rate at which the cannibalistic voracity of larvae increases with age; parameter c_r was increased tenfold. The bifurcation portrait for this case appears in Fig. 3.9b. It is subcritical with a bifurcation at $B^* = 135$. This is still well below the age-independent value of $B^* = 424$. Stable oscillations are first detected at $B = 60$ and the unstable orbit persists for a significant range of parameter values. The shape of the bifurcation portrait in Fig. 3.9b is qualitatively similar to the one depicted in Fig. 3.7.

What are the implications of these results? Clearly, age-dependent cannibalism has a profound effect on the dynamics of the egg-larval submodel. When cannibalism is age-independent, any level of cannibalism will make the Hopf bifurcation subcritical. In other words, the qualitative behavior of the model is independent of the cannibalism rate. This is no longer true for age-dependent cannibalism. The quantitative values of the cannibalism parameters will determine whether or not the bifurcation is subcritical. Age-dependence does not rule out a subcritical bifurcation *a priori*, but neither is it guaranteed. This is unfortunate, since parameter estimates must be obtained and simulations conducted before any predictions can be made.

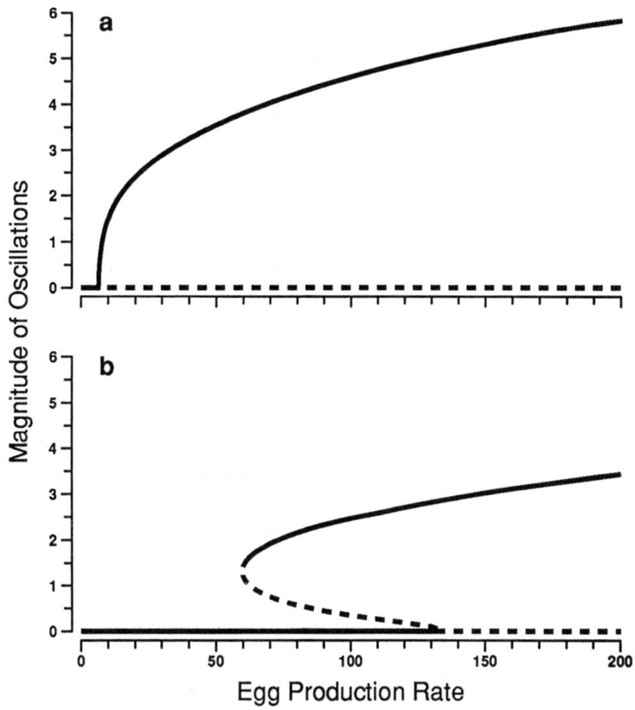

Fig. 3.9. Computed bifurcation portraits for the egg-larval submodel with age-dependent cannibalism. The solid lines are stable attractors and the dashed lines are unstable solutions. In case (a), the cannibalism parameters from Fig. 3.8 were used. The bifurcation is supercritical and occurs at $B^* = 6.5$ eggs per day. In case (b), the rate of increase of cannibalistic voracity with age (c_r) is ten times larger than in case (a). The bifurcation is subcritical and occurs at $B^* = 135$ eggs per day. The stable oscillations begin at $B = 60$ eggs per day. See text for parameter values.

3.4.2 Complete Life Cycle

Another key assumption of the egg-larval submodel is that adult numbers remain constant. While the arguments for separate fast and slow time scales makes this assumption seem reasonable, further insight can be obtained from an examination of the complete life cycle. In this section we embed the egg-larval submodel into a full model with all four age classes, but without all of the biological details described in Chapter 2. Specifically, we include adult cannibalism of pupae, since this has an important damping effect on adult recruitment, and omit adult cannibalism of eggs, age-dependent fecundity, and the effects of crowding on fecundity. We will also assume no natural mortality among the immature age classes. Since our main results come from numerical work, we focus on the difference equation model. Equations for the corresponding continous-time model can be found in Hastings and Costantino (1987).

We begin with the difference equations. Let δ, E, and L be defined as in §3.3.1, and let $P=\{D_E+D_L+\delta, D_E+D_L+2\delta, D_E+D_L+3\delta, \ldots, D_I=D_E+D_L+D_P\}$. No upper limit will be placed on adult age, so that $A=\{D_I+\delta, D_I+2\delta, D_I+3\delta, \ldots\}$. The model is given by

$$\begin{aligned} n(\delta,t+\delta) &= b\delta N_A(t)\,, & & \\ n(x+\delta,t+\delta) &= n(x,t)\exp[-\delta c_{EL}N_L(t)]\,, & x &\in E\,, \\ n(x+\delta,t+\delta) &= n(x,t)\,, & x &\in L\,, \\ n(x+\delta,t+\delta) &= n(x,t)\exp[-\delta c_{PA}N_L(t)]\,, & x &\in P\,, \\ n(x+\delta,t+\delta) &= n(x,t)\exp[-\delta\mu_A]\,, & x &\in A\,, \end{aligned} \tag{3.41}$$

where b is the per capita fecundity of adults (one-half the female fecundity), c_{EL} is the rate of larval cannibalism of eggs, c_{PA} is the rate of adult cannibalism of pupae, and μ_A is the adult mortality rate. As $\delta\to 0$, these equations approximate the continuous-time model of Hastings and Costantino (1987).

The equilibrium is found in the usual way by setting $n(x,t)=n^*(x)$. This gives

$$\begin{aligned} n^*(\delta) &= b\delta N_A^*\,, & & \\ n^*(x+\delta) &= n^*(x)\exp[-\delta c_{EL}N_L^*]\,, & x &\in E\,, \\ n^*(x+\delta) &= n^*(x)\,, & x &\in L\,, \\ n^*(x+\delta) &= n^*(x)\exp[-\delta c_{PA}N_L^*]\,, & x &\in P\,, \\ n^*(x+\delta) &= n^*(x)\exp[-\delta\mu_A]\,, & x &\in A\,, \end{aligned} \tag{3.42}$$

Summing (3.42) over $x=D_E, D_E+\delta, D_E+2\delta, \ldots, D_E+D_L-\delta$ and $x\ge D_I$ gives

$$\begin{aligned} N_L^* &= (D_L/\delta)\,n^*(D_E)\exp(-\delta c_{EL}N_L^*)\,, \\ N_A^* &= n^*(D_I)\exp(-\delta c_{PA}N_A^*)+N_A^*\exp(-\delta\mu_A)\,, \end{aligned} \tag{3.43}$$

for the equilibrium number of larvae and adults, respectively. Using (3.42) it is easy to show that

$$\begin{aligned} n^*(D_E) &= b\delta N_A^*\exp[-(D_E-\delta)c_{EL}N_L^*]\,, \\ n^*(D_I) &= b\delta N_A^*\exp[-D_E c_{EL}N_L^*-(D_P-\delta)c_{PA}N_A^*]\,. \end{aligned} \tag{3.44}$$

Substituting (3.44) into (3.43) gives two equations, the first of which can be rearranged to yield

$$N_A^* = (N_L^*/D_L b)\exp(D_E c_{EL}N_L^*)\,. \tag{3.45}$$

This can be substituted into the second equation to obtain

$$b\,\delta \exp\{-N_L^* [D_E c_{EL} + (D_P c_{PA} / D_L b) \exp(D_E c_{EL} N_L^*)]\} + \exp(-\delta\mu_A) = 1\,, \tag{3.46}$$

which has a unique real root N_L^*. Given this root, one can compute the remaining equilibrium densities using (3.45) and (3.42).

Let us assume that the stability boundaries from the egg-larval submodel also apply to the full model with $B = bN_A^*$ and $\gamma = c_{EL}N_L^*/D_L$. Using (3.46), $\gamma < \gamma^*$ implies

$$b\,\delta \exp\{-\gamma^* [D_E D_L + (D_P c_{PA} / c_{EL} b) \exp(\gamma^* D_E D_L)]\} < 1 - \exp(-\delta\mu_A)\,. \tag{3.47}$$

The critical value of the bifurcation parameter γ^* is computed using (3.34). This stability boundary is equivalent to (3.25).

The inequality (3.47) can be used to make inferences about stability in the full model. Since the two cannibalism rates enter as a ratio, it is their relative magnitude which influences stability. An increase in the adult cannibalism rate is stabilizing, while an increase in the larval cannibalism rate is destabilizing. Increases in the duration of the pupal stage or the adult mortality rate also have a stabilizing effect. An increase in the fecundity rate is destabilizing. The qualitative relationship among the fecundity rate and the duration of the egg and larval stages is similar to the curves presented in Fig. 3.3 (Hastings and Costantino 1987).

Numerical simulations of the difference equations were used to obtain a better picture of the dynamics of the full model. To facilitate comparisons with the numerical work described earlier, we used the same values for the parameters that are common to the full model and the submodel, i.e., $\mu_E = \mu_L = 0$, $D_E = 4$, $D_L = 18$, $c_{EL} = 0.025$, and $\delta = 0.125$. For the rate of adult cannibalism of pupae, we simply used the value $c_{PA} = 0.0056$ from Table 2.1. For the remaining parameters, we consulted the data of Park et al. (1965) for values appropriate for the cI strain of *T. castaneum*. We used $D_P = 6$ days for the duration of the pupal stage and $\mu_A = 0.003$ per day (Table 3.1) for the adult mortality rate. The fecundity rate b was varied.

One set of simulations was conducted to locate the bifurcation point. With the parameter values given above, we computed the equilibrium densities and multiplied these by 0.99 to obtain an initial condition in the neighborhood of the equilibrium point. We then iterated the difference equations to see if the population trajectory approached the equilbrium or moved away from it. In this way we estimated a bifurcation point of $b^* = 6.05$. Using (3.47), one predicts a bifurcation point of $b^* = 5.65$. This suggests that (3.47) is best viewed as qualitative statement about the way in which each parameter affects stability, although it does provide a rough approximation for the stability boundary in the full model.

A second set of simulations was conducted to test for multiple attractors. Using the same parameter values, we initiated simulations with no immatures and with adult densities at their equilibrium values. We then increased the fecundity parameter b until sustained oscillations were attained. If the bifurcation point were supercritical, small amplitude sustained oscillations should appear at the bifurcation point $b^* = 6.05$. Instead, we first obtained sustained oscillations at $b = 1.81$, and the magnitude of the oscillations were quite large relative to the equilibrium value (using (3.35), magnitude = 1.46). This is completely consistent with the hypothesis of a subcritical Hopf bifurcation (Fig. 3.7). It appears that the egg-larval submodel does a reasonably good job of approximating the dynamics of immatures in the full model.

3.4.3 Populations in Parameter Space

Although there are no experimental data that can be used to test directly the egg-larval submodel, there does exist a large body of information which can be used to estimate the parameters of the model. In particular, we focus again on the genetic strains cI–cIV of *T. castaneum* and bI–bIV of *T. confusum*. Our goal is to see what type of dynamical behavior the egg-larval submodel predicts.

The relevant biological data appear in Table 3.2. The papers by Park et al. (1961, 1964, 1965) should be consulted for detailed information on the experimental methods. The means and standard errors for the durations of the egg and larval stages were obtained directly from Park et al. (1961, Table 6). In Table 3 of the same paper, three-day fecundity values for females over a ninety day period are given. We divided these values by six to obtain rates per individual (female and male) per day (Table 3.2). In Park et al. (1964, Table 12), the means, sample sizes, and variances are provided for adult densities in experimental populations with eight grams of media after the populations began fluctuating around an asymptotic value. Using this information we computed standard errors. These data also appear in Table 3.2. Finally, Park et al. (1965, Table 10) report the percentage of 100 eggs eaten by groups of fifty larvae after forty-eight hours in eight grams of media. We converted these numbers to cannibalism rates per larvae per day by subtracting the proportion eaten from one, taking the negative of the natural logarithm, and dividing by 100 (50 larvae $\times$ 2 days). We then averaged over all ages of larvae to obtain the means and standard errors in Table 3.2. Of course, as with the egg-larval submodel, this latter step ignores the strong dependence of larval cannibalism rates on age.

Table 3.2. Parameter estimates for the egg-larval submodel for eight genetic strains of *Tribolium*

Genetic strain	Duration of egg stage (days)	Duration of larval stage (days)	Fecundity rate (per adult per day)	Mean Adult Density	Egg cannibalism rate (per larva per day)
cI	$4.09 \pm .012$	$17.58 \pm .439$	$7.62 \pm .303$	220.2 ± 3.88	$0.01807 \pm .00427$
cII	$4.17 \pm .028$	$25.11 \pm .488$	$3.43 \pm .202$	81.2 ± 1.66	$0.01054 \pm .00295$
cIII	$4.32 \pm .022$	$19.02 \pm .237$	$5.03 \pm .253$	102.7 ± 2.28	$0.00980 \pm .00321$
cIV	$4.19 \pm .014$	$20.93 \pm .295$	$6.15 \pm .310$	43.9 ± 0.54	$0.01389 \pm .00376$
bI	$5.37 \pm .028$	$18.96 \pm .248$	$4.77 \pm .232$	275.8 ± 2.90	$0.00057 \pm .00014$
bII	$5.43 \pm .020$	$20.11 \pm .477$	$4.51 \pm .165$	217.1 ± 2.71	$0.00170 \pm .00077$
bIII	$5.63 \pm .037$	$18.89 \pm .120$	$4.02 \pm .360$	104.3 ± 1.19	$0.00240 \pm .00092$
bIV	$5.60 \pm .023$	$19.92 \pm .412$	$5.20 \pm .165$	69.0 ± 0.71	$0.00429 \pm .00173$

The difference equations (3.27) were used to make predictions from the egg-larval submodel for each genetic strain. To accomodate the precision of the duration estimates, we let $\delta = 0.01$. First, we took the product of the per capita fecundities and the mean adult densities in Table 3.2 as a rough estimate of the actual egg production rate for each genetic strain. These numbers are reported in Table 3.3. The onset of the stable cycle was obtained by initiating simulations with no eggs or larvae and iterating the equations until a stable point or cycle was attained. The smallest value of B which yields a stable cycle marks the begininng of the multiple attractor region (the boundary of regions I and II in Fig. 3.5b). These values of B were obtained for each genetic strain and also appear in Table 3.3. The bifurcation point for the egg production rate was predicted using D_E and D_L from Table 3.2 in (3.24), (3.25), and (3.34). The predictions for the continuous and discrete models were nearly identical and are also given in Table 3.3. With these data, we can see where the populations fall in parameter space relative to the dynamical behavior of the model.

The predicted asymptotic behavior of each genetic strain is given in the last column of Table 3.3. Only strain cI of *T. castaneum* has an unstable equilibrium point. In fact, the estimated egg production rate is well above the bifurcation point (Table 3.3). Apparently, the short durations of the egg and larval stages coupled with "extremely high cannibalism of eggs by larvae" (Park et al. 1964, p. 113) makes this genetic strain qualitatively different from the others. Of the seven remaining strains, three fall into the range where only a stable point exists and four are in the range where multiple attractors are possible (Table 3.3). Although the cannibalism rates for *T. confusum* are significantly smaller than those of *T. castaneum* (Table 3.2), both species have strains in these two regions of parameter space. Together, these eight genetic strains span the entire range of possibilities, from a single stable point attractor to a single stable orbit.

Table 3.3. Predictions from egg-larval submodel for eight genetic strains of *Tribolium*

Genetic strain	Estimated egg production rate (eggs per day)	Onset of stable cycle (eggs per day)	Bifurcation point (eggs per day)	Predicted asymptotic behavior
cI	1678	173	471	stable orbit
cII	279	338	3655	stable point
cIII	517	289	852	multiple attractors
cIV	270	236	1077	multiple attractors
bI	1316	2814	4567	stable point
bII	981	952	1722	multiple attractors
bIII	419	588	859	stable point
bIV	359	346	570	multiple attractors

3.4.4 Experimental Possibilities

Many of the biological parameters in experimental populations can be altered by changing the environmental conditions. It is well known that developmental, fecundity, and mortality rates in *Tribolium* are functions of environmental conditions such as type of flour, temperature, and relative humidity (Sokoloff 1974). Under the assumption of random contacts between "predator" and "prey", the rates of cannibalism will be a function of flour volume. Even the geometry of the container cannot be overlooked (Stanley 1932). This plasticity allows for experimental manipulations that could be used to "move" populations around in parameter space.

One of the most exciting aspects of the egg-larval submodel is that it suggests a new range of possible experiments. Environmental conditions can be varied to position populations at varying distances from the bifurcation point. Initial conditions can be varied to test for the existence of multiple attractors. Many interesting questions can be asked. Considering the stochastic nature of real populations, can populations spontaneously "jump" from one stable attractor to another? Can populations remain "balanced" for a time on an unstable orbit? In the region of the unstable orbit, will the asymptotic dynamics be indeterminate? Could some well-documented phenomena such as spontaneous population "outbreaks" and "crashes" (Mertz 1969) be due to the existence of multiple attractors? These and other questions deserve experimental study.

Empirical investigations will not be easy. Even in the simplified framework of the egg-larval submodel, six parameters must be considered: the durations of the egg and larval stages, the rate of egg production, the rate of larval cannibalism of eggs, and the rates of natural mortality for eggs and larvae. Experimental conditions must be carefully controlled. Potential confounding effects such as genetic variability must also be considered. These are substantial but not insurmountable demands.

3.4.5 Theoretical Questions and Extensions

Despite its simple form, the egg-larval submodel leaves many theoretical questions unanswered. Is there some way of precisely predicting the period of the oscillations in the neighborhood of the bifurcation point? Which parameters affect the magnitude of the oscillations? Can it be shown that at most two stable attractors exist? Recall that an assumption of no larval mortality ($\mu_L = 0$) was necessary for the analysis of the model. Although Hastings (1987) has shown that this assumption can be viewed as a first order approximation for small values of μ_L, the question of how larval mortality affects the model's dynamics is still open. In particular, is the bifurcation always subcritical when larval mortality is high? If a mathematical analysis proves intractable, it should at least be possible to investigate these questions numerically.

More questions surround the problem of age-dependent larval cannibalism. The numerical work described in §3.4.1 shows that age-dependent cannibalism can change the nature of the bifurcation from subcritical to supercritical. It would interesting to know the conditions which favor one type of bifurcation over the other. For example, the results in Fig. 3.9 suggest that a quick increase in the cannibalistic voracity with larval age favors subcritical bifurcations, while a slow increase of voracity with age favors a supercritical bifurcation. Is this true in general? How do the durations of the egg and larval stages affect this result? What is the role for the asymptotic level of cannibalism? Although these sorts of questions can be investigated with numerical studies, a mathematical analysis could provide a more complete answer.

In §3.4.2, we presented a difference equation model of the complete life cycle which included the egg-larval submodel as a component. Our analytical results went no further than to prove the existence of a unique equilbrium. The same is true for the continuous-time version of Hastings and Costantino (1987). It might be possible to apply the techniques of local stability analysis which were used on the submodel to the full model. Perhaps with some effort, it might even be possible to complete an analysis of the direction of the bifurcation. These are extremely difficult problems, but they deserve further study.

Finally, there are many possible extensions of the egg-larval submodel. A stochastic version of the egg-larval submodel could be used to study the possibility of spontaneous transitions from the point attractor to the stable cycle. This type of model could also be used to investigate the question of an indeterminate outcome for the asymptotic dynamics. Genetics could be incorporated into the egg-larval submodel. Is it possible for the multiple attractor for population densities to induce multiple attractors for gene frequencies? Can natural selection have different effects depending on initial population size? Species interactions could also be included, with both intra- and interspecific sources of egg eating. Given the fact that simple competition models can also exhibit multiple attractors with respect to competitve exclusion, four or more stable attractors might be possible, with species survival and oscillatory behavior being two characteristics of the asymptotic behavior. These and other exciting possibilities are open to investigation.

3.5 Summary and Conclusions

In the present Chapter, we have continued our analysis of demographic dynamics in *Tribolium* populations by focusing on the biological mechanism for demographic oscillations – the cannibalism of eggs by larvae. We began by suggesting that the overall population dynamics could be viewed on different time scales, with adult population densities varying on the slow time scale and egg-larval interactions occuring on the fast time scale. With this scheme, we assumed a constant adult number and considered an egg-larval submodel for population dynamics. This model was analytically tractable and yielded predictions for the equilibrium, stability, and direction of the bifurcation. In particular, the model predicts the existence of a subcritical Hopf bifurcation which implies multiple attractors for a range of parameter values. Numerical simulations revealed the coexistence of a stable point and a stable cycle. We extended the numerical analysis to include age-dependent rates of cannibalism and the complete life cycle. Life history data were used to locate several real populations in parameter space. Lastly, we discussed unresolved theoretical questions and the implications for further experimental work.

Taken together, Chapters 2 and 3 provide an intriguing picture of population dynamics in *Tribolium*. On one hand, the cannibalism of eggs and pupae by adults, together with density-dependent fecundity, regulates population densities and helps stabilize population growth. On the other hand, the cannibalism of eggs by larvae can drive huge oscillations in the immature age classes. Even so, because of adult longevity and the cannibalism of pupae, these violent oscillations become mere ripples in the density of the adult age class. But adult numbers, although oscillating only slightly, may maintain densities well above the equilibrium level. Finally, we learn that oscillations need not imply an unstable equilibrium – a stable cycle and a stable equilibrium can coexist. Ironically, all this dynamic complexity is from a technically simple experimental system that has been under study for over sixty years!

We close this chapter by pointing out that our study of cannibalistic interactions, although specific to *Tribolium* in its details, may have relevance to the demography of many natural populations. In reviews of intraspecific predation, Fox (1975) and Polis (1981) cite over forty examples where cannibalism is suggested as a regulator of population size. In fish populations, cannibalism by older or larger individuals can cause skewed age or size distributions and can sometimes result in violent fluctuations in recruitment (Ricker 1954; Radovich 1962; Forney 1976; Holcik 1977; Popova and Sytina 1977; DeAngelis et al. 1979). Naturally occurring population cycles have been attributed to age-specific cannibalistic interactions in perch (Menshutkin 1964) and in the Dungeness crab *Cancer magister* (Botsford and Wickham 1978; McKelvey et al. 1980). Our results with *Tribolium* may exemplify a more general class of demographic phenomena.

3.6 Appendix

In this appendix we derive the characteristic equation and stability boundaries for the system of difference equations (3.26).

Consider the stability matrix $\mathbf{Z}$ defined by (3.29)–(3.31). If λ is an eigenvalue of $\mathbf{Z}$ and $\mathbf{w} = \{w(\delta), w(2\delta), \ldots, w(D_E + D_L)\}^T$ is the corresponding right eigenvector, then $\lambda\mathbf{w} = \mathbf{Z}\mathbf{w}$ which implies $w(\delta) = 0$ and

$$\lambda w(x+\delta) = \begin{cases} \rho w(x) - \delta^2 c B \rho^{x/\delta} \sum\limits_{y \in L} w(y), & x \in E, \\ w(x), & x \in L,\ x \neq D_E + D_L, \end{cases} \tag{3.48}$$

or, equivalently,

$$w(x) = \begin{cases} \rho^{-1} w(x+\delta) + \delta^2 c B \rho^{x/\delta - 1} \sum\limits_{y \in L} w(y), & x \in E, \\ \lambda w(x+\delta), & x \in L,\ x \neq D_E + D_L. \end{cases} \tag{3.49}$$

For $x \in L$, this simple difference equation gives

$$w(x) = w(D_E + D_L)\lambda^{(D_E + D_L - x)/\delta}, \qquad x \in L. \tag{3.50}$$

From this one obtains

$$\sum_{y \in L} w(y) = w(D_E + D_L)\left(\frac{1 - \lambda^{D_L/\delta}}{1 - \lambda}\right). \tag{3.51}$$

Since the eigenvector $\mathbf{w}$ can be scaled by any constant factor, we can set $w(D_E + D_L) = 1$. This gives

$$w(D_E + \delta) = \lambda^{D_L/\delta - 1}. \tag{3.52}$$

Using (3.51) and (3.52) in (3.49), one can show by induction that

$$w(x) = \rho^{(x - D_E)/\delta - 1}\lambda^{(D_E + D_L - x)/\delta} + \delta^2 cB\rho^{(x/\delta)-1}\left(\frac{1 - \lambda^{D_L/\delta}}{1 - \lambda}\right)\left(\frac{1 - \lambda^{1 + (D_E - x)/\delta}}{1 - \lambda}\right), \tag{3.53}$$

for $x \in E$. Using the fact that $w(\delta) = 0$, we can let $x = \delta$ in (3.53) to obtain the characteristic equation (3.33).

For the stability boundary, let $\lambda = \exp(\omega i)$. Multiplying (3.33) by $(1-\lambda)^2$ gives

$$\exp[\omega i((D_E + D_L)/\delta - 1)]\ \{1 - \exp[\omega i]\}^2$$
$$+ \delta^2\gamma\{1 - \exp[\omega i D_E/\delta]\}\ \{1 - \exp[\omega i D_L/\delta]\} = 0. \tag{3.54}$$

The complex exponentials can be expressed in their trigonometric forms. After much simplification using trigonometric identities, one arrives at the following pair of equations for the real and imaginary parts:

$$2\{1 - \cos(\omega)\}\cos[\omega(D_E + D_L)/\delta]$$
$$+ \delta^2\gamma\{1 + \cos[\omega(D_E + D_L)/\delta] - \cos[\omega D_E/\delta] - \cos[\omega D_L/\delta]\} = 0,$$

(3.55)

$$2\{1 - \cos(\omega)\}\sin[\omega(D_E + D_L)/\delta]$$
$$+ \delta^2\gamma\{\sin[\omega(D_E + D_L)/\delta] - \sin[\omega D_E/\delta] - \sin[\omega D_L/\delta]\} = 0.$$

These equations yield the following expressions:

$$\delta^2\gamma = \frac{2\{1-\cos(\omega)\}\cos[\omega(D_E+D_L)/\delta]}{1+\cos[\omega(D_E+D_L)/\delta]-\cos[\omega D_E/\delta]-\cos[\omega D_L/\delta]},$$

$$\delta^2\gamma = \frac{2\{1-\cos(\omega)\}\sin[\omega(D_E+D_L)/\delta]}{\sin[\omega(D_E+D_L)/\delta]-\sin[\omega D_E/\delta]-\sin[\omega D_L/\delta]}. \tag{3.56}$$

Note that $\gamma > 0$ implies $\omega \neq 2\pi j$ for $j = 1, 2, 3, \ldots$. Equating these solutions, dividing by $2[1-\cos(\omega)]$, and using trigonometric identities, one obtains

$$\sin[\omega(D_E+D_L)/2\delta]\,\sin[\omega D_E/2\delta]\,\sin[\omega D_L/2\delta] = 0. \tag{3.57}$$

Possible solutions are

$$\omega = \frac{2\pi\delta j}{D_E+D_L}, \qquad j = 1, 2, 3, \ldots \tag{3.58}$$

$$\omega = \frac{2\pi\delta j}{D_E}, \qquad j = 1, 2, 3, \ldots \tag{3.59}$$

$$\omega = \frac{2\pi\delta j}{D_L}, \qquad j = 1, 2, 3, \ldots \tag{3.60}$$

However, using (3.59) in (3.56) implies $\cos(2\pi D_L/D_E) = \sin(2\pi D_L/D_E) = 0$, a contradiction. Similarly, using (3.60) in (3.56) implies $\cos(2\pi D_E/D_L) = \sin(2\pi D_E/D_L) = 0$, another contradiction. Thus, the only remaining solutions are given by (3.58). Use of (3.58) in (3.56) yields (3.34).

Chapter 4 Continuous Stochastic Models

Even under carefully controlled laboratory conditions, uncertainty is an inherent feature of *Tribolium* populations. Replicate populations which are initiated with the same initial age structures and maintained in the same incubator will exhibit variation in the densities of individuals at any given time. The densities in a single population will also fluctuate over time in an apparently random way. Deterministic models are incapable of describing this type of behavior. Variability must be described using a *stochastic* model.

Stochastic population models are widely used as theoretical constructs in population ecology. Applications include studies of population fluctuations (Kendall 1949, Leslie and Gower 1958, Bartlett et al. 1960, Goodman 1968, Levins 1969, Lewontin and Cohen 1969, May 1973, Capocelli and Ricciardi 1977, Hanson and Tuckwell 1981, Aagaard-Hansen and Yeo 1984, Ellner 1984, 1985, Arrigoni and Steiner 1985), extinction (Tuckwell 1974, Keiding 1975, Turelli 1977, Dennis 1981, Tier and Hanson 1981, Strebel 1985, Hallam and Ma 1986, Goodman 1987), and age structure (Murphy 1968, Cohen 1976, 1977a,b, 1979a,b, Tuljapurkar and Orzack 1980, Slade and Levenson 1982, Tuljapurkar 1982, 1984, 1985, 1986, Cohen et al. 1983, Heyde and Cohen 1985, Orzack 1985). Other applications and references are contained in books by Bartlett (1960a), Goel and Richter-Dyn (1974), Ludwig (1974), May (1974a), Pielou (1977), Roughgarden (1979), and Nisbet and Gurney (1982).

The present chapter and the next one are about probabilities; specifically, these two chapters deal with the probability densities that arise in populations of *Tribolium* which contain some inherent features of uncertainty. We begin by formulating a simple detertministic model for the changes in adult numbers. Next we consider a stochastic analogue of this model: a stochastic differential equation (SDE). In the SDE model the stochastic process is continuous in both time and population number. In Chapter 5 we examine a "birth-death" process in which, again, the population changes continuously with time, but its size is recorded as a nonnegative integer, i.e., the state space is discrete. In both cases the overall goal is to associate biological attributes such as birth, death, and cannibalism rates with the stochastic dynamics of adult number. Variability is no longer dismissed as random error; it is an additional source of information about the population.

4.1 Deterministic Model for Adult Numbers

Among the insights obtained from the demographic analyses of Chapters 2 and 3 is the recognition that the interactions between adults and immatures occur on a "slow" time scale relative to the "fast" time scale for the interactions within the immature age classes.

As a consequence of this observation, which was also emphasized by Lloyd (1968, p. 257), we have an opportunity to obtain some information about the dynamics of adult numbers without specifying the enormously complex interactions among all the beetle life stages.

The model attempts to explain changes in the size of the adult beetle population based on the association between the number of adults and the number of the progeny produced by these adults. If we let c be the *per capita* rate at which adults prevent a potential recruit (pupa or incompletely sclerotized adult) from entering the adult population and if we assume that the adults act independently, then the proportion of recruits which survive to adulthood is $\exp[-cN(t)]$, where $N(t)$ is the total number of adults (Dennis 1981). Coupling this density dependent function together with the density independent rates of reproduction, b, and adult mortality, μ, we have

$$\frac{dN}{dt} = N(t)\,(b\exp[-c\,N(t)] - \mu) \tag{4.1}$$

as an ordinary differential equation for the rate of change in adult numbers. We shall refer to (4.1) as the *adult recruitment* model. Although this equation is a gross oversimplification of the true complexity inherent in *Tribolium* populations, it has served as a useful paradigm in many studies on beetle population dynamics as reviewed in Chapter 1 as well as in fisheries biology (Ricker 1954, Levin and Goodyear 1980, May 1980, McKelvey et al. 1980) and in general population theoretical studies (Moran 1950, May 1974b, 1976, Maynard Smith 1968a, 1974, Hoppensteadt 1975, Oster 1976, May and Oster 1976, Guckenheimer et al. 1977, Thieme 1979, Hunt 1980, Fisher et al. 1979, Cull 1981, Desharnais and Costantino 1983).

The dynamics of the population equation (4.1) are straightforward. If $b > \mu$, the equilibrium number of adults given by

$$N^* = c^{-1}\log\left(\frac{b}{\mu}\right) \tag{4.2}$$

is globally stable. If $b < \mu$, then the population will become extinct. In the neighborhood of N^*, the rate of approach to equilibrium is given by the eigenvalue, $-\lambda$, where

$$\lambda = \mu\log\left(\frac{b}{\mu}\right). \tag{4.3}$$

Since $N^* > 0$ implies $b > \mu$, $\lambda > 0$; the equilibrium is always stable.

4.2 Stochastic Model for Adult Numbers

4.2.1 Stochastic Differential Equation

In the stochastic version of the population model (4.1), infinitesimally small changes in adult number $dN(t)$ are assumed to be the result of both deterministic and unspecified "random" forces. The deterministic forces are given by the balance between recruitment and mortality rates. For the random forces, we follow the procedure of Dennis and Patil (1984) which involves the addition of a random variable, $d\omega(t)$, known as a *Wiener increment*, to the *per capita* growth rate. The model becomes

$$dN(t) = N(t)\,(b\exp[-c\,N(t)] - \mu)\,dt + \sigma N(t)\,d\omega(t)\,, \tag{4.4}$$

where the Wiener increment has the following properties

$$\begin{aligned} E[d\omega(t)] &= 0\,, & t &\geq 0\,, \\ E[d\omega(t)^2] &= dt\,, & t &\geq 0\,, \\ E[d\omega(t_1)d\omega(t_2)] &= 0\,, & t_1 &\neq t_2\,. \end{aligned} \tag{4.5}$$

The stochastic function $d\omega(t)$ is the continuous-time equivalent of a discrete random variable with no serial autocorrelations. The parameter σ is a measure of the amplitude of the random fluctuations. Dividing (4.4) by dt gives the stochastic differential equation (SDE)

$$\frac{dN(t)}{dt} = N(t)\,(b\exp[-c\,N(t)] - \mu + \sigma\gamma(t))\,, \tag{4.6}$$

where the derivative

$$\gamma(t) = \frac{d\omega(t)}{dt} \tag{4.7}$$

is known as *Gaussian white noise*. Nisbet and Gurney (1982) provide a lucid discussion of the assumptions involved in the SDE approach.

4.2.2 Stationary Distribution

SDE's such as (4.6) can be connected with a probability distribution for population size through the use of the *Kolomogorov diffusion equation*

$$\frac{\partial f(N,t)}{\partial t} = -\frac{\partial}{\partial N}[m(N)f(N,t)] + \frac{1}{2}\frac{\partial^2}{\partial N^2}[v(N)f(N,t)]\,, \tag{4.8}$$

where $f(N,t)$ is the probability density function for population size N at time t. The functions $m(N)$ and $v(N)$ are, loosely speaking, the infinitesimal mean and variance of the rate of change in population size. The population is analogous to a particle diffusing along the single dimension N, with $m(N)$ representing the "flow" and $v(N)$ representing the "drift." Both $m(N)$ and $v(N)$ are computed from the SDE. See Roughgarden (1979, p. 69), Crow and Kimura (1970, p. 371), and May (1973, p. 114) for applications of (4.8) in population biology.

Unfortunately, two methods of stochastic integration have been developed to obtain $m(N)$: the Ito and Stratonovich rules (see Soong 1973, Karlin and Taylor 1981, Horsthemke and Lefever 1984, Mangel 1985). To present the analyses for both methods, we imitate Dennis and Patil (1984) and define the following "indicator variable"

$$\zeta = \begin{cases} 1 & \text{for Ito calculus,} \\ 0 & \text{for Stratonovich calculus.} \end{cases} \tag{4.9}$$

The SDE (4.6) gives

$$\begin{aligned} m(N) &= N\,[b\exp(-cN) - \mu + (1-\zeta)(\sigma^2/2)]\,, \\ v(N) &= \sigma^2 N^2 \end{aligned} \tag{4.10}$$

for the functions measuring flow and drift. Note that with the Ito calculus the flow is the same as the growth rate in the deterministic model (4.1), but with the Stratonovich calculus the stochastic term affects $m(N)$. Fortunately, for our purposes, the Ito and Stratonovich methods yield qualitatively similar results.

In principle, one can substitute (4.10) into (4.8) to obtain the the time-dependent probability distribution $f(N,t)$ for adult numbers. In practice, the resulting partial differential equation can rarely be solved when the original model is nonlinear. However, if the deterministic and random forces eventually come to a balance, then the resulting *stationary* or *steady-state distribution* of population size will be independent of time. Setting $\partial f/\partial t = 0$ in (4.8) gives

$$f^*(N) = \left(\frac{k_0}{v(N)}\right)\exp\left[2\int\left(\frac{m(N)}{v(N)}\right)dN\right], \tag{4.11}$$

where k_0 is a constant which makes the integrated probability equal to one. Substituting (4.10) into (4.11) gives

$$f^*(N) = k\,N^{[2(b-\mu)/\sigma^2]-\zeta-1}\exp\left[\left(\frac{2b}{\sigma^2}\right)\sum_{i=1}^{\infty}\frac{(-cN)^i}{i(i!)}\right] \tag{4.12}$$

for the stationary distribution of adult numbers, where $k = k_0/\sigma^2$.

The infinite series in (4.12) is rather unwieldy. Costantino and Desharnais (1981) used a linear approximation to this series which is accurate near the deterministic equilibrium N^*. This yielded a *gamma* probability distribution for the steady state. An alternative derivation of the gamma distribution was suggested by Dennis and Patil (1984) and applied to *Tribolium* by Dennis and Costantino (1988). These authors returned to the original stochastic population model (4.6) and expanded the *per capita* growth rate in a Taylor series around the deterministic equilibrium point N^* given by (4.2). The approximating stochastic model is

$$\frac{dN}{dt} = N(t)\,[\lambda - c\mu N(t) + \sigma\gamma(t)]\ , \tag{4.13}$$

where λ is given by (4.3). Using (4.13) gives

$$\begin{aligned} m(N) &= N\,[\lambda - c\mu N + (1-\zeta)(\sigma^2/2)]\ , \\ v(N) &= \sigma^2 N^2, \end{aligned} \tag{4.14}$$

which, upon substitution into (4.11), yields

$$f^*(N) = k\,N^{\alpha-1}\exp(-\beta N)\,, \qquad 0 \le N \le \infty\,, \tag{4.15}$$

where

$$\begin{aligned} \alpha &= \frac{2\lambda}{\sigma^2} - \zeta\,, \\ \beta &= \frac{2c\mu}{\sigma^2}\,. \end{aligned} \tag{4.16}$$

The integration constant k is given by

$$k = \frac{\beta^\alpha}{\Gamma(\alpha)}\,, \tag{4.17}$$

where $\Gamma(\cdot)$ is the well-known gamma function (Abramowitz and Stegun 1964, p. 255). The gamma distribution (4.15) is defined for values of $\alpha > 0$ and $\beta > 0$.

4.2.3 Properties of the Stationary Distribution

A major advantage of the gamma approximation is that the moments and other properties of the distribution are known. The mean and variance of the gamma are given by

$$\begin{aligned} \mathrm{E}[N] &= \alpha\beta^{-1}, \\ \mathrm{Var}[N] &= \alpha\beta^{-2}, \end{aligned} \tag{4.18}$$

respectively. When $\alpha > 1$, the gamma has a single mode at

$$\text{mode} = (\alpha - 1)\beta^{-1} \tag{4.19}$$

and points of inflection which are equidistant from the mode at

$$\text{inflection points} = \beta^{-1}[(\alpha - 1) \pm \sqrt{\alpha - 1}]\,. \tag{4.20}$$

The distribution function of the gamma density is given by

$$F^*(N) = \int_0^N f^*(\xi)\,\mathrm{d}\xi\,, \tag{4.21}$$

for which there is no simple closed form solution. The median of the gamma distribution is defined as the solution to $F^*(\text{median}) = 0.5$.

A typical example of the gamma density function is given in Fig. 4.1. Parameter values of $\alpha = 5$ and $\beta = 0.1$ were used. This distribution has a mode at 40 adults and a mean at 50 adults. The median equals 46.709 adults. The shaded areas represent $F^*(15) = 0.0186$ and $1 - F^*(100) = 0.0293$. In this example, we would expect the population to consist of less than 15 adults 1.86% of the time and more than 100 adults 2.93% of the time.

The shape parameter of the gamma distribution is α and the scaling parameter is β. While a typical shape of the gamma is as given in Fig. 4.1, it can undergo marked alterations. The exponential distribution is a special case of the gamma when $\alpha = 1$. The chi-square distribution with υ degrees of freedom is also a gamma with $\alpha = \upsilon/2$ and $\beta = 0.5$. As α gets large, $f^*(N)$ approaches a Gaussian normal distribution (Johnson and Kotz 1970). The model is simple; yet it is flexible enough to have a variety of applications (Dennis and Patil 1984).

The most important property of $f^*(N)$ is that we can associate the gamma density function for adult numbers with the biological entities b, μ, c, and σ of the stochastic model (4.6). Substituting (4.16) into (4.18), we have

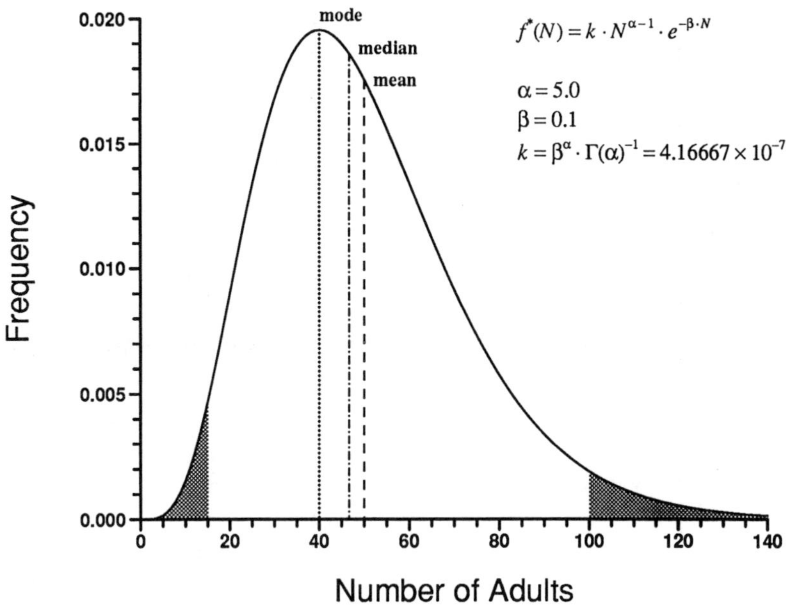

Fig. 4.1. Typical gamma probability density function with $\alpha = 5$ and $\beta = 0.1$. The mode, median, and mean are shown. The shaded areas represent the probabilities that adult numbers fall below 15 and above 100.

$$\begin{aligned} \mathrm{E}[N] &= N^* - \zeta\left(\frac{\sigma^2}{2c\mu}\right), \\ \mathrm{Var}[N] &= \left(\frac{\sigma^2}{2c\mu}\right)\left[N^* - \zeta\left(\frac{\sigma^2}{2c\mu}\right)\right] \end{aligned} \tag{4.22}$$

for the mean and variance of adult numbers, where N^* is the deterministic equilibrium point (4.2). With these expressions it is possible to predict the effects of a change in one of the biological parameters on the stationary distribution of adult numbers.

Several stationary distributions are sketched in Fig. 4.2. In each of the four panels, the solid curve is a common gamma with the following parameter values: $b = 0.40$, $\mu = 0.028$, $c = 0.044$, $\sigma^2 = 0.025$, and $\zeta = 1$ (Ito calculus). Associated with this curve are two distributions in which one of the parameters (either b, μ, c, or σ^2) was first reduced by half and then increased two-fold (except for σ^2). In Fig. 4.2a, the three gammas represent changes in b from 0.40 to 0.20 and from 0.4 to 0.80 with the other parameters the same as those in the common gamma. An increase in b resulted in an increase in both the mean and variance of the stationary distribution. In Fig. 4.2b, the three gammas are for μ = 0.014, 0.028 and 0.056. An increase in the adult death rate decreased both the mean and variance of $f^*(N)$. In Fig. 4.2c the curves are for $c = 0.022$, 0.044 and 0.088. An increase in the adult inhibition of recruitment reduced quite markedly the mean and

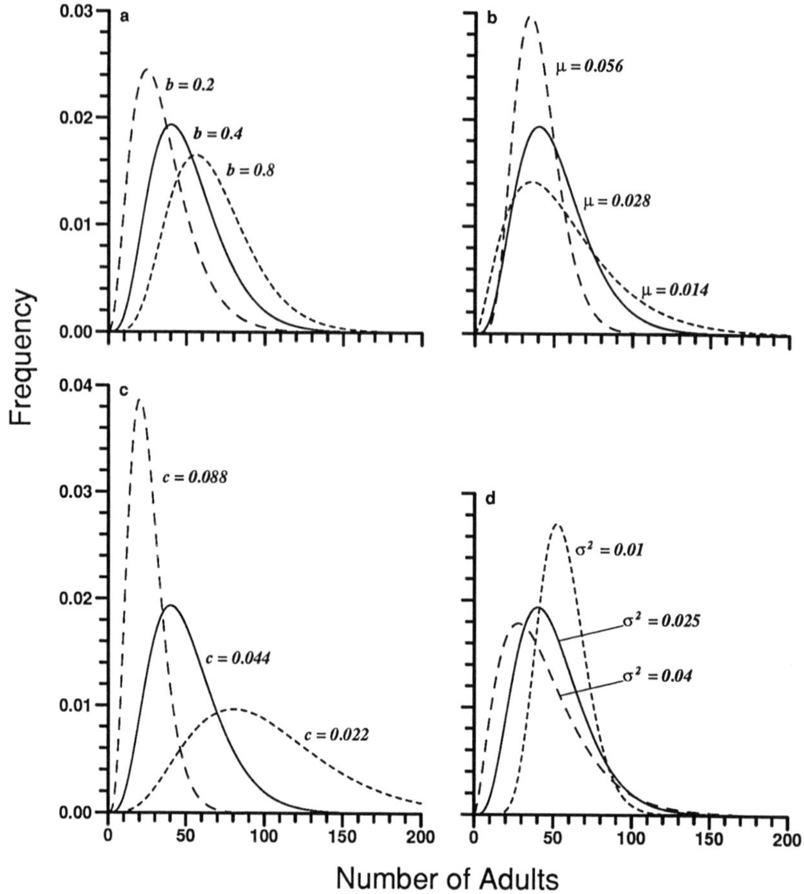

Fig. 4.2. Effects of changes in the parameters b, μ, c, and σ^2 on the stationary gamma density function. The solid curve is a gamma which is common to all four panels.

variance of the steady state. In Fig. 4.2d, we let $\sigma^2 = 0.01$, 0.025 and 0.04. As the amplitude of the stochastic fluctuations increased so also did the variance of $f^*(N)$. However, the expected number of adults was reduced slightly. (For the Stratonovich calculus the expected number of adults is unchanged.) The magnitude of the positive constant σ in the SDE model (4.13) can also result in a qualitative shift in the shape of the gamma. When

$$\sigma^2 = \frac{2\lambda}{1+\zeta}, \tag{4.23}$$

then $\alpha = 1$ and there is a "noise-induced transition" (Hamada 1981, Horsthemke and Lefever 1984) to an exponential stationary distribution. In this numerical example, $\sigma^2 = 0.07446$ is the critical value.

4.3 Statistical Procedures

4.3.1 Data Organization

The observations considered in this chapter are the number of adults recorded in many independent replicate *Tribolium* cultures. For the moment, our analysis is restricted to data collected in the region of the steady state. In §4.6, when we look at the transient or time-dependent distribution, rather than just the limiting gamma distribution, we will include all of the observations. We begin with a discussion of how the data are organized to obtain observed distributions of adult number.

We will consider data on the Chicago strain of *T. castaneum* (Lloyd 1965) as a representative example. Twelve individual replicates of the Chicago strain were initiated with 13 very small larvae (4–5 days old), 10 small larvae (13–14 days old), 8 medium larvae (22–23 days old), 7 large larvae (about 32 days old), 4 sexually immature adults (0–1 day after ecolsion, sexes equal) and 4 mature adults (9–10 days after ecolsion, sexes equal). The cultures were maintained for 36 weeks in incubators at 24±0.6° C and 31±5% relative humidity. At weekly intervals the cultures were removed from the incubator and censused. All of the life stages were counted and then all living stages were returned to fresh medium.

These data clearly are time series observations so it is worth mentioning some of the statistical difficulties associated with the analysis of time series observations on replicate populations. On the one hand, the stationary distribution of population size is an abstraction which is technically applicable only to a very large (infinite) ensemble of independent populations. On the other hand, the ergodicity of the stochastic process will yield a gamma distribution if the population is sampled at equal time intervals for a very long (infinite) period of time. Of course, any realistic experiment will be intermediate to these extremes. Our naive approach has been to pool observations among replicates of the same treatment to perform our analyses; consequently, the adult numbers for all of the replicates of the Chicago strain were used to establish the grouped frequency distribution for this data set. A more sophisticated approach would take into account observations within and between replicates. We discuss this problem in §4.3.5.

We plotted the census data on adult numbers for the Chicago strain in Fig. 4.3. In the upper graph (Fig. 4.3a), we present adult numbers as a function of time for each of the twelve replicates. The lower graph (Fig. 4.3b) is a series of grouped frequency histograms for each census as a function of time. Two important observations can be made. First, the data show clearly the variability in adult numbers within and between populations. Second, these populations went through a transient phase of sigmoidal growth and then fluctuated around an upper asymptote. It is the latter phase that we will characterize as a time-independent "steady-state."

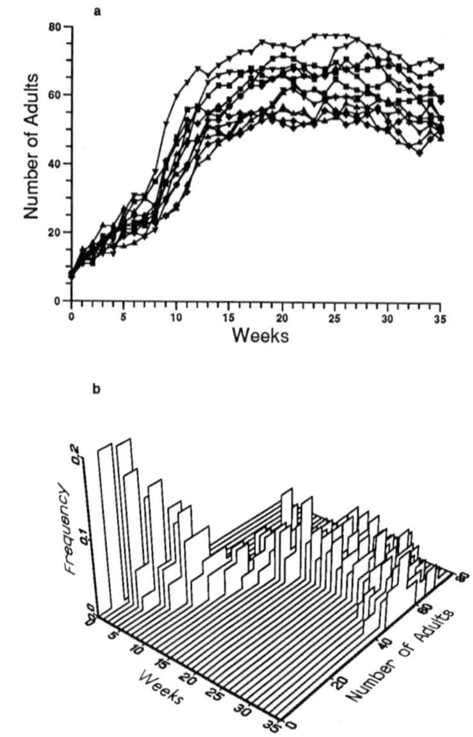

Fig. 4.3. *T. castaneum*, Chicago strain. The upper graph (a) is the number of adults recorded at weekly intervals in each of 12 replicate cultures. The lower graph (b) is a three-dimensional plot of histograms of adult number over time. The class intervals are 0–4, 5–9, 10–14, etc. Each histogram involves only 12 replicates, but the general idea of a *time-dependent* distribution is represented. Data are from Lloyd (1965).

Somewhat arbitrarily, we chose the 17 week interval from week 19 to the end of the experiment at week 35 as the steady-state. Class intervals $(s_0=0, s_1]$, $(s_1, s_2]$, $(s_2, s_3]$, ..., $(s_{m-2}, s_{m-1}]$, $(s_{m-1}, s_m=\infty]$ were selected to partition the entire sample space of adult numbers. The quantities $y_1, y_2, y_3, \ldots, y_{m-1}, y_m$ represent the counts of the observations grouped into the m interval classes. For the Chicago data set, we chose $m=7$ class intervals: $(0,48]$, $(48,53]$, $(53,58]$, $(58,63]$, $(63,68]$, $(68,73]$, and $(73,\infty]$. From this we obtained $y_1=7, y_2=34, y_3=45, y_4=42, y_5=38, y_6=23$, and $y_7=15$ for a total of $n=204$ observations. The values $y_1/n, y_2/n, \ldots, y_m/n$ form the observed stationary probability distribution.

4.3.2 Parameter Estimation

With the data organized into a grouped frequency distribution, we can use the procedure described by Dennis and Costantino (1988) to obtain estimates of the parameters α and β which specify the gamma distribution.

Let p_j denote the probability that an observation falls within the interval $(s_{j-1}, s_j]$. Under the gamma hypothesis,

$$p_j(\alpha, \beta) = F^*(s_j) - F^*(s_{j-1}), \tag{4.24}$$

where $F^*(s)$ is the cumulative distribution function given by (4.21). If the original observations are independent, then the probability of obtaining the grouped data set y_1, $y_2, \ldots, y_m$ is given by the multinomial likelihood function

$$L(\alpha, \beta) = n! \prod_{j=1}^{m} \left(\frac{p_j(\alpha, \beta)^{y_j}}{y_j!} \right). \tag{4.25}$$

The maximum likelihood estimates of α and β are the values which maximize the probability (4.25). They can be found by solving for the roots of the equations:

$$\begin{aligned} \frac{\partial \log L}{\partial \alpha} &= \sum_{j=1}^{m} \left(\frac{y_j}{p_j} \right) \left(\frac{\partial p_j}{\partial \alpha} \right) = 0, \\ \frac{\partial \log L}{\partial \beta} &= \sum_{j=1}^{m} \left(\frac{y_j}{p_j} \right) \left(\frac{\partial p_j}{\partial \beta} \right) = 0. \end{aligned} \tag{4.26}$$

These equations can be solved by performing a nonlinear least squares regression (Jennrich and Moore 1975). The regression uses the values $y_1, y_2, \ldots, y_m$ as observations on the dependent variable, and $np_1(\alpha, \beta), np_2(\alpha, \beta), \ldots, np_m(\alpha, \beta)$ – the expected values of the y_j's – become the model to be fit. The nonlinear regression must be "iteratively reweighted;" that is, weights of $[np_j(\alpha, \beta)]^{-1}$ must be computed at every iteration. The resulting least squares estimates of α and β are the maximum likelihood estimates.

The maximum likelihood estimates have a asymptotic multivariate normal distribution as $n \to \infty$ (see, for example, Bishop et al. 1975, p. 509). The variance-covariance matrix is

$$\Sigma = \begin{pmatrix} \dfrac{\theta_1}{\theta_1\theta_3 - \theta_2^2} & \dfrac{-\theta_2}{\theta_1\theta_3 - \theta_2^2} \\ \\ \dfrac{-\theta_2}{\theta_1\theta_3 - \theta_2^2} & \dfrac{\theta_3}{\theta_1\theta_3 - \theta_2^2} \end{pmatrix}, \tag{4.27}$$

where

$$
\begin{aligned}
\theta_1 &= n \sum_{j=1}^{m} p_j^{-1} \left(\frac{\partial p_j}{\partial \beta}\right)^2, \\
\theta_2 &= n \sum_{j=1}^{m} p_j^{-1} \left(\frac{\partial p_j}{\partial \alpha}\right)\left(\frac{\partial p_j}{\partial \beta}\right), \qquad (4.28) \\
\theta_3 &= n \sum_{j=1}^{m} p_j^{-1} \left(\frac{\partial p_j}{\partial \alpha}\right)^2,
\end{aligned}
$$

with $p_j = p_j(\alpha, \beta)$ as defined in (4.24). The elements in Σ are estimated by substituting the maximum likelihood estimates of α and β; the derivatives in (4.28) can be computed numerically. The square roots of the diagonal elements in Σ provide standard errors for α and β.

For the Chicago data set, $\alpha = 54.717 \pm 6.089$, $\beta = 0.90476 \pm 0.10116$, and

$$
\Sigma = \begin{pmatrix} 37.073 & 0.61362 \\ 0.61362 & 0.010233 \end{pmatrix}.
$$

4.3.3 Testing the Gamma Hypothesis

A test for the goodness of fit of the gamma distribution is accomplished using the maximum likelihood estimates. After convergence of the parameter estimates is achieved, the final value of the weighted residual sum of squares is

$$
\chi^2 = \sum_{j=1}^{m} \frac{[y_j - n\, p_j(\alpha, \beta)]^2}{n\, p_j(\alpha, \beta)}, \qquad (4.29)
$$

which is the Pearson chi-square statistic. Under the null hypothesis that the gamma model fits, χ^2 has a large sample chi-square distribution with $m-3$ degrees of freedom.

The observed and expected distributions for the Chicago strain are presented in Fig. 4.4. The critical chi-square value which must be exceeded to reject the gamma at the 0.01 level of probability with 4 degrees of freedom is 13.3. The calculated value of 5.26 allows us to accept the hypothesis of characterizing the steady state probability distribution of adult numbers as a gamma distribution.

A significant overall goodness of fit test may be followed with tests for significant deviations of the model in individual abundance classes. The following quantity can be regarded as a generalized residual for the jth class:

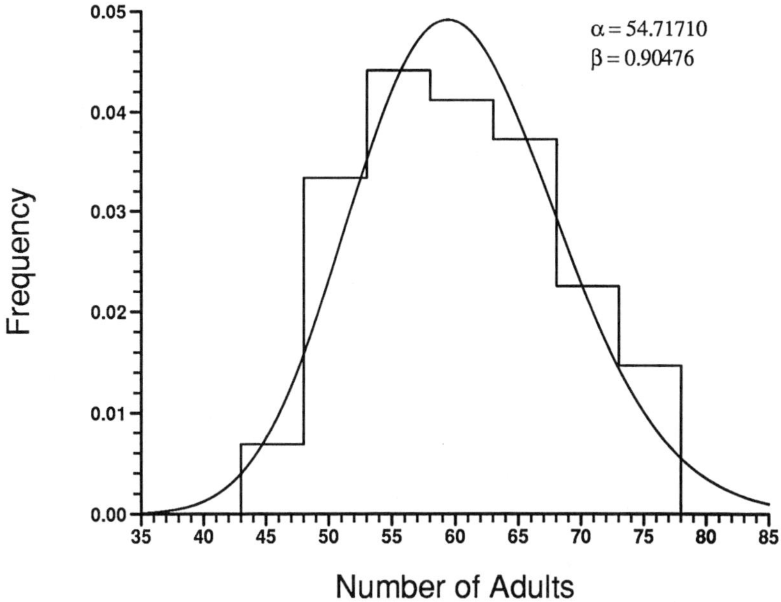

Fig. 4.4. Steady-state distribution of the Chicago strain of *T. castaneum*. The histogram is the observed grouped frequency distribution and the smooth curve is the fitted gamma probability density function.

$$\delta_j = \frac{[y_j - n\, p_j(\alpha, \beta)]}{\sqrt{n\, p_j(\alpha, \beta)}}. \tag{4.30}$$

The Pearson chi-square statistic (4.29) is the sum of the squared values of δ_j. The δ_j's have a large sample normal distribution, under the null hypothesis that the model fits, with a mean of zero and a variance of

$$\nu_j = 1 - p_j - (n\, p_j^{-1})\, q_j^{\mathrm{T}} \Sigma q_j\,, \tag{4.31}$$

where q_j is the vector of partial derivatives given by $q_j^{\mathrm{T}} = [\partial p_j / \partial \alpha, \partial p_j / \partial \beta]$ (e.g. Rao 1973, p. 394). This variance may be estimated by substituting the maximum likelihood estimates of α and β. The partial derivatives in q_j can be computed numerically. The test statistic for the jth class is given by

$$Z_j = \frac{\delta_j}{\sqrt{\nu_j}}. \tag{4.32}$$

Z_j has a large sample standard normal distribution under the null hypothesis that the model fits the jth class.

4.3.4 Comparing Two Distributions

The multivariate normal distribution of the maximum likelihood estimates may be used for comparing parameters from two gamma distributions.

Let α_1 and β_1 represent the parameter estimates from the first gamma distribution and let α_2 and β_2 be the estimates from the second distribution. The null hypothesis is

$$\mathrm{H}_0: \begin{pmatrix} \alpha_1 \\ \beta_1 \end{pmatrix} = \begin{pmatrix} \alpha_2 \\ \beta_2 \end{pmatrix};$$

i.e., both gammas have identical parameters. Since (α_1, β_1) and (α_2, β_2) both converge in distribution to multivariate normals, the difference

$$\Delta = \begin{pmatrix} \alpha_1 - \alpha_2 \\ \beta_1 - \beta_2 \end{pmatrix} \tag{4.33}$$

will also converge to a multivariate normal with a zero mean vector and a variance-covariance matrix of Σ if H_0 is true. The test statistic

$$D^2 = \Delta^{\mathrm{T}} (\Sigma_1 + \Sigma_2)^{-1} \Delta \tag{4.34}$$

can be used to test H_0, where Δ, Σ_1, and Σ_2 are computed using the maximum likelihood estimates. Under the null hypothesis, D^2 has a (large sample) chi-square distribution with 2 degrees of freedom.

4.3.5 The Problem of Time-Dependence

An underlying assumption for the maximum likelihood estimates of α and β is that the frequency counts $y_1, y_2, \ldots, y_m$ form a sample from a multinomial distribution. The assumption is not strictly valid if the underlying observed population sizes are time series data. For time series data, (4.25) is not the true likelihood function for the frequency counts. However, as the time interval between observations becomes large, the likelihood function for the counts approaches (4.25) asymptotically. Thus, the question becomes whether the observations are far enough apart to justify using (4.25) as a likelihood function.

This question can be partly addressed by examining the empirical autocorrelation structure of the data using standard time series methods. If the autocorrelation between observations is small, then it is reasonable to compute maximum likelihood estimates for the stationary gamma with the methods described. In §4.5, we use a linearization of the stochastic model (4.6) to derive an expected theoretical autocorrelation function.

These maximum likelihood estimates may be valid even when the interval between observations is small, due to the ergodic nature of the process $N(t)$. As t becomes large, the proportion of time that the process spends in the abundance interval $(s_{j-1}, s_j]$ approaches the proportion $p_j(\alpha, \beta)$ given by (4.24). If the process was sampled at equal time intervals, the proportion y_j/n of samples found in the abundance interval $(s_{j-1}, s_j]$ should approach $p_j(\alpha, \beta)$ asymptotically. Thus, estimating α and β by maximizing (4.25) appears to be a reasonable procedure, even when the observations of the process $N(t)$ are closely spaced. Two qualifications are: (i) the process must be observed for a long time, and (ii) the process should be observed at equally spaced intervals.

Recent results by Gleser and Moore (1985) indicate the chi-square goodness of fit test (4.29) will reject the null hypothesis too often when the observations are closely spaced. Thus, if the hypothesis of goodness of fit is not rejected by the chi-square test, one can be fairly confident that H_0 would also not be rejected using the true distribution of the test statistic. The effects of autocorrelation on the other statistical inferences described here, unfortunately, remain unknown at this time.

4.4 Applications

Stochastic models provide a framework for making probabilistic statements about the dynamics of a population and allow an analysis of population abundance variability. Although these are obvious advantages to the theoretician, they make experimental evaluations of stochastic population theory extremely difficult. Large numbers of observations are required to test probabilistic statements. Highly accurate estimates of population abundance must be used to eliminate sampling error as a source of variation. Experiments must contain many replicates and consist of long time series. For these reasons, there is a paucity of real data which can be used to evaluate stochastic models (Nisbet and Gurney 1982, p. 186). Data on the flour beetle are available.

In this section we present several observed stationary distributions. The latter were obtained from different laboratories and our descriptions of these experiments are necessarily brief; readers should consult the original sources for more details.

4.4.1 Different Environments

One possible application of the gamma distribution is the comparison of populations under different environmental regimes. As an example, we examine two species of *Tribolium* cultured at two different temperatures.

Thomas Park provided P. H. Leslie with data on *T. confusum* and *T. castaneum* for climate III (29° C, 70% RH) and climate V (24° C, 70% RH) of his studies on species competition (Park 1954). The observed frequency distributions of Park together with the predictions from Leslie's model are given in Table 9 of Leslie (1962). The four grouped frequency distributions were based on adult numbers recorded at day 360 onward to day 720 at 30 day intervals.

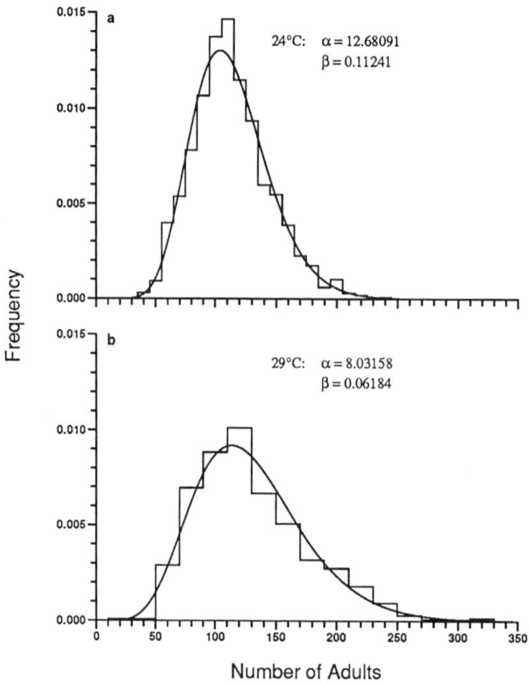

Fig. 4.5. Steady-state distributions for *T. confusum* cultured at (a) 24° C and at (b) 29° C. The histograms are the observed grouped frequency distributions and the smooth curves are the fitted gamma probability density functions. [Modified from R. F. Costantino and R. A. Desharnais (1981), Gamma distributions of adult numbers for *Tribolium* populations in the region of their steady states. *J. Anim. Ecol.* **50**: 667–681.]

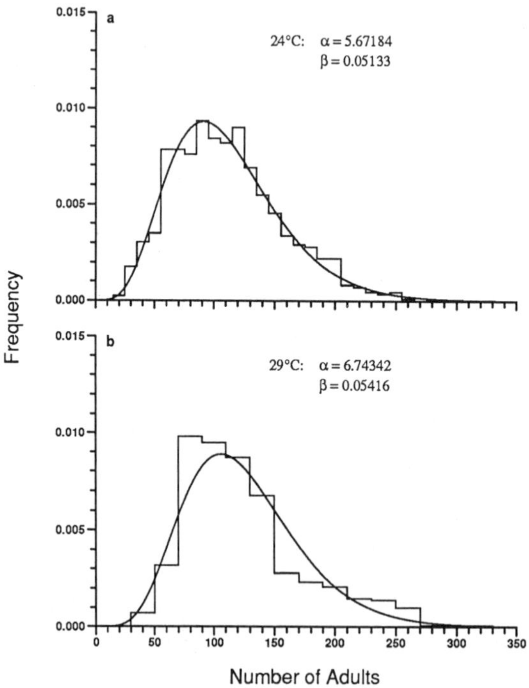

Fig. 4.6. Steady-state distributions for *T. castaneum* cultured at (a) 24° C and at (b) 29° C. The histograms are the observed grouped frequency distributions and the smooth curves are the fitted gamma probability density functions. [Modified from R. F. Costantino and R. A. Desharnais (1981), Gamma distributions of adult numbers for *Tribolium* populations in the region of their steady states. *J. Anim. Ecol.* **50**: 667–681.]

The observed frequency histograms for *T. confusum* cultured at 24° C and 29° C are presented in Fig. 4.5. The smooth curves are the gamma distributions with the parameters set equal to their maximum likelihood estimated values. In both data sets, the hypothesis that the data fit a gamma distribution was accepted at the 0.01 level of probability. For *T. castaneum* cultured at 24° C the hypothesis was also accepted, however, at 29° C the hypothesis of a gamma distribution was rejected (Fig. 4.6).

The computed gamma distributions reveal that for both species the mean adult number was larger at 29° C than at 24° C and that the variance of the distributions was larger at the higher temperature. Referring to the mean and variance of adult numbers (4.22) given in terms of the biological parameters b, μ, and c, the model suggests several explanations: (i) a decrease in the probability of adult inhibition of the immatures, (ii) a decrease in the mortality rate among adults, or (iii) an increase in the rate that pupae are are available for recruitment into the adult population. Leslie (1962, p. 8) pointed out that the mean length of time from egg to adult is approximately twice as long at 24° C than at 29° C. Although other possibilities cannot be ruled out, this observation is consistent with the third explanation.

4.4.2 Genetic Strains

Another application of the gamma distribution is the comparison of different genetic strains and species. As an example, we consider the genetic strains bI–bIV of *T. confusum* and the strains cI–cIV of *T. castaneum*. Our source of data is the grouped frequency distributions of the adult numbers from day 360 onward which are presented in Appendix Table N in Park et al. (1964).

The observed and gamma expected distributions for the four strains of *T. confusum* are presented in Fig. 4.7. For strains bII and bIV, the gamma was accepted at the 0.01 level of probability. For strains bI and bIII the gamma hypothesis was rejected.

The observed gamma steady-state distributions reflect the considerable differences that exist among these genetic strains. Park et al. (1964, p. 105) seem quite certain that cannibalism, c, is an important characteristic for differentiating these strains. In general, strains bIII and bIV were much more cannibalistic than strains bI and bII (Park et al. 1964, p. 150). Our interpretation of their data is as follows: the rates of cannibalism among the four strains are $c_4 > c_3 \gg c_2 > c_1$. The model predicts that the expected number of adults will rank $\mathrm{E}[N_1] > \mathrm{E}[N_2] \gg \mathrm{E}[N_3] > \mathrm{E}[N_4]$ and that the variances of adult numbers will rank $\mathrm{Var}[N_1] > \mathrm{Var}[N_2] \gg \mathrm{Var}[N_3] > \mathrm{Var}[N_4]$. These qualitative statements agree with the experimental data.

The observed and gamma expected distributions for the four strains of *T. castaneum* are presented in Fig. 4.8. For strains, cIII and cIV the gamma hypothesis was accepted. In the cases of strains cI and cII, this hypothesis was rejected.

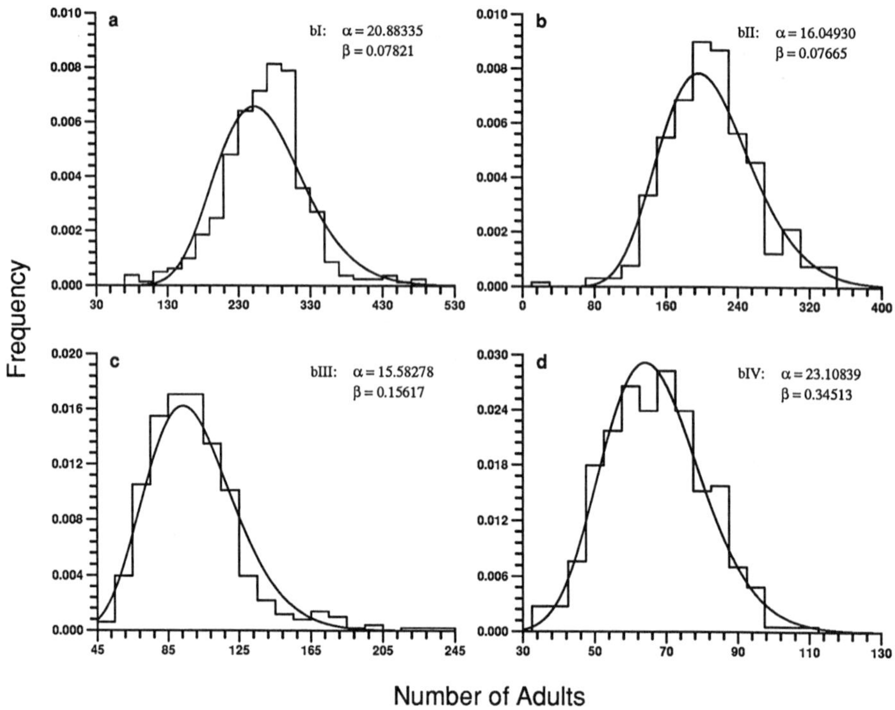

Fig. 4.7. Steady-state distributions for four genetic strains of *T. confusum*: (a) bI, (b) bII, (c) bIII, and (d) bIV. The histograms are the observed grouped frequency distributions and the smooth curves are the fitted gamma probability density functions. [Modified from R. F. Costantino and R. A. Desharnais (1981), Gamma distributions of adult numbers for *Tribolium* populations in the region of their steady states. *J. Anim. Ecol.* **50**: 667–681.]

The equilibrium distributions for the *T. castaneum* data can be interpreted as follows. From Park et al. (1964, p. 150), the rates of cannibalism among the strains are $c_1 > c_4 > c_2 > c_3$. For the mean number of adults, the model predicts that $E[N_3] > E[N_2] > E[N_4] > E[N_1]$. The qualitative agreement between the theory and the data is correct for strains cIII, cII, and cIV but is completely wrong for strain cI which had the largest mean number of adults and not the smallest as predicted. The model also predicts that the variances of adult numbers will have the following ranking: $\mathrm{Var}[N_3] > \mathrm{Var}[N_2] > \mathrm{Var}[N_4] > \mathrm{Var}[N_1]$. Strain cI did not have the smallest observed variance but the largest followed by strains cII, cIII and cIV in that order.

Figure 4.9 allows a comparison of the fitted gamma distributions for all eight genetic strains. The large differences among the steady-states for these strains are quite apparent. Qualitatively, the stochastic model does a reasonable job of describing these differences. The important exception is strain cI, which is poorly fit by a gamma (Fig. 4.8a) and has an extremely large mean and variance.

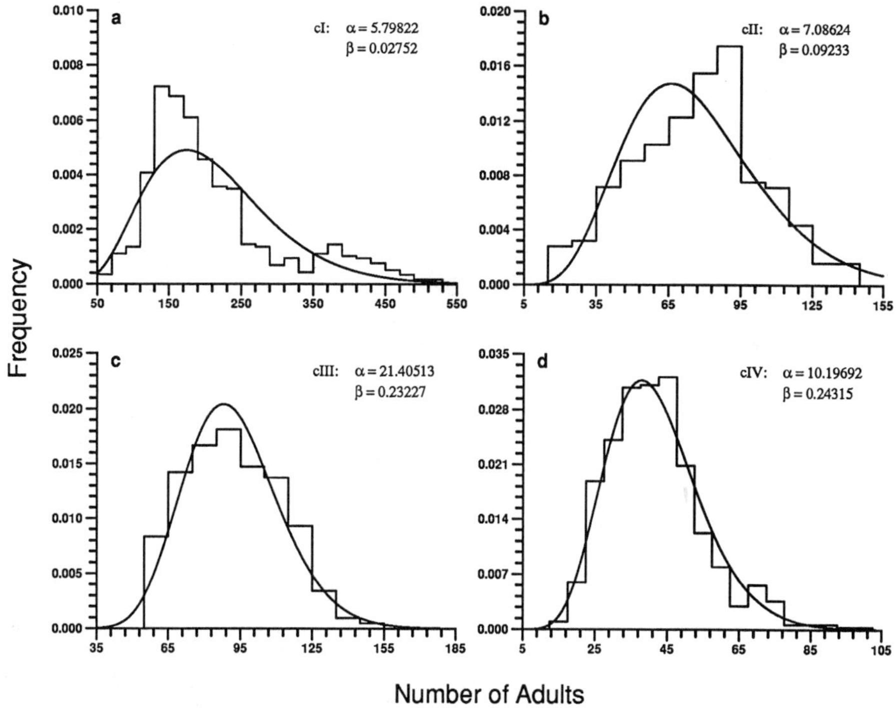

Fig. 4.8. Steady-state distributions for four genetic strains of *T. castaneum*: (a) cI, (b) cII, (c) cIII, and (d) cIV. The histograms are the observed grouped frequency distributions and the smooth curves are the fitted gamma probability density functions. [Modified from R. F. Costantino and R. A. Desharnais (1981), Gamma distributions of adult numbers for *Tribolium* populations in the region of their steady states. *J. Anim. Ecol.* **50**: 667–681.]

It is important to remember that we are interpreting these data with a stochastic version of the demographically reduced adult recruitment model. We are restricted to discussing adult cannibalism on eggs and pupae and not, for example, larval cannibalism on eggs.

In Chapter 3 we considered an egg-larval cannibalistic model. From our analysis we concluded that strain cI was the only population in the parameter region where the sole dynamic outcome was sustained oscillations in egg and larval numbers. The equilibrium distribution of adults for strain cI may be inadequately characterized by the adult recruitment model. A demographically more complete stochastic model which incorporates the short duration of the egg and larval stages and high fecundity coupled with "moderately high cannibalism of pupae by adults and extremely high cannibalism of eggs by larvae" (Park et al. 1964, p. 113) may be necessary for this strain.

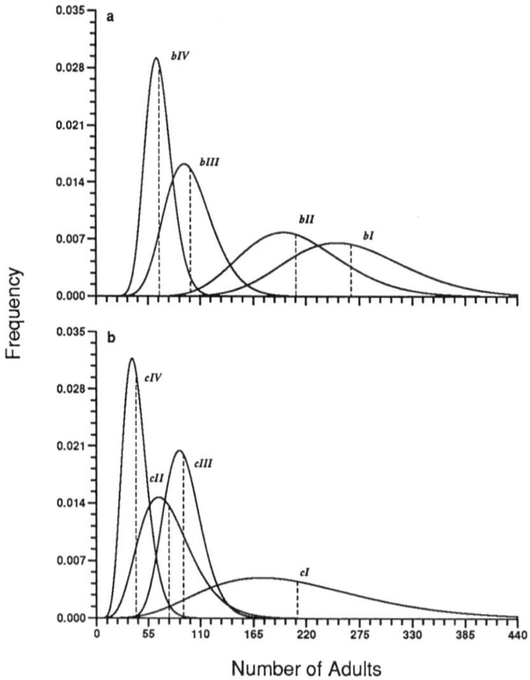

Fig. 4.9. The theoretical gamma stationary distributions for the four genetic strains of (a) *T. confusum* and (b) *T. castaneum*. The curves are plotted on the same graph to facilitate comparisons among strains. [Modified from R. F. Costantino and R. A. Desharnais (1981), Gamma distributions of adult numbers for *Tribolium* populations in the region of their steady states. *J. Anim. Ecol.* **50**: 667–681.]

4.4.3 Geographical Strains

The gamma stationary distribution can be used to detect differences between populations. In this section, we use the gamma to compare two *Tribolium* populations obtained from different geographical areas.

Lloyd (1965, p. 91) discussed the origin of the strains called Chicago and Brazil:

> In 1953 [the Chicago] strain had been cultured in the laboratory for about 20 years, or maybe even longer. A seeding was initially received in 1934 from Dr. N. E. Good of the United States Department of Agriculture, where the stock had been maintained for an unknown period. Strain B (sometimes called "Brazil") was originally obtained from Professor A. de Costa Lime of the Escola Nacional de Agronomia, Rio de Janeiro, in 1946; its history before that time is not known.

Lloyd (1965, 1968) examined the growth pattern of these two strains of *T. castaneum*. (The data from the Chicago strain was presented previously in Figs 4.3 and 4.4.) The observed stationary distributions (Fig. 4.10) were generated in the usual way beginning with the adult counts at week 19. The region of the steady-state extended over 17 weeks and gave a total of 204 observations on each strain.

The gamma hypothesis was accepted for both the Chicago and Brazil strains. Using the D^2 statistic (4.34), we conclude that these populations have statistically different gamma distributions. Interestingly, Lloyd (1968) used the Mann-Whitney U-test to

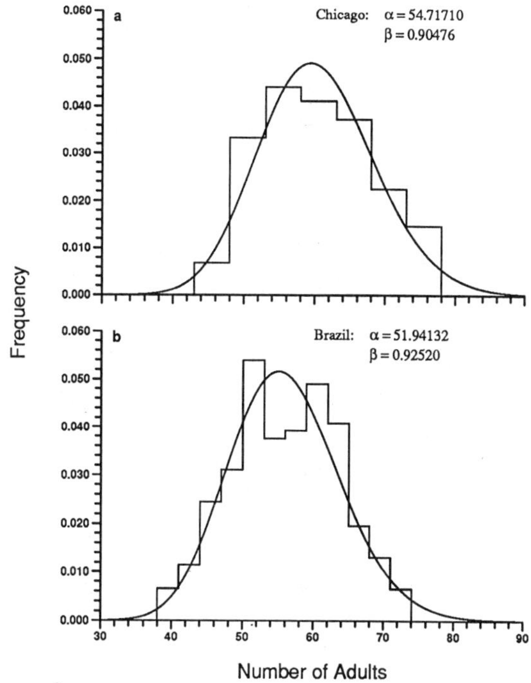

Fig. 4.10. Steady-state distributions for *T. castaneum* from (a) Chicago and (b) Brazil. The histograms are the observed grouped frequency distributions and the smooth curves are the fitted gamma probability density functions. Data are from Lloyd (1965). [Modified from B. Dennis and R. F. Costantino (1988), Analysis of steady-state populations with the gamma abundance model and its application to *Tribolium*. *Ecology* **69**: 1200–1213. Copyright 1988 by the Ecological Society of America.]

evaluate the average number of adults in these two strains and concluded that the levels were essentially the same. Clearly, the differences are not large (Fig. 4.10), but the use of the gamma distribution does allow a more precise characterization of the steady-state.

4.4.4 Populations at Different Times

Since they are continuously maintained populations, *Tribolium* stocks can evolve significant changes in their biological attributes after many generations of laboratory culture. For two of the genetic strains described in §4.4.2, we show how these biological changes are reflected as differences in the stationary gamma distributions.

T. castaneum strain cIV was one of eight genetic strains studied by Thomas Park and his colleagues in 1964 (Park et al. 1964). Four years later, the same strain, which was relabelled cIV-a, was reexamined (Leslie et al. 1968). The observed and fitted gamma stationary distributions of cIV and cIV-a are given in Fig. 4.11. For the 1964 data set the gamma was accepted but for the 1968 data this was not the case. Using the test statistic (4.32), the gamma significantly overpredicted the first cell and significantly underpredicted the fourth cell of the 1968 data, though it captured the essential shape of the histogram. Comparing these two populations using the D^2 test (4.34) is not appropriate, since the gamma does not fit one of them. However, it is clear that the distributions are not the same.

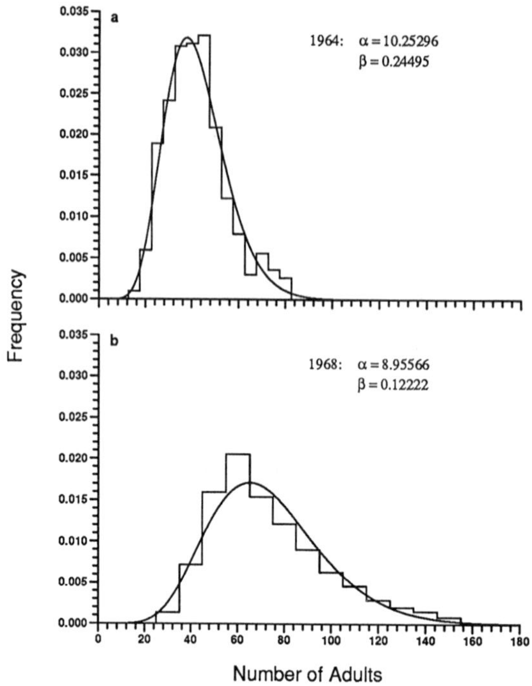

Fig. 4.11. Steady-state distributions for *T. castaneum* strain cIV in (a) 1964 and (b) 1968 (strain cIV-a). The histograms are the observed grouped frequency distributions and the smooth curves are the fitted gamma probability density functions. Data are from Park et al. (1964) and Leslie et al. (1968). [Modified from B. Dennis and R. F. Costantino (1988), Analysis of steady-state populations with the gamma abundance model and its application to *Tribolium*. *Ecology* **69**: 1200–1213. Copyright 1988 by the Ecological Society of America.]

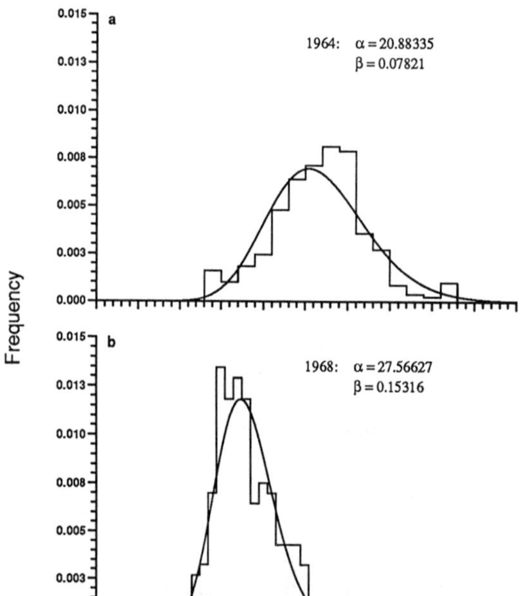

Fig. 4.12. Steady-state distributions for *T. confusum* strain bI in (a) 1964 and (b) 1968 (strain bI-a). The histograms are the observed grouped frequency distributions and the smooth curves are the fitted gamma probability density functions. Data are from Park et al. (1964) and Leslie et al. (1968). [Modified from B. Dennis and R. F. Costantino (1988), Analysis of steady-state populations with the gamma abundance model and its application to *Tribolium*. *Ecology* **69**: 1200–1213. Copyright 1988 by the Ecological Society of America.]

In their analysis of strain cIV-a, Leslie et al. (1968, p. 15) discussed three changes in this stock: (i) the average *per capita* adult death rate per 30 days was reduced from 0.1982 to 0.1571; (ii) the average fertility of eggs was increased from 66.5% to 85.4%; and (iii) the average duration of the immature life stages was shortened from 31.09 days to 27.81 days. It is evident from Fig. 4.11 that both the mean and variance of the stationary distribution increased in 1968 as compared with 1964. Given that the biological changes (i)–(iii) imply a decrease in μ and an increase in b, the observed increases in $E[N]$ and $Var[N]$ are entirely consistent with the predictions of the gamma density function (see Fig. 4.2).

In a similar manner, strain bI of *T. confusum* was studied by these same authors in 1964 and again in 1968. The gamma model fits the 1968 data, but not the 1964 data (Fig. 4.12). The observed distributions for these two years indicate a reduction in both the mean and variance of the stationary density. What had happened? Leslie et al. (1968, p. 15) wrote, "the principle change in this strain appears to lie in an increase of its cannibalistic powers." The gamma prediction is in agreement with their statement: an increase in the rate of cannibalism, c, would reduce both the mean and variance of the gamma distribution (see Fig. 4.2).

4.4.5 Population Outbreaks and Crashes

Tribolium populations sometimes exhibit extremely large or small numbers of adults. These occurrences are often called "outbreaks" and "crashes," respectively. In this section we show how the tails of the stationary distribution can be used to predict the probability of these events.

Mertz (1969) investigated the effects of nine sets of conditions defined by initial densities and ages of the beetles using strain cIV-a of *T. castaneum*. Ten cultures were started in each of the nine treatments and were sampled every thirty days, though three of the cultures were lost at the outset. By day 360, most of the transient effects of initial conditions, such as age-structure imbalances, had damped out in the cultures. The 87 populations from day 360 through day 600 provided a total of 783 observations on equilibrium adult numbers.

The equilibrium frequency distribution appears unimodal and positively skewed (Fig. 4.13). A gamma distribution does not fit very well, due to significant deviations in the first, second, fourth, and fourteenth cells.

Mertz labeled certain observations as representing "outbreaks" in population levels. An outbreak was defined as a surge in population size of 25% or more within two sample periods. A stationary probability density function such as the gamma provides a different definition of outbreak which may be useful in biological monitoring. An outbreak can be defined as the population exceeding some specified level, N. For the gamma model the probability of that event is $1-F^*(N)$, where $F^*(N)$ is the gamma cumulative density function (4.21). An estimate of the outbreak probability is obtained by substituting the maximum likelihood estimates α and β in $F^*(N)$. For instance, using a threshold level of $N = 132.5$, the estimated outbreak in Fig. 4.13 is about 0.04, roughly in accord with the observed frequency of about 0.05.

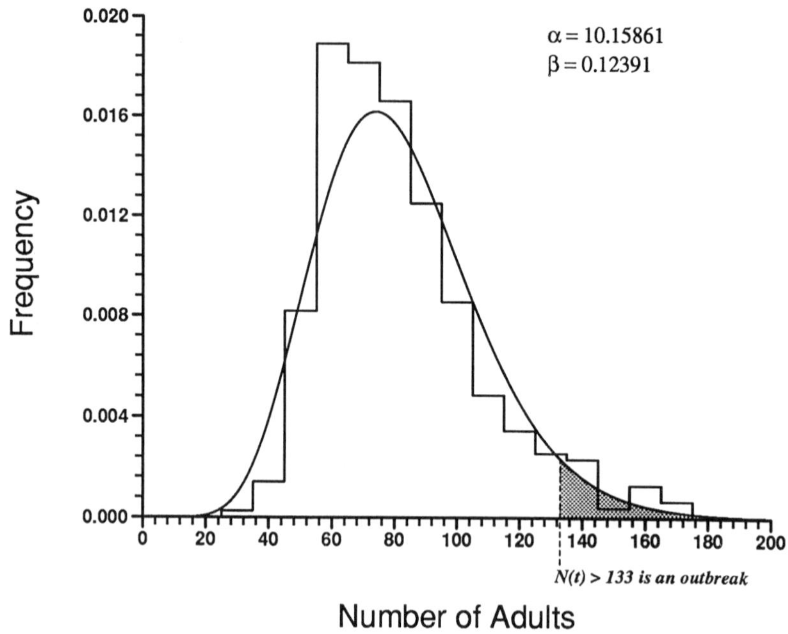

Fig. 4.13. Steady-state distribution for *T. castaneum* strain cIV-a. The histogram is the observed grouped frequency distribution and the smooth curve is the fitted gamma probability density function. Estimated proportion of time spent at outbreak levels ($N > 133$) is represented by the shaded area. Data are from Mertz (1969). [Modified from B. Dennis and R. F. Costantino (1988), Analysis of steady-state populations with the gamma abundance model and its application to *Tribolium*. *Ecology* **69**: 1200–1213. Copyright 1988 by the Ecological Society of America.]

Interestingly, a lognormal distribution fits this data set somewhat better than the gamma (though still not acceptably at the 0.01 level). A stochastic differential equation (SDE) model which predicts a lognormal stationary distribution has a convex *per capita* growth rate as a function of population size (Dennis and Patil 1984, 1988). The original SDE flour beetle model (4.6) also has a convex form; perhaps the quadratic approximation (4.13) implicit in the gamma model has some shortcomings for this data set. The possibility that details of curvature in the *per capita* growth rate might be detected through statistical analysis of stationary distributions deserves further study.

4.5 Stationary Time Series

In our analysis of the stationary distribution, we ignored time-series aspects of the stochastic process. In this section we use a linear approximation of the stochastic model to obtain predictions for the stationary time-series of adult numbers.

4.5.1 Linear Approximation for the Stochastic Model

If the amplitudes of the random fluctuations are small, then we can obtain a first approximation to the autocorrelation function of adult numbers through time (Nisbet and Gurney 1982). We begin by linearizing the stochastic differential equation (4.6) around the point $N(t)=N^*$ and $\gamma(t)=0$:

$$\frac{dN}{dt} \approx \lambda[N^* - N(t)] + \sigma N^* \gamma(t), \tag{4.35}$$

where λ is the absolute value of the eigenvalue from the deterministic model (4.1) and N^* is the deterministic equilibrium point (4.2). For the population equation (4.35), the "flow" $m(N)$ and "drift" $v(N)$ of the diffusion equation are given by

$$\begin{aligned} m(N) &= \lambda(N^* - N), \\ v(N) &= (\sigma N^*)^2. \end{aligned} \tag{4.36}$$

Substituting these expressions into (4.11) gives the steady-state probability distribution

$$f^*(N) = \left(\frac{\sqrt{\lambda}}{\sigma N^* \sqrt{\pi}}\right) \exp\left[-\frac{\lambda(N - N^*)^2}{(\sigma N^*)^2}\right]. \tag{4.37}$$

This is a Gaussian normal distribution with a mean and variance

$$\begin{aligned} \mathrm{E}[N] &= N^*, \\ \mathrm{Var}[N] &= \frac{\sigma^2 N^*}{2c\mu}. \end{aligned} \tag{4.38}$$

The normal distribution (4.37) can be viewed as an approximation to the gamma (4.15) when the fluctuations are small. In fact, as $\sigma \to 0$, we have $\alpha \to \infty$ for the shape parameter of the gamma, and, from (4.22), the mean and variance of the gamma converge to (4.38). In the case of the Stratonovich calculus ($\zeta = 0$), (4.22) and (4.38) are identical for all admissible values of σ.

4.5.2 Autocorrelation Function

Using standard Fourier transform techniques, Desharnais and Costantino (1985) used (4.35) to derive the following theoretical function for the autocorrelation of $N(t)$ with $N(t+\tau)$:

$$\rho_N(\tau) = \exp(-\lambda|\tau|). \tag{4.39}$$

With this expression, one can obtain information on the demographic stability of the population by estimating λ from time-series data at the steady state.

Until now, we have assumed that the random variable $\gamma(t)$ represents uncorrelated white noise. As an alternative model, we will assume that the fluctuations are themselves autocorrelated. Specifically, we now assume that $\gamma(t)$ represents "pink noise," that is, the correlation of $\gamma(t)$ with $\gamma(t+\tau)$ is

$$\rho_\gamma(\tau) = \exp(-\phi|\tau|), \tag{4.40}$$

where $\phi > 0$ is a measure of the rate of decay of the autocorrelations (Nisbet and Gurney 1982). With this modified definition of $\gamma(t)$ in (4.35), the expected autocorrelation function of adult numbers is

$$\rho_N(\tau) = \begin{cases} \dfrac{\lambda\exp(-\phi|\tau|) - \phi\exp(-\lambda|\tau|)}{\lambda - \phi}, & \lambda \neq \phi, \\[2ex] (1+\lambda|\tau|)\exp(-\lambda|\tau|), & \lambda = \phi. \end{cases} \tag{4.41}$$

By comparing equations (4.40) and (4.41) with experimental data, we can evaluate the assumption that the time scale for the random fluctuations is much shorter than the time scale for changes in population size.

Finally, it is possible to obtain a confidence interval on the expected autocorrelation function. If n is the number of paired observations used to compute the autocorrelation and the parent series is normally distributed, then Bartlett (1946) has shown that the variance of the estimated autocorrelation is given by

$$\begin{aligned} \mathrm{Var}[\rho_N(\tau)] = n^{-1} \sum_{j=-\infty}^{+\infty} [&\rho_N(j)^2 + \rho_N(j-\tau)\rho_N(j+\tau) \\ &- 4\rho_N(\tau)\rho_N(j)\rho_N(j+\tau) + 2\rho_N(j)^2\rho_N(\tau)^2], \end{aligned} \tag{4.42}$$

where $\rho_N(\tau)$ is the expected autocorrelation. The substitution of equation (4.39) into (4.42) gives

$$\mathrm{Var}[\rho_N(\tau)] = n^{-1}\left(\frac{[1-\exp(-2\lambda\tau)]\,[1+\exp(-2\lambda\tau)]}{[1-\exp(-2\lambda)]} - 2\tau\exp(-2\lambda\tau)\right) \tag{4.43}$$

for the variance of the estimated autocorrelations under the assumptions of the stochastic white noise model.

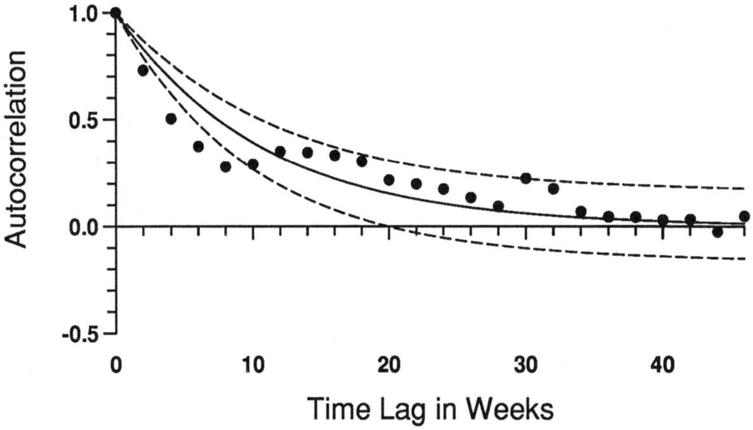

Fig. 4.14. Stationary autocorrelations of adult numbers versus time. The circles are the observed autocorrelations and the solid curve is the fitted theoretical function (4.39). The dashed curves form confidence bands of ± 2 standard errors. Data are from Moffa and Costantino (1977).

As an illustrative example, we have applied the autocorrelation functions to the data of Moffa and Costantino (1977). The number of adults were recorded in biweekly censuses on 32 independent replicate cultures of *T. castaneum* maintained in the region of their steady states. The analysis covers a period of 50 weeks (25 biweekly censuses from week 20 to week 68) yielding 800 data points.

The autocorrelation of order τ was computed by pairing the adult census number at week t with week $t+\tau$ for each replicate. These paired observations were pooled among replicates and an overall correlation coefficient was computed in the usual way. This procedure was repeated for values of τ from 2 through 44 weeks at intervals of two weeks. The estimated correlation coefficients were corrected for bias by multiplying by $(1-\tau/n)$, where n is the total number of paired data points used in the computation (Bloomfield 1976, p. 184). These corrected estimates are plotted in Fig. 4.14.

The estimated autocorrelations were used to evaluate the "white noise" and "pink noise" models. A nonlinear least squares technique was used to fit the observed autocorrelations to the theoretical functions (4.39) and (4.41). For the pink noise model, the fit continued to improve as ϕ increased, yielding the white noise model in the limit. Thus, there is no evidence in these data to support an autocorrelation in the noise function $\gamma(t)$. The parameter estimate was $\lambda = 0.09361 \pm 0.00806$ (Fig. 4.14). The variance bounds of ± 2 standard errors (dashed curves in Fig. 4.14) were computed assuming a theoretical sample size of $n = 800$. Although the observed autocorrelations deviate significantly from the simple exponential prediction, the autocorrelations remain positive and do eventually decay to zero (Fig. 4.14).

4.6 Time-Dependent Gamma Distribution

The stationary probability density function, $f^*(N)$, was obtained from the SDE model (4.6) through the use of the diffusion equation (4.8). As we discussed in §4.2.2, one can, in principle, obtain a *time-dependent* probability distribution $f(N,t)$ for adult numbers. Unfortunately, the resulting partial differential equation for $f(N,t)$ cannot be solved. In the case of the quadratic SDE (4.13), Prajneshu (1980) solved for $f(N,t)$, but this solution involves integrals of functions of complex variables and is difficult to apply. Dennis (1988) has proposed an approximation to the time-dependent distribution which builds on the quadratic SDE (4.13). We apply his results.

Let $u_1(N_0,t)$ and $u_2(N_0,t)$ represent the first two moments of the stochastic process $N(t)$ given by (4.13) with $N(0)=N_0$. Using singular perturbation methods, the first and second moments of $N(t)$ can be approximated by

$$u_1(N_0,t) \approx \left[\frac{N^*}{1+\kappa_0 e^{-\lambda t}}\right] + \left[\frac{\sigma^2 N^*}{2\lambda(1+\kappa_0 e^{-\lambda t})^3}\right]$$

$$\times[(2\kappa_0+2)e^{-\lambda t} - (2\kappa_0+1)e^{-2\lambda t} - 1 - 2\lambda\kappa_0 t\, e^{-\lambda t}] \,, \tag{4.44}$$

$$u_2(N_0,t) \approx \left[\frac{N^*}{1+\kappa_0 e^{-\lambda t}}\right]^2 + \left[\frac{(\sigma N^*)^2}{\lambda(1+\kappa_0 e^{-\lambda t})^4}\right]$$

$$\times\left[(4\kappa_0+2)e^{-\lambda t} - \left(4\kappa_0+\frac{3}{2}\right)e^{-2\lambda t} - \frac{1}{2} + \lambda(\kappa_0 e^{-\lambda t} - 2)\kappa_0 t\, e^{-\lambda t}\right], \tag{4.45}$$

respectively, where $\kappa_0=(N^*-N_0)/N_0$ (Dennis 1988). The transient distribution is approximated with a gamma with the first two moments given by (4.44) and (4.45); i.e.,

$$f(N,t\,|\,N(0)=N_0) = \beta^\alpha \Gamma(\alpha)^{-1} N^{\alpha-1} e^{-\beta N} \,, \qquad 0 \le N \le \infty \,, \tag{4.46}$$

where

$$\alpha = \alpha(N_0,t) = \frac{u_1(N_0,t)^2}{u_2(N_0,t) - u_1(N_0,t)^2} \,,$$

$$\beta = \beta(N_0,t) = \frac{u_1(N_0,t)}{u_2(N_0,t) - u_1(N_0,t)^2} \,. \tag{4.47}$$

The mean and variance of adult number are given by

$$\mathrm{E}[N(t)\,|\,N(0)=N_0] = \alpha(N_0,t)\,\beta(N_0,t)^{-1} = u_1(N_0,t) \,,$$

$$\mathrm{Var}[N(t)\,|\,N(0)=N_0] = \alpha(N_0,t)\,\beta(N_0,t)^{-2} = u_2(N_0,t) - u_1(N_0,t)^2 \,. \tag{4.48}$$

As $t \to \infty, f(N,t) \to f^*(N)$, where $f^*(N)$ is the stationary distribution given by (4.15) with $\zeta = 1$ (Ito calculus).

The maximum likelihood method can also be used to estimate the parameters of the time-dependent distribution. Let $N(t_0)$, $N(t_1)$, ..., $N(t_n)$ represent a sequence of observations on a single population. Since $N(t)$ is a Markov process, each transition in population number is independent of the past transitions. The probability of the entire sequence is given by

$$L(\lambda, N^*, \sigma) = \prod_{j=1}^{n} f(N(t_j), (t_j - t_{j-1}) \mid N(0) = N(t_{j-1})) . \tag{4.49}$$

In effect, one uses (4.46) with the initial time and density reset to the previous observation as j increases. Maximizing the logarithm of (4.49) with respect to the parameters λ, N^*, and σ yields the maximum likelihood estimates. The negative inverse of the matrix of second derivatives of $\log L$ with respect to the parameters provides a variance-covariance matrix for the estimates (Basawa and Prakasa-Rao 1980). These derivatives can be evaluated numerically at the maximum likelihood point. If data are available for two or more independent replicate populations, then the overall likelihood function is simply the product of the likelihood functions for each population.

In §4.3 we used data on the Chicago strain of *T. castaneum* as an illustrative example of the statistical procedures for the stationary distribution. In Fig. 4.3b we presented the data for twelve replicate populations as a series of histograms which change over time. We now use these data to compute the maximum likelihood estimates of the parameters for the time-dependent gamma distribution.

For the Chicago strain, the number of adults, $N(0), N(1), \ldots, N(35)$, were recorded each week for 35 weeks. We computed the time-dependent maximum likelihood estimates by maximizing (4.49) numerically. The following values were obtained: $\lambda = 0.277410 \pm 0.008669$, $N^* = 61.824 \pm 2.200$, $\sigma^2 = 0.009129 \pm 0.000445$. These estimates have the advantage of being computed from the entire data set, not just the steady-state. The standard errors were computed from the diagonal elements of the variance-covariance matrix.

The mean and variance of the time-dependent gamma distribution along with the parameters α and β of the gamma are presented in Fig. 4.15a. Both the mean and the variance quickly increase. The trajectories of the mean and variance follow a sigmoidal pattern, although the variance slightly overshoots its asymptotic value of 61.86. Other numerical examples showed that the variance can be significantly larger than its asymptotic value during the transient phase of population growth. At week 19 the mean is within 98% of its equilibrium value and the variance is essentially stationary. The parameters α and β are undefined at $t = 0$ because the variance, which is in the denominator of each, is initially zero. The shape parameter α is over 100 at week one, then drops to 29.11 at week seven and slowly increases again to its equilibrium value of $\alpha^* = 59.78$. This implies that the distribution becomes more skewed during the growth phase of the population. The scale parameter β declines rapidly to a minimum of 0.7879 at week 10 and then increases slightly to its stationary value of $\beta^* = 0.9830$.

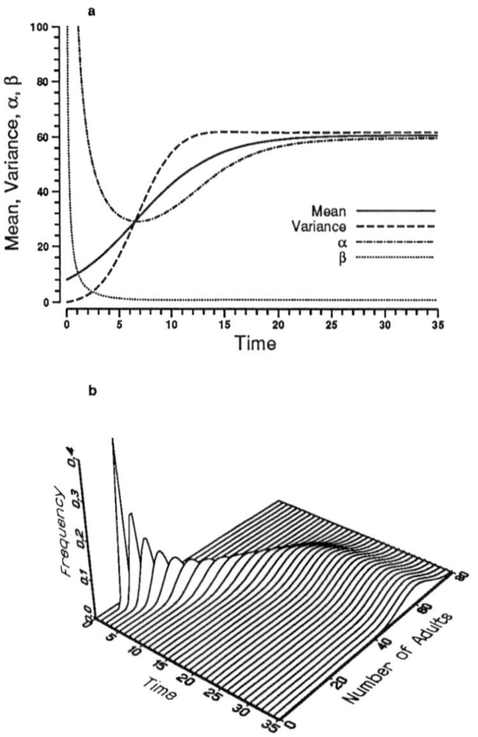

Fig. 4.15. Results for the time-dependent gamma distribution with parameters estimated from data on the Chicago strain of *T. castaneum*. In (a), the mean and variance of adult numbers are plotted as functions of time together with the parameters $\alpha(N_0,t)$ and $\beta(N_0,t)$ of the gamma. In (b), a three-dimensional plot of the time-dependent gamma is given.

A three-dimensional plot of the time-dependent gamma is presented in Fig. 4.15b. The distribution is peaked at $t=1$; however, after the interval of rapid population growth (approximately by week 19) the distribution assumes its steady-state shape. This plot is the theoretical counterpart to Fig. 4.3b.

The time-dependent analysis provides an estimate of the positive constant σ which is a measure of the magnitude of the stochastic forces. From (4.23) we know that the distribution will undergo a "noise-induced transition" to an exponential-like density function when $\sigma^2 > \lambda$. Since the estimates of the parameters are $\lambda = 0.277410$ and $\sigma^2 = 0.009129$, this strain appears to be far removed from the critical value for a qualitative change in the shape of the distribution.

The gamma abundance model was applied to the observations of the Chicago strain in two ways. First, the analysis was restricted to the data in the steady-state (week 19 to week 35). Pooling observations, we found $\alpha = 54.72$ and $\beta = 0.9048$ so that $E[N] = 60.48$ and $\mathrm{Var}[N] = 66.84$. In the second case, *all* of the data were used to obtain asymptotic predictions of $\alpha^* = 59.78$ and $\beta^* = 0.9830$ as $t \to \infty$, which give similar population mean and variance values of $E[N] = 60.81$ and $\mathrm{Var}[N] = 61.86$. This concordance suggests that our statistical methods are fairly robust.

4.7 Summary and Conclusions

The purpose of a stochastic population model is to make predictions about variability in population size. In this Chapter, we dealt with two categories of predictions. First, we used stochastic models to obtain a statistical description of the variation in population size. Gamma stationary and transient distributions of adult number were derived together with expressions for the means and variances of the distributions. Functions for the serial autocorrelations in adult number were also derived. Statistical procedures for parameter estimation and hypothesis testing were also developed. These results were used to characterize the data from experimental populations of *Tribolium*. The second category of predictions involves the relationship between the biological attributes of a population and its stochastic dynamics. Differences in the stationary distributions of adult numbers were associated with biological entities such as birth, death, and cannibalism rates. Comparisons were made among populations with different environmental conditions, genetic backgrounds, geographical origins, and at different points in time. In most cases there was a good qualitative agreement between the observed differences in the means and variances of the populations and known differences in the biology of the beetles. These applications utilized variation in population number as an additional source of information.

The gamma stationary distribution did not do a perfect job of describing the experimental results. In many cases, we were forced to reject the gamma hypothesis at the 0.01 probability level. In fact, throughout this monograph, we apply the gamma distribution to 31 data sets (17 in this chapter, 13 in Chapter 6, and *T. brevicornis* in Chapter 9), and we reject the gamma in 7 of those cases. This error rate of 22.6% is far above the expected 1% probability level. In most cases where the gamma was rejected, the overall shape of the distribution conformed well to the gamma, but in a few cells the deviations were too large. In a few cases, such as strain cI of *T. castaneum* (Fig. 4.8a), the gamma did a very poor job. One possible source of error is discussed in §4.3.5; when applied to time-series data, the gamma will be rejected too often. Of course, another likely reason for the failures of the gamma is that the adult recruitment model (4.1) is an inadequate description of the dynamics of the adult population. Indeed, considering the true complexity of *Tribolium* demographic dynamics as discussed in Chapters 2 and 3, we might also ask how the gamma managed to do so well! In using the gamma, we have sacrificed realism for simplicity and generality. Considering what we have gained, the compromise was a good one.

Although all of the theoretical predictions are in terms of the biological parameters b, c, μ, and σ, our results do not allow us to estimate these parameters. Fitting the gamma stationary distribution involves the estimation of α and β. Although α and β are functions of the biological parameters, the relationship involves two known quantities and four unknown quantities – a unique correspondence does not exist. This means that we cannot estimate all the biological parameters from data on the steady-state. In the case of the time-dependent distribution, three parameters are estimated: λ, N^*, and σ. Again, we are unable to specify uniquely b, c, and μ, but we do have an estimate of σ. This is significant, because there is no way to estimate σ from other sources of biological data such as assays of fecundity or cannibalism. A more complex approximation might allow the estimation of all four biological parameters. This problem deserves further research.

What is the nature of the stochasticity? On one hand, it could be argued that changes in environmental conditions induce fluctuations in demographic rates such as fecundity and adult mortality. On the other hand, one could argue that the probabilistic nature of individual behaviors and physiological events gives rise to variation in the timing of reproduction and mortality even under constant environmental conditions. The latter explanation requires a consideration of population dynamics in terms of life history events – births and deaths at the very least. We look at *Tribolium* population growth from this perspective in the next chapter.

Chapter 5 Discrete Stochastic Models

In the deterministic and stochastic models of the previous chapters, population size was treated as a continuous variable. But animals come in discrete units. For large populations, a continuous variable may be a reasonable approximation; a model which predicts a population size of 11,924.38 is not much different from one that predicts a population of 11,924. But in small populations the difference may be important; a prediction of 5 animals is more reasonable than a prediction of 5.42. This is especially true when one considers the question of extinction, where the dynamics of small populations are of fundamental interest. A more realistic model should consider population size as a discrete variable restricted to the nonnegative integers.

In this chapter, population growth is viewed as a stochastic ***birth-death process***. Time is continuous, but the population changes occur in discrete "jumps" of +1 or −1; these correspond to isolated "birth" or "death" events. We begin with a general description of birth-death processes and then specialize the model using recruitment and mortality rates appropriate to *Tribolium*. Discrete "quasi-steady state" probability distributions are obtained for adult numbers and the problem of extinction is considered. Since exact solutions are unwieldy, simple approximations are derived and illustrated. An approximate linear birth-death model is used to obtain a time-dependent probability distribution for population size. These results are applied to laboratory data on *T. brevicornis*. The chapter summary includes a discussion of the merits of stochastic models, a comparison of the continuous and discrete stochastic formulations, and comments on the problem of environmental versus demographic stochasticity.

5.1 Population Growth as a Birth-Death Process

Markov chains are often used as stochastic models where the dependent variables are discrete. The system is described using a finite or countably infinite number of discrete *states*, and the dynamics are specified using *transition probabilities* between states. In a *discrete-time Markov chain*, the transitions occur at discrete time intervals in a manner similar to difference equations; the transition probabilities are what define the model. In a *continuous-time Markov chain* (sometimes called a *Markov process*), there is an exponentially-distributed waiting time between transitions; the mean waiting time between transitions and the transition probabilities themselves depend on *transition rates* which are defined for each state of the model. Both types of models are *memoryless*; transitions depend only on the current state of the system.

A *birth-death process* is a special type of Markov chain where the states represent integer numbers and transitions are only allowed to neighboring states. A "birth" represents a transition from state j to state $j+1$; a "death" is a transition from state j to state $j-1$. These transitions need not represent actual births or deaths; they could be people entering or leaving a queue or species colonizing or becoming extinct on an island. For population growth, these transitions represent recruitment and mortality.

In this section we introduce the mathematical machinery for a general birth-death process and define several important concepts such as the time spent in a state and the quasi-steady state. Detailed derivations are omitted; expository presentations can be found in many textbooks such as those by Ross (1972), Rosenblatt (1974), Karlin and Taylor (1975), and Iosifescu (1980).

5.1.1 Kolmogorov Forward Equations

Let $N(t)$ be a discrete random variable that represents the size of a population at time $t \geq 0$, where time is continuous. We are interested in the probability distribution of $N(t)$ given any initial condition $N(0)$. This distribution will be represented by $P(t,i,j)$, where

$$P(t,i,j) = \text{Prob}\{N(t)=j \mid N(0)=i\} . \tag{5.1}$$

In other words, $P(t,i,j)$ is the probability that $N(t)=j$ given $N(0)=i$.

Consider an infinitesimal time interval from t to $t+\Delta t$ and assume $N(t+\Delta t)=j$ at the end of the interval. If $N(t)$ is governed by a birth-death process, three possible transitions could have occurred (Fig. 5.1): (i) $N(t)$ could have gone from $j-1$ to j with probability $P(t,i,j-1)\,r(j-1)\Delta t$; (ii) $N(t)$ could have gone from $j+1$ to j with probability $P(t,i,j+1)\,m(j+1)\Delta t$; (iii) $N(t)$ could have remained in state j during the interval with probability $P(t,i,j)[1-r(j)\Delta t - m(j)\Delta t]$. If Δt is sufficiently small, the probability of two or more transitions, which is of order $(\Delta t)^2$, can be ignored. The functions $r(j)$ and $m(j)$ represent the transition rates for "births" and "deaths," respectively, while the system is in state j. The probability of $N(t+\Delta t)=j$ is given by the sum of these three mutually exclusive events:

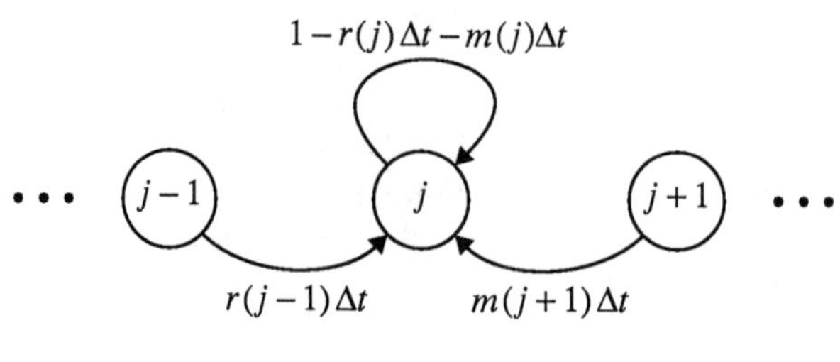

Fig. 5.1. Possible transitions into state j during a small time interval Δt. The functions $r(j)$ and $m(j)$ are the transition rates for "births" and "deaths," respectively, given the process is in state j.

$$P(t+\Delta t,i,j) = P(t,i,j-1)r(j-1)\Delta t + P(t,i,j+1)r(j+1)\Delta t$$
$$+ P(t,i,j)[1-r(j)\Delta t - m(j)\Delta t] \ . \tag{5.2}$$

Subtracting $P(t,i,j)$ from both sides, dividing through by Δt, and taking the limit as $\Delta t \to 0$ gives

$$\frac{dP(t,i,j)}{dt} = P(t,i,j-1)r(j-1) + P(t,i,j+1)m(j+1)$$
$$- P(t,i,j)[r(j)+m(j)] \tag{5.3}$$

for $j>0$, $i \geq 0$, and $t \geq 0$. We can use (5.3) for $j=0$ if we define $r(-1)=0$ and $m(0)=0$. This system of ordinary differential equations is called the *Kolmogorov forward equations*.

5.1.2 Stationary Probabilities

We begin with the case where the state 0 is a "reflecting boundary." This means that $r(j)>0$ for $j \geq 0$, $m(0)=0$, and $m(j)>0$ for $j \geq 1$. In particular, $r(0)>0$ implies immigration into the population which prevents extinction. As $t \to \infty$, the probabilities have limiting values which are independent of the initial state i:

$$\lim_{t \to \infty} P(t,i,j) = P^*(j) \ . \tag{5.4}$$

These limiting probabilities must solve the system (5.3) with the time derivatives set equal to zero:

$$\begin{aligned} P^*(1)m(1) - P^*(0)m(0) &= 0 \ , \\ P^*(j-1)r(j-1) + P^*(j+1)m(j+1) - P^*(j)[r(j)+m(j)] &= 0 \ , \quad j \geq 1 \ . \end{aligned} \tag{5.5}$$

The first equation can be used to obtain $P^*(1)$ in terms of $P^*(0)$. This expression for $P^*(1)$ can then be substituted into the equation for $j=1$ to obtain $P^*(2)$ as a function of $P^*(0)$. Proceeding in this fashion one obtains

$$P^*(j) = P^*(0) \prod_{k=0}^{j-1} \frac{r(k)}{m(k+1)} , \qquad j \geq 1. \tag{5.6}$$

For the sequence $P^*(0)$, $P^*(1)$, ... to be a probability distribution, the sum must add to one. This gives the following expression for $P^*(0)$:

$$P^*(0) = \left(1 + \sum_{j=1}^{\infty} \prod_{k=0}^{j-1} \frac{r(k)}{m(k+1)}\right)^{-1}. \tag{5.7}$$

If the summation in (5.7) is finite, then the stationary probability distribution exists. Otherwise, $P^*(j) = 0$ for $j \geq 0$ and there is no steady-state distribution.

As a concrete example, consider a *linear birth-death process with immigration.* The transition rates are given by

$$\begin{aligned} r(j) &= \alpha + \beta j\,, \\ m(j) &= \mu j\,, \end{aligned} \qquad j \geq 0\,, \tag{5.8}$$

where α is the immigration rate, β is the *per capita* birth rate, and μ is the mortality rate. Making use of the following identities,

$$\prod_{k=0}^{j-1} \frac{r(k)}{m(k+1)} = \prod_{k=0}^{j-1} \frac{\alpha + \beta k}{\mu(k+1)} = \binom{\theta + j - 1}{j} \phi^j\,,$$

$$1 + \sum_{j=1}^{\infty} \prod_{k=0}^{j-1} \frac{r(k)}{m(k+1)} = 1 + \sum_{j=1}^{\infty} \binom{\theta + j - 1}{j} \phi^j = (1 - \phi)^{-\theta}\,,$$

where $\theta = \alpha / \beta$, $\phi = \beta / \mu$, and

$$\binom{\theta + j - 1}{j} = \frac{\Gamma(\theta + j)}{\Gamma(\theta)\, j!}\,,$$

the limiting stationary distribution is given by

$$P^*(j) = \binom{\theta + j - 1}{j} \phi^j (1 - \phi)^\theta\,, \qquad j \geq 0\,. \tag{5.9}$$

This distribution exists provided $\beta < \mu$. The probabilities (5.9) follow a *negative binomial distribution* with mean $\theta(1-\phi)\phi^{-1}$ and variance $\theta(1-\phi)\phi^{-2}$. We will encounter this distribution again in §5.3.

We now consider the case where the state 0 is an "absorbing state." This occurs when $r(0) = 0$; i.e. there is no immigration to save the population from extinction. The existence of an absorbing does not guarantee extinction since population size may drift indefinitely or go to infinity. Karlin and Taylor (1975) show that the probability of extinction when $N(0) = i$ is given by

$$\Omega(i) = \begin{cases} \dfrac{\omega(i)}{1+\omega(1)}, & \text{if } \omega(1) < \infty, \\ 1, & \text{if } \omega(1) = \infty, \end{cases} \tag{5.10}$$

where

$$\omega(i) = \sum_{j=i}^{\infty} \prod_{k=1}^{j} \frac{m(k)}{r(k)}, \qquad i \geq 1. \tag{5.11}$$

For example, if we use the linear rates $r(j)$ and $m(j)$ from (5.8) with $\alpha=0$, then $\Omega(i)=1$ if $\beta \leq \mu$ and $\Omega(i)=(\beta/\mu)^i$ if $\beta>\mu$.

5.1.3 Extinction Times and the Quasi-Steady State

Assume $\Omega(i)=1$ in (5.10). Although the population goes extinct with a probability of one, it may take a very long time for extinction to occur. Given that extinction has not yet occurred and the population size remains bounded, we can view the fluctuations in population size as a sort of "quasi-steady state." We investigate this problem by considering the amount of time $N(t)=j$ for $j \geq 1$ prior to extinction.

Let $\delta(t,i,j)$ represent an indicator variable. Given $N(0)=i$,

$$\delta(t,i,j) = \begin{cases} 1 & \text{if } N(t) = j. \\ 0 & \text{if } N(t) \neq j. \end{cases} \tag{5.12}$$

The expected value for $\delta(t,i,j)$ is given by

$$\mathrm{E}\{\delta(t,i,j)\} = 1 \cdot P(t,i,j) + 0 \cdot [1-P(t,i,j)] = P(t,i,j). \tag{5.13}$$

Let $\tau(i,j)$ represent the expected total amount of time that the "process" $N(t)$ spends in state j given it started in state i. Integrating (5.13) we obtain

$$\tau(i,j) = \mathrm{E}\left\{ \int_0^{\infty} \delta(t,i,j)\mathrm{d}t \right\} = \int_0^{\infty} \mathrm{E}\{\delta(t,i,j)\}\,\mathrm{d}t = \int_0^{\infty} P(t,i,j)\mathrm{d}t. \tag{5.14}$$

Therefore, we must integrate the Kolomogorov forward equations (5.3) to obtain the expected times for each state.

Consider the case where $i=0$. Since $N(0)=0$ and there is no immigration (i.e. $r(0)=0$), the process must remain in state 0. Hence $\tau(0,0)=\infty$ and $\tau(0,j)=0$ for $j \geq 1$.

Now consider the case where $i>0$. Integrating (5.3) from $t=0$ to ∞ gives

$$P(\infty,i,j)-P(0,i,j) = \begin{cases} \tau(i,1)m(1)\,, & j=0, \\ \tau(i,2)m(2)-\tau(i,1)\,[r(1)+m(1)]\,, & j=1, \\ \tau(i,j-1)r(j-1)+\tau(i,j+1)m(j+1) \\ \qquad -\tau(i,j)\,[r(j)+m(j)]\,, & j\geq 2. \end{cases} \tag{5.15}$$

Since the population eventually becomes extinct, $P(\infty,i,0)=1$ for $i\geq 1$ and $P(\infty,i,j)=0$ for $i,j\geq 1$. For the initial probabilities, $P(0,i,j)=1$ if $i=j$, otherwise $P(0,i,j)=0$. Using these facts in (5.15) one obtains

$$\tau(i,j) = \begin{cases} \left(\dfrac{1}{m(1)}\right), & 1=j\leq i, \\ \left(\dfrac{1}{m(j)}\right)\left(1+\sum\limits_{k=1}^{j-1}\prod\limits_{h=k}^{j-1}\dfrac{r(h)}{m(h)}\right), & 2\leq j\leq i\,, \\ \left(\dfrac{1}{m(j)}\right)\left(\sum\limits_{k=1}^{i}\prod\limits_{h=k}^{j-1}\dfrac{r(h)}{m(h)}\right), & 1\leq i<j\,. \end{cases} \tag{5.16}$$

This result is proved in the appendix. The expressions (5.16) give the expected amount of time the population will contain exactly j individuals given it started with i individuals initially.

How long will the population survive before becoming extinct? Since the $\tau(i,j)$ gives the expected time spent in state j, we need only sum over $j\geq 1$ to obtain the expected time to extinction:

$$\Psi(i) = \sum_{j=1}^{\infty}\tau(i,j)\,, \qquad i\geq 1\,. \tag{5.17}$$

However, we have implicitly assumed that this value is finite. It is possible for $\Psi(i)=\infty$ although the probability of extinction is equal to one. Specifically, the mean time to extinction is finite provided

$$\sum_{j=1}^{\infty}\left(\frac{1}{m(j)}\right)\left(\prod_{i=1}^{j-1}\frac{r(i)}{m(i)}\right) < \infty \tag{5.18}$$

(Karlin and Taylor 1975). The linear birth-death process (5.8) with $\alpha=0$ and $\beta=\mu$ is an example of a birth-death process where extinction is certain but the mean time to extinction is infinite. We will only consider cases where (5.18) holds.

Let us assume that $\Psi(i)$ is finite but very large relative to the lifetime of an individual. Observations of the population would reveal fluctuations which appear stationary. We wish to characterize this "quasi-stationary" distribution. One possibility is to use the expected times spent in a state as weights for the probability distribution. Thus the quasi-stationary probabilities are defined using

$$Q(i,j) = \frac{\tau(i,j)}{\Psi(i)}, \qquad i,j \geq 1 . \tag{5.19}$$

Although this distribution depends on the initial state i, numerical results suggest that this dependence is weak throughout most of the range of the distribution. Alternate distributions based on conditional probabilities are given by Darroch and Seneta (1967). Their distributions yield numerical results which are very similar to ours. We will use (5.19) as our definition of the quasi-steady state.

5.2 Nonlinear Birth-Death Process for *Tribolium*

The theory developed in §5.1 is for any general birth-death process. To apply these results, the "birth" and "death" functions $r(j)$ and $m(j)$ must be specified. In this section we derive results based on a simple model for adult numbers in *Tribolium.*

5.2.1 Recruitment and Mortality Rates

In §4.1 we derived a simple deterministic model for adult numbers in *Tribolium* based on the assumption that adults inhibit the recruitment of immatures into the adult population. The differential equation (4.1) has two terms: $Nb\exp(-cN)$ for recruitment and $N\mu$ for mortality. We will use these terms to develop predictions based on a stochastic birth-death process.

The transition rates for the birth-death process are given by the following recruitment and mortality functions:

$$\begin{aligned} r(j) &= bj\exp(-cj), \\ m(j) &= \mu j, \end{aligned} \qquad j \geq 0 . \tag{5.20}$$

The constant parameters b, μ, and c have the same meanings as before: b is a *per capita* reproductive rate, μ is a mortality rate, and c is the rate at which adults inhibit recruitment. However, the "birth" and "death" rates $r(j)$ and $m(j)$ are defined for discrete nonnegative integers only; j represents the number of adults in the population. As in the deterministic model, $r(j)<m(j)$ for $j<N^*$ and $r(j)>m(j)$ for $j>N^*$, where $N^*=c^{-1}\log(b/\mu)$. The transition rates still favor the deterministic equilibrium N^*.

5.2.2 Extinction Times and the Quasi-Steady State

Given explicit functions for $r(j)$ and $m(j)$, we can make use of the general results for a birth-death process provided in §5.1. First we note that

$$\begin{aligned}\prod_{h=k}^{j-1}\frac{r(h)}{m(h)} &= \left(\frac{b}{\mu}\right)^{j-k}\exp\left[-c\sum_{h=k}^{j-1}h\right] \\ &= \left(\frac{b}{\mu}\right)^{j-k}\exp\left[-c\left(\sum_{h=1}^{j-1}h\right)+c\left(\sum_{h=1}^{k-1}h\right)\right] \\ &= \left(\frac{b}{\mu}\right)^{j-k}\exp\left[-\left(\frac{c}{2}\right)j(j-1)+\left(\frac{c}{2}\right)k(k-1)\right] = \frac{u(j)}{u(k)},\end{aligned} \tag{5.21}$$

where the function $u(j)$ is defined using

$$u(j) = \left(\frac{b}{\mu}\right)^{j}\exp\left[-\left(\frac{c}{2}\right)j(j-1)\right]. \tag{5.22}$$

Use of this definition simplifies the notation.

It is clear from (5.20) that $r(0)=0$; hence the state zero acts as an absorbing state. But is extinction a certain outcome? Substituting the inverse of (5.21) for $k=1$ into (5.11) for $i=1$, we can see that

$$\begin{aligned}\omega(1) &= \sum_{j=1}^{\infty}u(j)^{-1} = \sum_{j=1}^{\infty}\left(\frac{\mu}{b}\right)^{j}\exp\left[\left(\frac{c}{2}\right)j(j+1)\right] \\ &= \sum_{j=1}^{\infty}\exp\left[\left(\frac{c}{2}\right)j(j+1)-j\log\left(\frac{b}{\mu}\right)\right] \\ &> \sum_{j=1}^{\infty}\exp\left[\left(\frac{c}{2}\right)j^{2}-j\log\left(\frac{b}{\mu}\right)\right] = \infty.\end{aligned} \tag{5.23}$$

From (5.10) we know that $\Omega(i)=1$ for all $i\geq 1$; extinction is guaranteed for all initial population sizes.

We now obtain the expected time to extinction. Let

$$w(j) = \sum_{k=1}^{j}u(k)^{-1}. \tag{5.24}$$

Substituting (5.22) and (5.24) into (5.16) we obtain

$$\tau(i,j) = \begin{cases} w(j)(\mu j)^{-1}u(j), & 1 \le j < i, \\ w(i)(\mu j)^{-1}u(j), & 1 \le i \le j, \end{cases} \tag{5.25}$$

for the expected amount of time that $N(t)=j$ given $N(0)=i$. Summing (5.25) over $j \ge 1$ gives

$$\Psi(i) = \sum_{j=1}^{i-1} w(j)(\mu j)^{-1}u(j) + w(i)\sum_{j=i}^{\infty}(\mu j)^{-1}u(j) \tag{5.26}$$

for the expected "lifetime" of the population. It is easy to see from (5.22) that $u(j) \to 0$ at a rate that is dominated by the term $\exp[-(c/2)j^2]$. Hence the infinite series in (5.26) converges and extinction occurs in a finite amount of time.

The quasi-stationary distribution was defined in (5.19) as the ratio of the expected time spent in a state to the expected time to extinction. For the explicit transition rates in (5.20),

$$Q(i,j) = \begin{cases} \dfrac{w(j)u(j)}{\mu j \Psi(i)}, & 1 \le j < i, \\[2ex] \dfrac{w(i)u(j)}{\mu j \Psi(i)}, & 1 \le i \le j. \end{cases} \tag{5.27}$$

This gives an exact formula for the quasi-steady state.

Intuitively, it should be the case that a population with a long lifetime should exhibit dynamical behavior which is relatively independent of its initial size. For example, a laboratory population of *Tribolium* which has fluctuated around some mean value for several years and shows no sign of deterioration should have a stationary distribution and expected time to extinction which is independent of whether the population was initiated with 50 beetles or 100. We now show that the results (5.25)–(5.27) are indeed insensitive to the initial state i.

Consider the series (5.24) for $w(j)$. Let $z(j,k)$ represent the ratio of the jth term in the series to the kth term; i.e. $z(j,k) = u(j)^{-1}/u(k)^{-1} = u(k)/u(j)$. The ratio of two adjacent terms is given by

$$z(j+1,j) = \left(\frac{\mu}{b}\right)\exp(cj). \tag{5.28}$$

For $j<N^*$, $z(j+1,j)<1$; the terms of the series are monotonic decreasing. For $j>N^*$, $z(j+1,j)>1$ and the terms begin to increase. Thus the sequence $u(j)^{-1}, j=1, 2, \ldots$ reaches a minimum near N^*. As j approaches N^*, the ratio of $u(j)^{-1}$ to the first term is given by

$$z(j,1) \approx \left(\frac{\mu}{b}\right)^{(N^*-1)/2} \qquad \text{for} \quad j \approx N^*. \tag{5.28}$$

Since $b > \mu$ and $c \ll 1$, N^* will be large and the ratio (5.29) will be a very small number. Thus we can characterize the series $w(j)$ as follows. As j increases, the terms decrease according to (5.28). When $j \ll c^{-1}$, $\exp(cj) \approx 1$ and the terms of $w(j)$ approximate a geometric series with common ratio μ/b. By the time the factor $\exp(cj)$ becomes significant, the magnitude of the sequence will be very small and additional terms will contribute negligibly to the series $w(j)$. Thus we can approximate $w(j)$ as the sum of a geometric sequence:

$$w(j) \approx \frac{1-(\mu/b)^j}{(b/\mu)-1} \qquad \text{for} \quad 1 \le j \le N^*. \tag{5.30}$$

In fact, for biologically reasonable values of $\mu < b$, this sequence will converge quickly, and $w(j)$ will be relatively insensitive to the value of j. As a further approximation, we will assume

$$w(j) \approx w_0 = \frac{\mu}{b-\mu}. \tag{5.31}$$

Using w_0 in place of $w(j)$ yields approximations for the expected time to extinction and the quasi-stationary distribution which are independent of the initial state i:

$$\Psi_0 = (b-\mu)^{-1} \sum_{j=1}^{\infty} j^{-1} u(j); \tag{5.32}$$

$$Q_0(j) = \frac{j^{-1} u(j)}{\sum_{k=1}^{\infty} k^{-1} u(k)}. \tag{5.33}$$

Numerical results suggest that these approximations are quite good.

5.2.3 Approximations Based on the Normal Distribution

The results (5.26) and (5.27) are complex and difficult to use. They involve an infinite series and one must specify an initial state i. Even the approximations $Q_0(j)$ and Ψ_0, which do not involve an initial state, require the computation of an infinite series. This is especially a problem for the quasi-stationary distribution, since we would like to obtain closed form expressions for its moments. In this section we obtain approximations which are based on the normal distribution.

We begin by noting that $j^{-1}u(j)$ can be written as

$$j^{-1}u(j) = \kappa_0 \exp\left[-\left(\frac{c}{2}\right)(j-N_q)^2 - \log(j)\right], \qquad j \ge 1, \tag{5.34}$$

where $N_q = N^* + 1/2$ and $\kappa_0 = \exp[(c/2)N_q^2]$. This function has a form which is similar to a Gaussian density function. All that remains is to approximate the factor $\log(j)$ with a constant. Since the function $u(j)$ has its mode near N_q, we expand $\log(j)$ around N_q:

$$\log(j) = \log(N_q) + \frac{(j-N_q)}{N_q} - \frac{(j-N_q)^2}{2N_q^2} + \frac{(j-N_q)^3}{3N_q^3} - \cdots . \tag{5.35}$$

As a first approximation, we set $\log(j)$ equal to $\log(N_q)$ and substitute into (5.34). This approximation, together with (5.31), gives

$$Q_1(j) = \kappa_1 \exp\left[-\left(\frac{c}{2}\right)(j-N_q)^2\right], \qquad j \ge 1, \tag{5.36}$$

as the quasi-stationary distribution. The normalization constant can be obtained by summing the terms in (5.36). Alternatively, one can use the fact that

$$\sum_{j=1}^{\infty} \exp\left[-\left(\frac{c}{2}\right)(j-N_q)^2\right] \approx \int_{-\infty}^{\infty} \exp\left[-\left(\frac{c}{2}\right)(x-N_q)^2\right] dx = \sqrt{\frac{2\pi}{c}} \tag{5.37}$$

and let

$$\kappa_1 = \sqrt{\frac{c}{2\pi}}. \tag{5.38}$$

With this discrete version of a normal distribution, the approximate mean and variance of population size are given by

$$\mathrm{E}[N] \approx \overline{N}_1 = N_q; \tag{5.39}$$

$$\mathrm{Var}[N] \approx V_1 = c^{-1}. \tag{5.40}$$

Similarly, we can use (5.31), (5.38) and the first term of (5.35) to obtain

$$\Psi_1 = \sqrt{\frac{2\pi \exp(cN_q^2)}{(b-\mu)^2 cN_q^2}} \qquad (5.41)$$

as a crude approximation for the expected time to extinction.

Better approximations can be obtained by including more terms of log(j) in (5.35). In fact, using the first three terms in (5.34) one still obtains a normal distribution:

$$Q_2(j) = \kappa_2 \exp\left[-\left(\frac{1}{2V_2}\right)(j-\overline{N}_2)^2\right], \qquad j \geq 1, \qquad (5.42)$$

where $\kappa_2 = (2\pi V_2)^{-1/2}$ and

$$\mathrm{E}[N] \approx \overline{N}_2 = N_q\left[1 - \frac{c^{-1}N_q^{-2}}{1-c^{-1}N_q^{-2}}\right], \qquad (5.43)$$

$$\mathrm{Var}[N] \approx V_2 = (c - N_q^{-2})^{-1}. \qquad (5.44)$$

The corresponding result for the expected time to extinction is

$$\Psi_2 = \sqrt{\frac{2\pi \exp(3 + V_2^{-1}\overline{N}_2^2)}{(b-\mu)^2 V_2^{-1} N_q^2}}. \qquad (5.45)$$

The results (5.43)–(5.45) are uglier but more accurate approximations to the true values of the population mean, variance and expected lifetime.

How good are these approximations? As a numerical example, we let $b=0.4152$, $\mu=0.2480$, $c=0.002485$, and $i=20$. These values are parameter estimates for *T. brevicornis* which we derive in §5.4. Using these values we compared the exact distribution $Q(i,j)$ with $Q_0(j)$, $Q_1(j)$, and $Q_2(j)$ in the range of ± 2 standard deviations from the mean. The distributions $Q(i,j)$ and $Q_0(j)$ were nearly identical; differences were on the order of magnitude of 10^{-15}. Thus the assumption (5.31) introduces little error into the quasi-stationary distribution. The probabilities obtained from $Q_1(j)$ differed from $Q(i,j)$ with a relative error of less than 21%; this error was due to the fact that the approximate distribution was shifted slightly to the right. The second approximation $Q_2(j)$ did much better; relative errors were less than 0.29%. In general, all three approximations are reasonably close to the exact distribution.

Using the same parameter estimates we also compared the mean, variance, and expected time to extinction for adult numbers. The results are presented in Table 5.1. For the mean and variance, the largest differences were for the crude approximation $Q_1(j)$; it predicted slightly higher mean and lower variance. The approximations for the expected time to extinction possess a slightly larger error. For all three approximations the differences occur in the third significant figure. It is worth noting that Ψ_0 and Ψ_2 are

Table 5.1. A comparison of approximations for the mean, variance, and expected time to extinction for adult numbers

Distribution	Mean	Variance	Expected time to extinction
$Q(i,j)$	$E[N]=205.903$	$Var[N]=406.389$	$\Psi(i)=3.06673\times10^{23}$
$Q_0(j)$	$\overline{N}_0=205.903$	$V_0=406.389$	$\Psi_0=3.03827\times10^{23}$
$Q_1(j)$	$\overline{N}_1=207.877$	$V_1=402.414$	$\Psi_1=3.00943\times10^{23}$
$Q_2(j)$	$\overline{N}_2=205.923$	$V_2=406.197$	$\Psi_2=3.03778\times10^{23}$

very similar; this suggests that most of the error in the expected time to extinction lies with the assumption (5.31). However, Ψ_0 and Ψ_2 are still slightly better approximations than Ψ_1 when compared to the exact value. All three approximations give reasonably accurate predictions for most purposes.

5.2.4 Cannibalism Rates and Extinction Times for Eight Genetic Strains

The approximations (5.39)–(5.41) can be used to obtain information on cannibalism rates and extinction times. To illustrate, we will use the data of Park et al. (1964) on four genetic strains of *T. confusum* and four strains of *T. castaneum*. Using age-specific life history data, they estimated density-independent intrinsic rates of increase, $b-\mu$, for each genetic strain (Park et al., 1964, Table 10). They also tabulated the mean and variance of adult numbers at steady-state from populations maintained on 8g of flour media (Park et al., 1964, Table 12). These results are summarized in Table 5.2. From these data we will derive estimates for cannibalism rates and extinction times.

Table 5.2. Estimated growth rates, observed means and variances of adult numbers, and expected times to extinction for eight genetic strains of *Tribolium*

Genetic strain	Growth rate, $b-\mu$ (days^{-1})[a]	Mean number of adults[b]	Variance in adult numbers[b]	Coefficient of variation	Expected time to extinction (days)
cI	0.1340	220.2	8862.3	0.4275	123
cII	0.0930	81.2	691.8	0.3239	1,025
cIII	0.0870	102.7	1115.6	0.3252	1,059
cIV	0.1120	43.9	179.0	0.3048	1,485
bI	0.1044	275.8	3423.1	0.2121	340,638
bII	0.0980	217.1	2414.5	0.2263	100,336
bIII	0.0990	104.3	716.3	0.2566	128,847
bIV	0.1036	69.0	186.3	0.1978	1,695,532

[a] Data from Park et al. (1964, Table 10).
[b] Data from Park et al. (1964, Table 12).

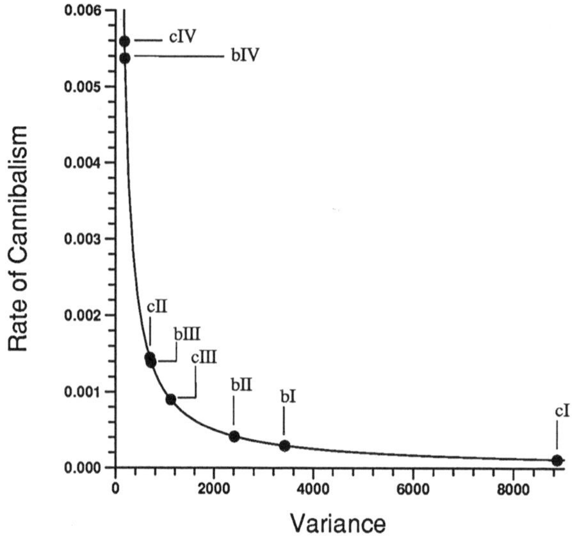

Fig. 5.2. The predicted relationship between cannibalism rate and the variance of the quasi-stationary distribution. Eight genetic strains of *Tribolium* have been positioned on this curve based on their observed steady-state variances in adult numbers.

The expression (5.40) predicts a rate of cannibalism which equals the inverse of the variance at steady-state: $c = V_1^{-1}$. This simple prediction is displayed graphically by the curve in Fig. 5.2. Using the observed variances of adult numbers (Table 5.2), we computed the predicted rates of cannibalism for the eight genetic strains (Fig. 5.2). Note that this figure does not imply a fit between the theory and the data; indeed, the eight data points are constrained to the curve. Nevertheless, there is a good qualitative agreement between these estimates of c and the conclusions of Park et al. (1964, p. 150) based on assays which determined relative rates of cannibalism. For *T. confusum* they ranked the four strains from most to least cannibalistic as follows: bIV > bIII > bII > bI. The model, based on the variance of the observed stationary state, makes exactly the same prediction. For *T. castaneum*, Park and his colleagues concluded that cI > cIV > cII > cIII. Our model predicts cIV > cII > cIII > cI. The match between the theory and the independent biological observations is correct for strains cII, cIII, and cIV, but completely opposite for strain cI. In Table 3.3 of Chapter 3, we noted that this strain was unique in that it fell into the region of parameter space where a stable orbit was the only attractor for the egg-larval submodel. It may be the case that the "fast scale" egg-larval oscillations invalidate the predictions from a model based on adult numbers.

The mean and variance of adult numbers at the steady state can also be used to predict expected time to extinction. Let C_V represent the coefficient of variation for adult numbers. From (5.39) and (5.40),

$$C_V \approx \frac{\sqrt{V_1}}{\overline{N}_1} = \sqrt{cN_q^2}. \qquad (5.46)$$

Using this result in (5.41) we obtain

$$\Psi_1 = \sqrt{\frac{2\pi\exp(C_V^{-2})}{(b-\mu)^2 C_V^{-2}}} . \tag{5.47}$$

In Table 5.2 we list the expected times to extinction for the eight genetic strains. Surprisingly, the predicted times to extinction for all of the *T. castaneum* strains are quite short. This is due to the large coefficients of variation for these four strains. Once again, the prediction for strain cI is the most questionable result. Unfortunately, we have no data which could be used to test these predictions.

5.3 Linear Birth-Death Approximation

In principle, the Kolmogorov equations (5.3) specify the entire probability distribution for population size as it changes through time. In practice, the differential system can rarely be solved explicitly for nonlinear transition rates. In fact, the theory is complex even for linear transition rates, but explicit solutions have been derived (Karlin and McGregor 1958). In this section we linearize the nonlinear transition rate $r(j)$ for *Tribolium* and apply the theory of linear birth-death processes to the dynamics of adult numbers.

5.3.1 Linearized Recruitment Rate

Our goal is to approximate the nonlinear recruitment function given in (5.20) with a linear function of the form given in (5.8). We begin by treating the transition rates $r(j)$ and $m(j)$ as functions of continuous variables:

$$\begin{aligned} r(x) &= bj\exp(-cx), \\ m(x) &= \mu x, \end{aligned} \qquad x \geq 0, x \in \Re . \tag{5.48}$$

The recruitment and mortality rates are equal when $x = N^*$. Taylor expanding $r(x)$ around the value N^* gives

$$r(x) = r(N^*) + (x - N^*)\mu\left[1 - \log\left(\frac{b}{\mu}\right)\right] + \sum_{k=2}^{\infty}\left[\frac{(x-N^*)^k}{k!}\right]\left[\frac{d^k r(x)}{dx^k}\right]_{x=N^*} . \tag{5.49}$$

Dropping the higher order terms which appear in the summation and returning to a domain of nonnegative integers, we obtain the linear recruitment approximation

$$r_L(j) = \alpha + \beta j , \qquad j \geq 0 , \tag{5.50}$$

where

$$\begin{aligned} \alpha &= \lambda N^*, \\ \beta &= \mu - \lambda, \end{aligned} \tag{5.51}$$

and $\lambda = \mu \log(b/\mu)$. The parameter λ was introduced earlier in §4.1 as the eigenvalue of the deterministic model for adult numbers. If the slope of the linear function $\beta > 0$, then the model is equivalent to a linear birth-death process with immigration. It was shown in §5.1.2 that this type of process has a stationary probability distribution provided $\beta < \mu$ which in this case implies $b > \mu$. However, if $\beta < 0$, then one must confine the analysis to $0 \le j \le -\alpha/\beta$; this requires the use of a finite state space. The condition $\beta < 0$ implies $b/\mu > e = 2.71828\ldots$. Thus the linear approximation is questionable if the reproductive rate exceeds the adult mortality rate by a large factor. In what follows we will assume $b/\mu < e$.

5.3.2 Time-Dependent Mean and Variance

With linear rates of recruitment and mortality, the mean of adult numbers can be computed directly from the Kolomogorov forward equations. The mean number of adults is defined using

$$\overline{N}(t,i) = \sum_{j=0}^{\infty} jP(t,i,j), \tag{5.52}$$

where i is the initial state. Taking the derivative of this expression we have

$$\frac{\mathrm{d}\overline{N}(t,i)}{\mathrm{d}t} = \frac{\mathrm{d}\left(\sum_{j=0}^{\infty} jP(t,i,j)\right)}{\mathrm{d}t} = \sum_{j=0}^{\infty} j\left(\frac{\mathrm{d}P(t,i,j)}{\mathrm{d}t}\right). \tag{5.53}$$

The last summation can be obtained by multiplying each of the equations (5.3) by its corresponding value of j and summing all of the equations. After some rearranging of terms we obtain

$$\frac{\mathrm{d}\overline{N}(t,i)}{\mathrm{d}t} = \sum_{j=0}^{\infty} P(t,i,j)[r(j) - m(j)] \tag{5.54}$$

as an ordinary differential equation for the mean number of adults. Substituting the linear functions $r_L(j)$ and $m(j)$ into (5.54) we have

$$\frac{d\overline{N}(t,i)}{dt} = \lambda[N^* - \overline{N}(t,i)] . \tag{5.55}$$

Applying the initial condition $\overline{N}(0,i) = i$, the closed form solution to this differential equation is given by

$$\overline{N}(t,i) = N^*[1 - a_0 \exp(-\lambda t)] , \tag{5.56}$$

where $a_0 = (i - N^*)/N^*$. As $t \to \infty$, the linear model predicts $\overline{N}(t,i) \to N^*$. The approach to equilibrium is asymptotic at a rate governed by the eigenvalue λ.

A similar procedure yields the variance of adult numbers. The variance is defined using

$$V(t,i) = \sum_{j=0}^{\infty} j^2 P(t,i,j) - \overline{N}(t,i)^2 . \tag{5.57}$$

The time-derivative of the variance is

$$\begin{aligned} \frac{dV(t,i)}{dt} &= \frac{d\left(\sum_{j=0}^{\infty} j^2 P(t,i,j)\right)}{dt} - 2\overline{N}(t,i)\left(\frac{d\overline{N}(t,i)}{dt}\right) \\ &= \sum_{j=0}^{\infty} j^2\left(\frac{dP(t,i,j)}{dt}\right) - 2\lambda\overline{N}(t,i)\,[N^* - \overline{N}(t,i)] . \end{aligned} \tag{5.58}$$

Let us first consider the summation. We multiply each of the equations (5.3) by j^2 and sum all the equations. After some algebraic simplification we obtain

$$\sum_{j=0}^{\infty} j^2\left(\frac{dP(t,i,j)}{dt}\right) = \sum_{j=0}^{\infty} P(t,i,j)\,[r(j)(2j+1) - m(j)(2j-1)] . \tag{5.59}$$

Substituting $r_L(j)$ and $m(j)$ into (5.56) and making use of (5.52) we have

$$\sum_{j=0}^{\infty} j^2\left(\frac{dP(t,i,j)}{dt}\right) = -2\lambda\left(\sum_{j=0}^{\infty} j^2 P(t,i,j)\right) + \lambda N^* + (2\lambda N^* + 2\mu - \lambda)\overline{N}(t,i) . \tag{5.60}$$

Substituting (5.60) into (5.58) and using (5.56) and (5.57), we arrive at a differential equation for the variance:

$$\frac{dV(t,i)}{dt} = 2\lambda[V^* - V(t,i)] + \lambda a_0(2V^* - N^*)\exp(-\lambda t) , \tag{5.61}$$

where $V^* = c^{-1}$. Upon integration this yields

$$V(t,i) = V^*[1 + a_1 \exp(-\lambda t) + a_2 \exp(-2\lambda t)] , \tag{5.62}$$

with constants $a_1 = a_0 c(2V^* - N^*)$ and $a_2 = -(a_1+1)$. Note that $V(0,i)=0$ and $V(t,i) \to V^* = c^{-1}$ as $t \to \infty$.

The asymptotic results for the linearized model are consistent with the results for the quasi-stationary distribution when the expected time to extinction is large. For the mean of adult numbers, the approximation (5.39) predicts $\mathrm{E}[N] \approx N_q = N^* + 1/2$. This agrees favorably with the prediction that $\overline{N}(t,i) \to N^*$. For the variance, both formulations are in agreement: $V_1 = V^* = c^{-1}$. Since the linearization (5.50) improves as $N(t)$ nears N^*, (5.56) and (5.62) provide accurate information on the asymptotic rate of approach to the steady state.

5.3.3 Transient and Stationary Probability Distribution

Karlin and McGregor (1958) conducted a thorough analysis of linear birth-death processes. For a linear birth-death process with immigration (process label F in their analysis), they derived a closed form solution for the time-dependent probability distribution given an arbitrary initial state. If we let $\rho = \exp(-\lambda t)$, $\theta = \alpha/\beta$, $\phi = \beta/\mu$, then the probability distribution is given by

$$P(t,i,j) = \left(\frac{1-\phi}{1-\phi\rho}\right)^{\theta} \left(\frac{\phi-\phi\rho}{1-\phi\rho}\right)^{j} \times \sum_{k=0}^{\min(i,j)} (-1)^k \binom{i}{k} \binom{\theta+i+j-k-1}{j-k} \left(\frac{1-\rho}{1-\phi\rho}\right)^{i-k} \left(\frac{1-\rho/\phi}{1-\rho}\right)^{k} . \tag{5.63}$$

(In several places in Karlin and McGregor's paper, the exponent corresponding to our θ appears with a negative sign; this is a typographical error.) The parameters α and β are specified in (5.51). Time dependence is through the parameter ρ. Given b, μ, c, and the initial state i, (5.63) can be used to predict the distribution of adult numbers for any $t \geq 0$. For $t=0$, $P(0,i,j)=0$ if $j \neq i$ and $P(0,i,j)=1$ if $j=i$. As $t \to \infty$, $P(t,i,j) \to P^*(j)$, where $P^*(j)$ is given by (5.9); i.e., the asymptotic distribution is a negative binomial with parameters θ and ϕ. This is consistent with the results obtained by setting the Kolmogorov forward equations equal to zero and solving for the steady state. The mean and variance for this distribution were derived in §5.3.2; they are specified in (5.56) and (5.62), respectively.

In the special case where $i=0$, (5.63) reduces to

$$P(t,0,j) = \binom{\theta+j-1}{j} \left(\frac{\phi-\phi\rho}{1-\phi\rho}\right)^{j} \left(\frac{1-\phi}{1-\phi\rho}\right)^{\theta}, \qquad j \geq 0 . \tag{5.64}$$

In this case population size has a negative binomial distribution for all values of $t>0$. Note that as $t \to \infty$, $\rho \to 0$ and (5.64) reduces to (5.9).

5.4 Application to *Tribolium brevicornis*

In this section we illustrate the results obtained from the birth-death formulation using data from four laboratory cultures of *Tribolium brevicornis*. This species seems particularly suited to the birth-death approach because it has rates of recruitment and mortality which are much less than those observed in *T. castaneum* and *T. confusum*. Fewer "birth" and "death" events occur between censuses, and there are no large amplitude cyclic fluctuations among the immature age classes. The mechanism of density regulation is different in *T. brevicornis*: adults inhibit the metamorphosis of large larvae into pupae (Jillson and Costantino 1980). This mechanism results in a constant pool of large larvae. These larvae cannibalize eggs and this inhibits the formation of demographic oscillations. A relatively constant age distribution is maintained; large waves of recruitment are not observed. The attributes seem to favor the birth-death paradigm.

The data were obtained from Desharnais and Costantino (1982). Four cultures of *T. brevicornis* were maintained for 72 weeks. Each culture was initiated with 20 young adults (10 of each sex) and kept in a one-half pint bottle with 20 grams of standard media (percentage composition: 95% wheat flour and 5% dried brewers yeast) and held in an unlighted incubator at 33±1°C and 60±5% relative humidity. All life stages (except eggs) were counted every two weeks and following the census the animals were placed in fresh media. We will focus on the adult census data.

The observed number of adults are presented in Fig. 5.3. Since each culture was started with adults only, no immatures were available for recruitment during the first two census periods. Therefore we have rescaled time so that $t=0$ corresponds to week 6 and one time unit equals 2 weeks. Since there was little mortality during this initial period, we will use $N(0)=i=20$ adults.

The values of N^* and λ in (5.56) were estimated using a general nonlinear least-squares regression of the mean number of adults versus time. Least squares estimates and standard errors of $N^*=207.4\pm3.4$ and $\lambda=0.1278\pm0.0077$ were obtained. The regression equation accounted for 86% of the variation in the observed numbers of adult beetles. The regression equation is plotted as a solid curve in Fig. 5.3.

Given the estimates of N^* and λ, we can compute the time required for the mean number of adults to reach any given proportion of its asymptotic value. Let $T(x,i)$ denote the time needed for $\overline{N}(t,i)/N^*=x$. From (5.56) we have

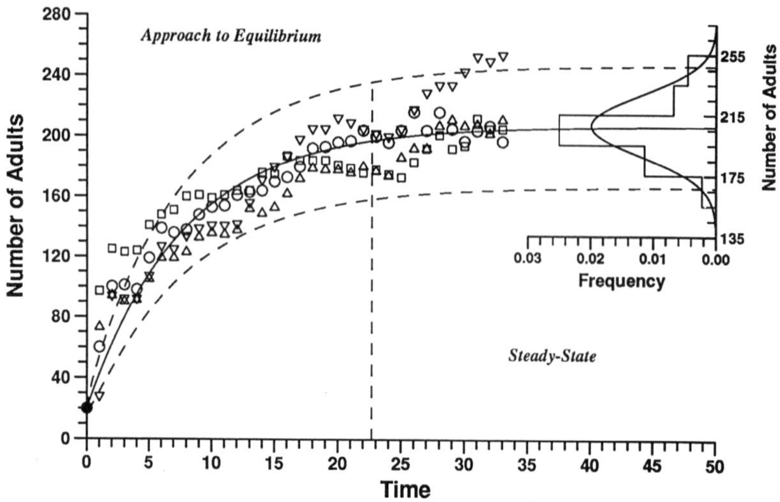

Fig. 5.3. Observed number of adults for four cultures of *T. brevicornis*. The solid curve is the mean number of adults and the dashed lines are the mean $\pm 2[\text{variance}]^{1/2}$. The vertical dashed line at $t=22.65$ corresponds to $T(0.95,20)$. The observed frequency data (histogram) and theoretical (smooth curve) steady-state probability distributions appear in the upper right of the figure. [From R. A. Desharnais and R. F. Costantino (1982), The approach to equilibrium and the steady-state probability distribution of adult numbers in Tribolium brevicornis. *Am. Nat.* **119**: 102-111.]

$$T(x,i) = \lambda^{-1}\log\left[\frac{N^*-i}{N^*(1-x)}\right]. \tag{5.65}$$

For *T. brevicornis*, $T(0.95,20)=22.65$. We will use this time value to demarcate the approach to equilibrium from the steady state. This demarcation is represented by the vertical dashed line in Fig. 5.3. Data falling to the right of this line form the quasi-stationary distribution.

An estimate of the variance in the vicinity of the equilibrium, $\hat{V}^*=402.48$, was computed from the adult data for $t=T(0.95,20)$ ($n=44$ observations). Since the expected quasi-stationary distribution is very nearly gaussian, the quantity $(n-1)\hat{V}^*/V^*$ follows a chi-square distribution with $(n-1)$ degrees of freedom (Snedecor and Cochran 1967, p 73). Using this fact, we found the 95% confidence interval on the variance of adult numbers at the steady-state to be $274.8<V^*<646.1$.

Crude estimates of the biological parameters c, μ and b can be obtained from the estimates of V^*, N^* and λ. To estimate c, the rate at which adults inhibit the recruitment of immatures, we made use of the fact that $c=1/V^*=0.002485$. Similarly, the 95% confidence interval on the parameter c can be approximated by inverting the lower and upper bounds on V^* to obtain $0.0016<c<0.0036$. An estimate of the per capita adult mortality rate was obtained by re-arranging (4.2) and (4.3) to read $\mu=\lambda/(cN^*)=0.2480$. Finally, an estimate of the rate at which potential recruits are produced per adult per time unit was computed from the equation $b=\mu\exp(cN^*)=0.4152$.

The parameter estimates can be used to predict the variance over time. Substituting the estimates of b, μ, c, and $i=20$ into (5.62) gives a variance which increases from 0 to c^{-1}. This variance is represented by the dashed curves in Fig. 5.3 which form an interval of $\overline{N}(t,i)\pm 2\,V(t,i)^{1/2}$. Most of the data points fall within this interval for $t>5$.

The quasi-stationary distribution was computed using the "normal" approximation (5.36). The hypothesis that the observed grouped frequency data (histogram in upper right of Fig. 5.3) is distributed according to this normal density was accepted using a chi-square test for goodness-of-fit at the 0.05 level of probability with 2 degrees of freedom. The mean number of adults computed directly from the 44 "steady-state" observations is 205.6. Interestingly, this estimate is very close to the exact value listed in Table 5.1 which was based on the parameter estimates obtained above.

The estimate for the expected time to extinction and its various approximations have already been presented in Table 5.1 for the parameter values of *T. brevicornis*. Using the exact value, $\Psi(20)=3.06673\times 10^{23}$ time units or, equivalently, $\Psi(20)=1.17547\times 10^{22}$ years. Although extinction is certain, it is unlikely to be observed unless there are changes in the parameters of the system. Thus it is reasonable to assume the existence of a quasi-stationary probability distribution.

Finally, we can use the linear birth-death approximation to predict the transient probability distribution of adult numbers. The parameter estimates $b=0.4152$, $\mu=0.2480$, and $c=0.002485$ were used in (5.63) to produce the time-varying probability distribution illustrated in Fig. 5.4. The distribution at $t=1$ is nearly a spike. As time proceeds both the mean and the variance of the distribution increase. Asymptotically, the probability distribution approaches a negative binomial which closely approximates a discrete normal distribution.

5.5 Summary and Conclusions

In this chapter and the previous one, we have focused our attention on simple stochastic models of population dynamics in *Tribolium*. We conclude this analysis with some comments on the merits of stochastic models, a comparison of the discrete and continuous formulations, and some remarks on the nature of stochasticity in population dynamics.

5.5.1 Deterministic versus Stochastic Models

In deterministic models, population fluctuations can only be viewed in a statistical sense as "random error." For example, in our analysis of life stage interactions in Chapter 2, the lack of regular cyclical fluctuations in adult number (Fig. 2.3) was attributed to the swamping effects of stochastic noise. Although the data forces us to acknowledge the

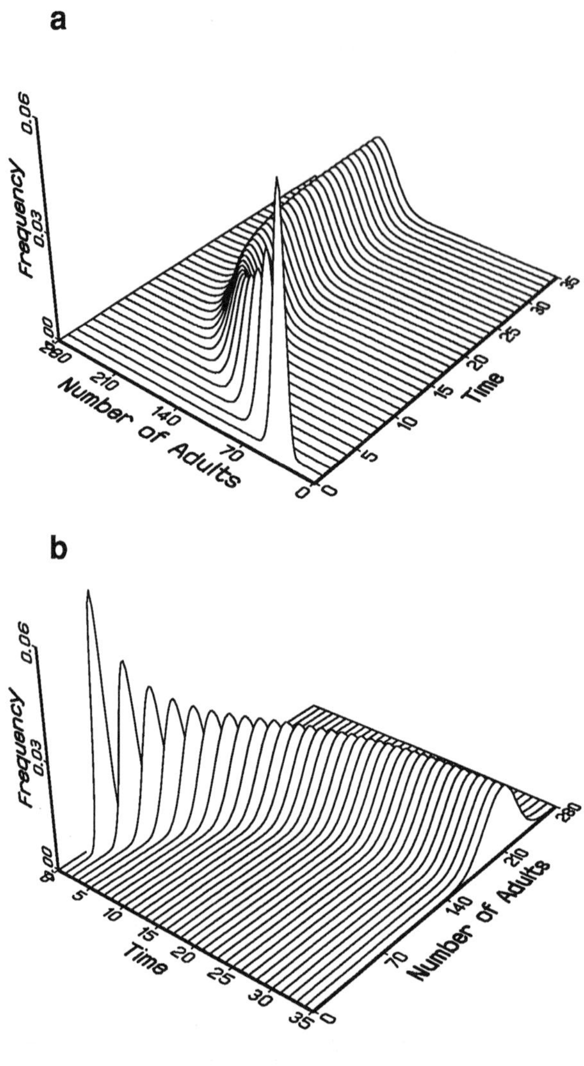

Fig. 5.4. Two views of the transient distribution (5.63) for the parameter estimates from *T. brevicornis*.

existence of population fluctuations, these fluctuations are treated as hindrances to our understanding of population dynamics. Indeed, we usually design our experiments to minimize these random effects.

Stochastic models, on the other hand, incorporate fluctuations as a fundamental component of the populations dynamics. Variability in population numbers is one of the model's predictions; it is an additional source of information on the population. For example, in our analysis of *T. brevicornis* in §5.4, we used a least squares technique to estimate two of the parameters for the change in the mean number of adults. In a deterministic formulation, we would be forced to stop there. However, variability in adult numbers provided us with a third parameter: $c = 1/V^*$. With this third piece of information, we were able to estimate all three parameters in the original biological model. We were presented a similar opportunity in §4.6 using the time-dependent gamma distribution. In this context, variability becomes an asset to our understanding of population dynamics.

Why should we bother with deterministic models at all? Stochastic models have a price. Underlying the analyses of Chapters 4 and 5 is a very simple model for adult numbers. However, the resulting stochastic diffusion equations and Kolmogorov forward equations are anything but simple; it was necessary to resort to approximations. It is easy to imagine that stochastic analogues of the complex deterministic demographic models of Chapters 2 and 3 would be hopelessly intractable. Stochastic models are desirable but not always practical. We must utilize both deterministic and stochastic models in our efforts to understand population dynamics.

5.5.2 Continuous versus Discrete Stochastic Models

Chapters 4 and 5 represent two alternative approaches to stochastic modeling: continuous and discrete state spaces. In the continuous case, one begins with a stochastic differential equation which must be integrated using Ito or Stratanovich calculus. The result is a diffusion equation that yields an asymptotic probability function for population size. In the discrete case one begins with an infinite number of linear differential equations: the Kolomogorov forward equations. These equations can be integrated to find the stationary or quasi-stationary discrete probability distribution and expected time to extinction. Ultimately, both formulations provide probabilistic predictions for the long term dynamics of population size. Which formulation should be used? Do these alternative approaches have different underlying biological assumptions?

In mathematical terms, the diffusion and birth-death paradigms are both representative of the same type of phenomenon: a continuous-time Markovian process. In fact, mathematical techniques exist for approximating one type of process with the other. The state space of a continuous process can be discretized to yield a discrete state space in much the same way a histogram can be used to represent a continuous distribution. Conversely, a discrete process which wanders over a large number of states can be "smoothed" into an approximating diffusion process. In fact, one can approximate a birth-death process with transition rates $r(N)$ and $m(N)$ using the following stochastic differential equation (Nisbet and Gurney 1982):

$$\frac{dN}{dt} = [r(N)-m(N)] + [r(N)+m(N)]^{1/2}\gamma(t)\,, \tag{5.66}$$

where $\gamma(t)$ is a "white noise" random variable as discussed in §4.2.1. Using the transition rates (5.48) and linearizing this equation around the point $N(t) = N^*$ and $\gamma(t) = 0$, we have

$$\frac{dN}{dt} = \lambda[N^* - N(t)] + (2\mu N^*)^{1/2}\gamma(t)\,. \tag{5.67}$$

This linear stochastic differential equation can be connected with an asymptotic probability distribution for adult numbers using the diffusion equation (4.11). Using either Ito or Stratanovich calculus, the probability density function is

$$f^*(N) = \sqrt{\frac{c}{2\pi}}\,\exp\left[-\left(\frac{c}{2}\right)(N - N^*)^2\right]. \tag{5.68}$$

Using the diffusion paradigm, we have obtained a result which is essentially identical to the quasi-stationary distribution (5.36). The population will have a normal distribution with a mean near N^* and a variance of c^{-1}. Diffusion equations and birth-death processes yield compatible results when the same biological assumptions are applied. The choice of method should be based on practical considerations such as mathematical tractability and the type of questions one wishes to address. When in doubt, try both approaches.

There are some circumstances in which a continuous or discrete stochastic model has the advantage. For problems of extinction, the birth-death model is more straightforward and easier to understand. Diffusion approximations are notoriously problematical when one wishes to estimate the probability of extinction (Nisbet and Gurney 1982). On the other hand, our experience suggests that diffusion equations are more convenient for questions concerning the steady-state. In birth-death processes, one must contend with absorbing states and "quasi-stationary" distributions. On the other hand, the method of integration is not an issue for a birth-death process whereas one must choose between Ito and Stratanovich calculus when using stochastic differential equations. Perhaps most importantly, in the continuous stochastic differential equation one can let the parameter σ represent any unknown source of variability in the growth rate of the population. In the birth-death approach, any "external" variability (temperature, humidity, etc.) must be incorporated into the transition rates themselves by making them random functions of time. This dichotomy of "demographic" versus "environmental" stochasticity is discussed below. Nevertheless, it should be pointed out here that both continuous and the discrete stochastic models can allow for demographic and environmental effects.

5.5.3 Environmental versus Demographic Stochasticity

Consider the following experiment. Two *Tribolium* populations are initiated with identical age structures from the same genetic stock. One population is placed in an incubator at a constant temperature and humidity. The second population is placed in another incubator where the mean temperature is the same as the first, but the temperature is made to fluctuate wildly from one day to the next. Clearly, we would expect to see variation in adult numbers in both populations. However, we would also expect the variation in the second population to be higher than in the first. If we applied the methods of this chapter to a derivation of the cannibalism rate, from $c = 1/V^*$ we might conclude that the first population was more cannibalistic than the second. Would this conclusion be correct? Are the adults in the first population really eating more eggs and pupae than adults in the second population? The most likely answer is no. Variation in population abundances is dominated by demographic stochasticity in the first population and environmental stochasticity in the second.

Untangling these two sources of variability is a formidable challenge for population biologists. Both types of variability are always present. Furthermore, complex demographic dynamics such as those described in Chapters 2 and 3 may confound the effects of environmental and demographic stochasticity. For example, in the egg-larval submodel of Chapter 3, a stable equilibrium may coexist with a stable limit cycle. Two identical *Tribolium* populations under the same environmental conditions might exhibit very different patterns of fluctuation. If these populations were kept in separate incubators, it would be easy to blame environmental stochasticity for the differences in their variability. This fact must be kept in mind when using simple stochastic models.

On the other hand, simple stochastic models might prove themselves as an effective tool for handling complex dynamics in certain situations. The simple adult model of Chapters 4 and 5 is based on the assumption of a fast time scale for immatures and a slow time scale for adults. In this context, the variable bursts of recruitment due to egg-larval cycles might be treated as random variation in the "birth" rate of the adult age class. Among the observed distributions of Chapter 4, many if not all of the populations exhibit immature cycles. Yet the gamma distribution did an adequate job of describing the variability in adult numbers. At present, we will be liberal in our interpretation of "growth rate variability." But the problem of environmental versus demographic stochasticity deserves further research.

5.6 Appendix

In this appendix we prove the result (5.16).

Consider the equations (5.15). For $j=0$, we have $\tau(i,1)m(1)=1$ or

$$\tau(i,1) = \frac{1}{m(1)}, \qquad i \geq 1. \tag{5.69}$$

Substituting this into the equation for $j=1$ gives

$$\tau(i,2) = \begin{cases} \dfrac{r(1)}{m(1)m(2)}, & i=1, \\[2ex] \dfrac{r(1)+m(1)}{m(1)m(2)}, & i \geq 2. \end{cases} \tag{5.70}$$

Finally, we can substitute (5.69) and (5.70) into (5.15) for $j=2$ to obtain

$$\tau(i,3) = \begin{cases} \dfrac{r(1)r(2)}{m(1)m(2)m(3)}, & i=1, \\[2ex] \dfrac{r(1)r(2)}{m(1)m(2)m(3)}+\dfrac{r(2)}{m(2)m(3)}, & i=2, \\[2ex] \dfrac{r(1)r(2)}{m(1)m(2)m(3)}+\dfrac{r(2)}{m(2)m(3)}+\dfrac{1}{m(3)}, & i\geq 3. \end{cases} \tag{5.71}$$

It is easy to verify that (5.69) – (5.71) satisfy (5.16). For $j>3$ we complete the proof by induction.

Assume the (5.16) holds for $j=k-2$ and $j=k-1$, $k>3$. Let ζ(i, j) be an indicator variable such that $\zeta(i,j)=1$ if $i=j$ and $\zeta(i,j)=0$ if $i\neq j$. For $i\geq k-1$,

$$\begin{aligned}\tau(i,k) &= \left(\frac{1}{m(k)}\right)\left\{[r(k-1)+m(k-1)]\left(\frac{1}{m(k-1)}\right)\left[1+\sum_{s=1}^{k-2}\prod_{h=s}^{k-2}\frac{r(h)}{m(h)}\right]\right. \\ &\qquad \left. -r(k-2)\left(\frac{1}{m(k-2)}\right)\left[1+\sum_{s=1}^{k-3}\prod_{h=s}^{k-3}\frac{r(h)}{m(h)}\right]-\zeta(i,k-1)\right\} \\ &= \left(\frac{1}{m(k)}\right)\left\{\frac{r(k-1)}{m(k-1)}+\sum_{s=1}^{k-2}\prod_{h=s}^{k-2}\frac{r(h)}{m(h)}+\sum_{s=1}^{k-2}\prod_{h=s}^{k-1}\frac{r(h)}{m(h)}\right. \\ &\qquad \left. -\frac{r(k-2)}{m(k-2)}-\sum_{s=1}^{k-3}\prod_{h=s}^{k-2}\frac{r(h)}{m(h)}+[1-\zeta(i,k-1)]\right\} \\ &= \left(\frac{1}{m(k)}\right)\left\{\sum_{s=1}^{k-1}\prod_{h=s}^{k-1}\frac{r(h)}{m(h)}+[1-\zeta(i,k-1)]\right\} \end{aligned} \tag{5.72}$$

which agrees with (5.16). For $i\leq k-1$,

$$\begin{aligned}\tau(i,k) &= \left(\frac{1}{m(k)}\right)\left\{[r(k-1)+m(k-1)]\left(\frac{1}{m(k-1)}\right)\left[\sum_{s=1}^{i}\prod_{h=s}^{k-2}\frac{r(h)}{m(h)}\right]\right. \\ &\qquad \left. -\left(\frac{r(k-2)}{m(k-2)}\right)\left[\zeta(i,k-2)+\sum_{s=1}^{i-\zeta(i,k-2)}\prod_{h=s}^{k-3}\frac{r(h)}{m(h)}\right]\right\}\end{aligned}$$

$$= \left(\frac{1}{m(k)}\right)\left\{\sum_{s=1}^{i}\prod_{h=s}^{k-1}\frac{r(h)}{m(h)} + \sum_{s=1}^{i}\prod_{h=s}^{k-2}\frac{r(h)}{m(h)} - \sum_{s=1}^{i-\zeta(i,k-2)}\prod_{h=s}^{k-2}\frac{r(h)}{m(h)} - \zeta(i,k-2)\left(\frac{r(k-2)}{m(k-2)}\right)\right\}$$

$$= \left(\frac{1}{m(k)}\right)\sum_{s=1}^{i}\prod_{h=s}^{k-1}\frac{r(h)}{m(h)} \tag{5.73}$$

which also agrees with (5.16). Hence we have shown that if (5.16) holds for $j=k-1$ and $j=k-2$, $k>3$, it also holds for $j=k$. Since (5.69) – (5.71) show that (5.16) holds for $j=1$, 2, and 3, the general result is proved by induction.

Chapter 6 Natural Selection

Genetic variation due to the segregation of known Mendelian genes provides a way to study the synthesis of population growth and natural selection. One such allelic polymorphism in *Tribolium* is associated with the utilization of unsaturated fatty acids, where the metabolism or lack of metabolism is essentially due to a single autosomal locus with two alleles (Costantino et al. 1966, 1967, 1968). In a population genetic analysis, interest in this sort of fitness-related physiological mutant is focused on the dynamics of *both* allele frequency and population size. While selection in the context of Malthusian growth is the more traditional approach (Moran 1962, Crow and Kimura 1970), an extensive literature on non-Malthusian growth is available (MacArthur 1962, Anderson 1971, Roughgarden 1971, Charlesworth 1971, Clarke 1972, 1973, Ginzburg 1977a, Asmussen and Feldman 1977, Hastings 1978, Leon and Charlesworth 1978, Gregorius 1979, Slatkin and Maynard Smith 1979, Desharnais and Costantino 1983, Ginzburg 1983, Selgrade and Namkoong 1984). New questions arise. With selection and non-Malthusian growth, the equilibrium allele frequency no longer corresponds to the maximization of the population's growth rate but rather leads to the maximization of the equilibrium population size. In laboratory experiments with the flour beetle, is population size maximized? Do cultures with different initial allele frequencies all converge to a common allele frequency? Do they converge to a common population size? How is the concept of a *stochastic* equilibrium, as discussed in Chapters 4 and 5, to be viewed? In this chapter we look at the long term or stationary dynamics while in the next chapter we emphasize the dynamics of the approach to equilibrium.

We begin with an elementary genetic system coupled with the adult recruitment model introduced in Chapter 4. Our overall strategy is the simultaneous study of allele frequency and population size; nevertheless, our experimental emphasis is on population size for the following reason. We cannot measure allele frequency any more accurately or with less work in *Tribolium* than we can in other animals. Moreover, linkage, epistasis, mutation, etc. are also as difficult to handle. But the measurement of population size *is* different. Here, then, as we begin to address genetic variability our objective is to do so by exploiting the major advantage of the *Tribolium* experimental system: accurate measurement of population size.

6.1 Ecological-Genetic Perspective

In the genetic model of natural selection in a sexually reproducing diploid population with continuous time, we can describe the change in allele frequency at a single multi-allelic autosomal locus and the change in population number, respectively, as (Crow and Kimura, p. 191, 1970)

$$\frac{\mathrm{d}p_i(t)}{\mathrm{d}t} = p_i(t)\,(W_i - W)\,, \qquad i = 1, 2, 3, \ldots, n, \tag{6.1}$$

$$\frac{\mathrm{d}N(t)}{\mathrm{d}t} = N(t)\,W\,, \tag{6.2}$$

where p_i is the frequency of allele i, n is the number of alleles, and N is the population size. Here W_i represents the marginal fitness of the *ith* allele and W is the average population fitness. With the usual assumptions of random mating, weak selection, and Hardy-Weinberg genotypic frequencies, W_i and W are calculated using

$$W_i = \sum_{j=1}^{n} W_{ij}p_j\,, \qquad i = 1, 2, \ldots, n, \tag{6.3}$$

$$W = \sum_{i=1}^{n} W_i p_i = \sum_{i=1}^{n}\sum_{j=1}^{n} W_{ij}p_i p_j\,. \tag{6.4}$$

The genotype fitness W_{ij} is defined as the contribution of an individual with genotype ij to the total population growth rate. When only two alleles are segregating ($n=2$), we will drop the subscripts on allele frequencies and let $p(t)=p_1(t)$ and $1-p(t)=p_2(t)$.

The population growth equation (6.2) is an explicit part of the genetic model of natural selection. However, if we assume that the genotypic fitnesses, W_{ij}, are constant, then population size, N, does *not* appear in the equations for the change in allele frequency, the average fitness, or the equilibrium allele frequencies (Fig. 6.1). The growth equation is Malthusian and the average fitness is the population's growth rate. The model, in this form, is not appropriate for *Tribolium*. It is important to recognize that the constancy

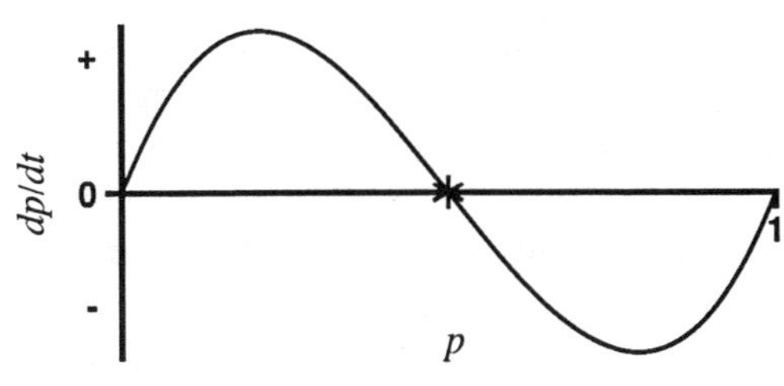

Fig. 6.1. A genetic perspective. The change in allele frequency, dp/dt, as a function of allele frequency, p, for the single locus model of natural selection (6.1) with the heterozygote superior. The asterisk marks a globally stable polymorphic equilibrium. Population size does not affect the rate of genetic change.

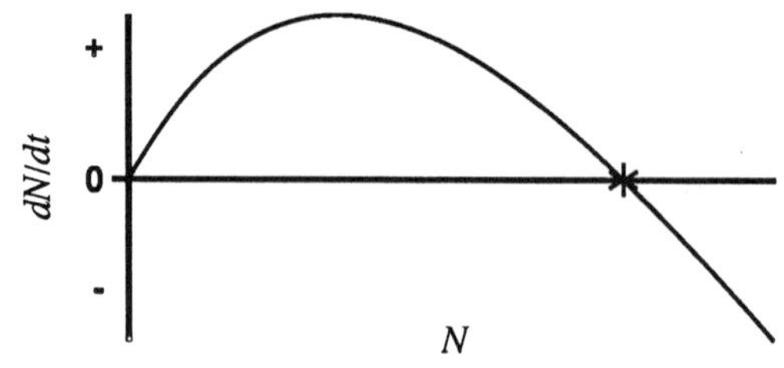

Fig. 6.2. A demographic perspective. The change in population size, dN/dt, as a function of population size, N, for the simple growth model (6.5). The asterisk marks a globally stable equilibrium. Without natural selection, allele frequency does not affect the rate of population growth.

assumption is not formally part of the genetic argument and that the W_{ij}'s may depend, for example, on the frequencies of the other genotypes or on the size of the population or both (Namkoong and Selgrade 1986).

In our discussion of the flour beetle, we have chosen to replace the linear growth rule with the nonlinear differential equation

$$\frac{dN(t)}{dt} = N(t)\,[b\,\exp(-cN(t)) - \mu]\,, \tag{6.5}$$

where b represents the rate at which potential recruits are produced per adult, c is the *per capita* rate at which adults cannibalize the immature life stages, and μ is the *per capita* adult mortality rate. This adult recruitment model has a property which is lacking in the Malthusian equation: a positive, globally stable equilibrium population size (Fig. 6.2). It is in the context of (6.5) that we will define the genotypic fitnesses.

6.2 Density-Regulated Natural Selection

6.2.1 Density-Dependent Fitness

The fitnesses of the different genotypes may be written as a function of population size. Our choice for the genetic fitness expression is

$$W_{ij} = b_{ij}\,r(N) - \mu_{ij}\,m(N)\,, \qquad i,j = 1, 2, \ldots, n, \tag{6.6}$$

where b_{ij} and μ_{ij} are the *per capita* density independent rates of recruitment and mortality, respectively, associated with genotype ij. We will assume that the density-dependent recruitment function, $r(N)$, and mortality function, $m(N)$, satisfy the following conditions: (i) $r(N)$ and $m(N)$ are continuous and differentiable; (ii) $r(0) = 1$ and $m(0) = 1$; (iii) $r(N)$ is nonincreasing ($dr/dN \leq 0$) and $m(N)$ is nondecreasing ($dm/dN \geq 0$); (iv) the function $R(N) = r(N)/m(N)$ is strictly monotonic decreasing ($dR/dN < 0$) with $R(N) \to 0$ as $N \to \infty$. We assume the condition (i) for convenience. Condition (ii) allows exponential population growth at low densities and conditions (iii) and (iv) assure density dependence throughout the range of population size. These restrictions on the fitness expression (6.6) satisfy the assumptions of the continuous time model of Ginzburg (1977a).

Substituting (6.6) into the equations of natural selection (6.1) and (6.2), the rates of change in allele frequency and population size become

$$\frac{dp_i}{dt} = p_i\,[(b_i - b)r(N) + (\mu - \mu_i)m(N)]\,, \qquad i = 1, 2, 3, \ldots, n, \tag{6.7}$$

$$\frac{dN}{dt} = N\,[b\,r(N) - \mu m(N)]\,, \tag{6.8}$$

where, for recruitment we have

$$b_i = \sum_{j=1}^{n} b_{ij}p_j \qquad \text{and} \qquad b = \sum_{i=1}^{n} b_i p_i\,, \tag{6.9}$$

and for mortality we have

$$\mu_i = \sum_{j=1}^{n} \mu_{ij}p_j \qquad \text{and} \qquad \mu = \sum_{i=1}^{n} \mu_i p_i\,. \tag{6.10}$$

Both the change in allele frequency (6.7) and population size (6.8) are altered when fitness, $W_{ij}(N)$, is explicitly stated in terms of genotypically determined birth and death rates together with the density-dependent expressions $r(N)$ and $m(N)$.

Genetic variation at a single locus, in the context of this model, may affect the rate of recruitment, or just the adult mortality rate or both. The implications of each of these cases are discussed in Desharnais and Costantino (1983). Here we will review the general results.

6.2.2 Selection on Recruitment

With genotype differences associated only with the density-independent rates of recruitment (6.6) becomes

$$W_{ij}(N) = b_{ij}\,r(N) - \mu m(N)\,. \tag{6.11}$$

The change in allele frequency and population size are given by

$$\frac{dp_i}{dt} = p_i(b_i - b)r(N), \qquad i = 1, 2, 3, \ldots, n, \tag{6.12}$$

$$\frac{dN}{dt} = N\,[b\,r(N) - \mu m(N)]\,, \tag{6.13}$$

where b_i and b are computed using (6.9). In this case, only the recruitment function $r(N)$ affects the rate of selection.

The equilibrium allele frequencies, p_i^*, are obtained in the usual way by setting equation (6.13) equal to zero and solving for p_i. The equilibria and the stability properties of p_i^* are the same as the classical model with constant fitnesses.

A single nontrivial equilibrium population size, $N^*>0$, is found from equation (6.13) when $R(N)=\mu/b^*$. The star* notation indicates that b is evaluated at $p_i=p_i^*$. Since the function R is strictly monotonic, the equilibrium is given by $N^*=R^{-1}(\mu/b^*)$ where R^{-1} is the inverse function of R, i.e., $R^{-1}(R(N))=N$. This equilibrium is locally stable provided $b>\mu$. Natural selection will result in N^* being a local maximum by maximizing b^*.

While inclusion of $r(N)$ does not alter the equilibrium allele frequencies nor the time-independent trajectory of frequencies, given by

$$\frac{dp_i}{dp_j} = \frac{p_i(b_i - b)}{p_j(b_j - b)}, \tag{6.14}$$

this density dependent recruitment function can influence the rate of approach to genetic equilibrium. Specifically, if $dr/dN<0$, then the magnitude of change in allele frequency will decrease with increases in population size, N. In other words, given a set of populations with the same fitness values, b_{ij}, and the same initial allele frequencies, $p_i(0)$, the population with the smallest initial density, $N(0)$, is predicted to approach genetic equilibrium at the quickest rate.

We now make an application of these ideas to *Tribolium*. Reasonable forms for the density-dependent recruitment and mortality functions are $r(N)=\exp(-cN)$ and $m(N)=1$, respectively. Anticipating the experimental data that are presented in the next section, we have chosen a numerical example with two alleles ($n=2$) and balancing selection ($b_{11}<b_{12}>b_{22}$). The rates of change for allele frequency and population size are plotted in Fig. 6.3. The equilibrium allele frequency is polymorphic, $0<p^*<1$, and globally stable. The magnitude of dp/dt is uniformly a decreasing function of population size (Fig. 6.3a). The sketch of the change in adult number reveals that for $p=p^*$, dN/dt is a maximum (Fig. 6.3b). All trajectories of the population converge to the globally stable equilibrium point and N^* is a maximum.

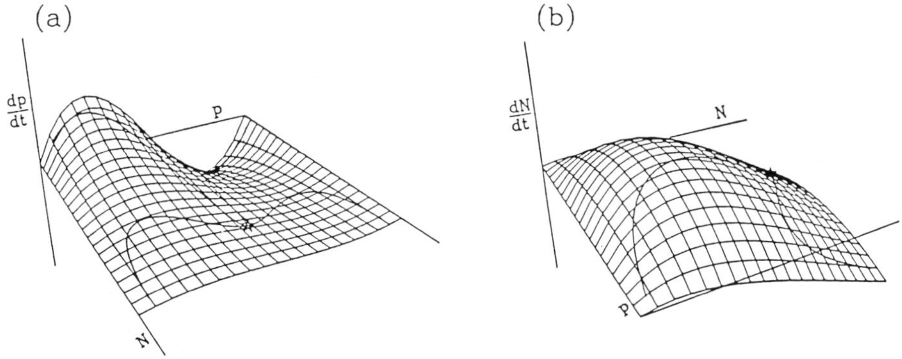

Fig. 6.3. Balancing selection on recruitment. (a) The change in allele frequency, dp/dt, and (b) the change in population size, dN/dt, as functions of allele frequency, p, and population size, N, for the model given by (6.12) and (6.13). The density-dependent functions are $r(N) = \exp(-cN)$ and $m(N) = 1$ and the following parameter values were used: $b_{12} = 3$, $b_{11} = b_{22} = 0.5$, $\mu = 0.25$, and $c = 0.02$. The stable equilibria are shown by stars. The curved lines are sample trajectories. [From R. A. Desharnais and R. F. Costantino (1983), Natural selection and density-dependent population growth. *Genetics* **105**: 1029–1040.]

6.2.3 Selection on Mortality

In this case, the genotypic differences are visualized in the *per capita* density-independent death rates, μ_{ij}, so that equation (6.6) is

$$W_{ij}(N) = b\, r(N) - \mu_{ij}\, m(N)\,. \tag{6.15}$$

The rates of change in allele frequency and population size are given by

$$\frac{dp_i}{dt} = p_i(\mu - \mu_i)m(N)\,, \qquad i = 1, 2, 3, \ldots, n, \tag{6.16}$$

$$\frac{dN}{dt} = N\,[b\, r(N) - \mu m(N)]\,, \tag{6.17}$$

where μ_i and μ are given by (6.10). In this case only the mortality function $m(N)$ affects the rate of selection.

As in the case of selection on recruitment, the equilibrium allele frequencies are independent of population size and the stability properties are the same as in the classical model. Also, the trajectory of allele frequency, dp_i/dp_j, is again independent of population size. The effect of gene frequency on the rates of change in adult number is qualitatively the same in both cases, i.e., for a given value of N, dN/dt is always maximum for a stable p^* and minimum when p^* is unstable.

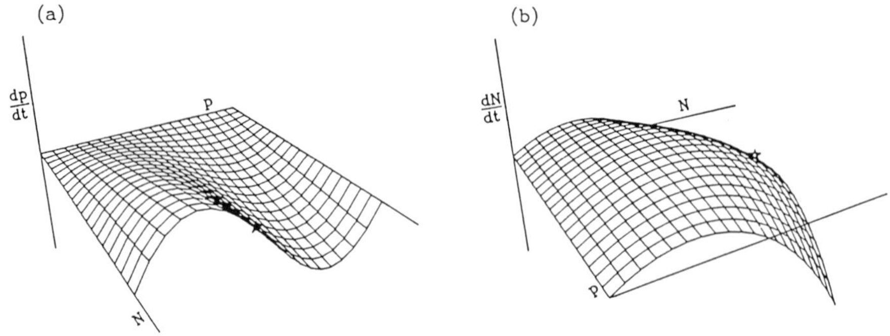

Fig. 6.4. Balancing selection on mortality. (a) The change in allele frequency, dp/dt, and (b) the change in population size, dN/dt, as functions of allele frequency, p, and population size, N, for the model given by (6.16) and (6.17). The density-dependent functions are $r(N)=1$ and $m(N)=N+1$ and the following parameter values were used: $b=2$, $\mu_{12}=0.0086$, and $\mu_{11}=\mu_{22}=0.0266$. The stable equilibria are shown by stars. [From R. A. Desharnais and R. F. Costantino (1983), Natural selection and density-dependent population growth. *Genetics* **105**: 1029–1040.]

The most striking comparison between recruitment and mortality selection is that with selection on mortality the magnitude of change in allele frequency is forecast to *increase* as N approaches N^* (Fig. 6.4). In addition, a high mortality represents a low fitness, consequently, the equilibrium population size $N^*=R^{-1}(\mu^*/b)$ will be a local maximum because natural selection will minimize μ.

6.2.4 Selection on Both Recruitment and Mortality

We now imagine that the genetic locus effects both the rate of recruitment and the rate of mortality. The rates of change in allele frequency and population size are given in (6.7) and (6.8).

The equilibria are found by setting the system of equations (6.7) and (6.8) equal to zero. The equilibrium population size is

$$N^* = R^{-1}(\mu^*/b^*)\,, \tag{6.18}$$

where $R(N)=r(N)/m(N)$. Setting equations (6.7) to zero we have

$$(b_i^* - b^*)R(N^*) - (\mu_i^* - \mu^*) = 0\,, \qquad i = 1, 2, \ldots, n. \tag{6.19}$$

From (6.18) we know that $R(N^*)=\mu^*/b^*$ and substituting this into (6.19) we obtain

$$b^* - (b_i^*/\mu_i^*)\mu^* = 0\,, \qquad i = 1, 2, \ldots, n. \tag{6.20}$$

In particular, note from equation (6.20) that the ratios

$$b_i^*/\mu_i^* = b^*/\mu^*, \qquad i = 1, 2, \ldots, n, \tag{6.21}$$

at the genetic equilibria. The local maxima of the ratio b/μ coincide exactly with the stable equilibria p_i^*. This suggests that natural selection will maximize N^* by maximizing the *ratio* of the density independent rates of recruitment to mortality, b/μ.

This conjecture can be demonstrated by using Ginzburg's (1977a) result which states that $N^* = R^{-1}(\mu^*/b^*)$ is maximized at a stable p^* and by recalling that R^{-1} is a continuous monotonic decreasing function. Alternatively, one can use equation (6.20) to show that the characteristic values, ζ, of the pencil of the quadratic form $b - \zeta\mu$ are $\zeta = \max(b/\mu)$ and that these maxima are assumed for the characteristic vectors, $\mathbf{p}^* = (p_1, p_2, \ldots, p_n)^{\mathrm{T}}$ of the pencil (Gantmacher 1959, pp. 317–323).

An important property of the rates of change in allele frequency for selection imposed on either recruitment or mortality is that the trajectory of allele frequency, dp_i/dp_j, is independent of N. With selection on both recruitment and mortality this time-independent trajectory is now a function of population size. For example, when N is small (i.e., $N \ll N^*$ such that $r(N) \approx 1$ and $m(N) \approx 1$) the genetic differences in fitness (6.6) are determined by $b_{ij} - \mu_{ij}$ and selection will change allele frequencies so as to maximize the exponential rate of population growth $b - \mu$. Alternatively, when N is large (i.e., $N \gg N^*$ such that $R(N) = r(N)/m(N) \approx 0$) differences among the μ_{ij}'s dominate and selection will alter allele frequencies so as to minimize the mean mortality rate μ. But, as we have already shown, selection will ultimately maximize the ratio b/μ when population density has reached its equilibrium. Since the allele frequencies corresponding to a maximum of $b - \mu$, a minimum of μ, and a maximum of b/μ may be quite different, qualitative changes in the direction of selection can occur.

To illustrate what happens with genetic differences in both recruitment and mortality we let $r(N) = \exp(-cN)$ and $m(N) = \exp(aN)$. At low populations densities the change in allele frequency (Fig. 6.5a) is uniformly positive and natural selection will maximize the difference between the mean recruitment and the mean mortality rates. At these low densities, allele frequency will move toward $p^* = 1$. However, as the population continues to grow natural selection will no longer maximize the difference $b - \mu$ but will maximize the ratio b/μ. The direction of dp/dt will become negative and the population will then move toward the equilibrium allele frequency of $0 < p^* < 1$. The equilibrium population density is $N^* = [\log(b^*/\mu^*)]/(c + a)$. The graphical representation of the change in population size (Fig. 6.5b) reveals that, for a given N, dN/dt is not always maximum at p^*.

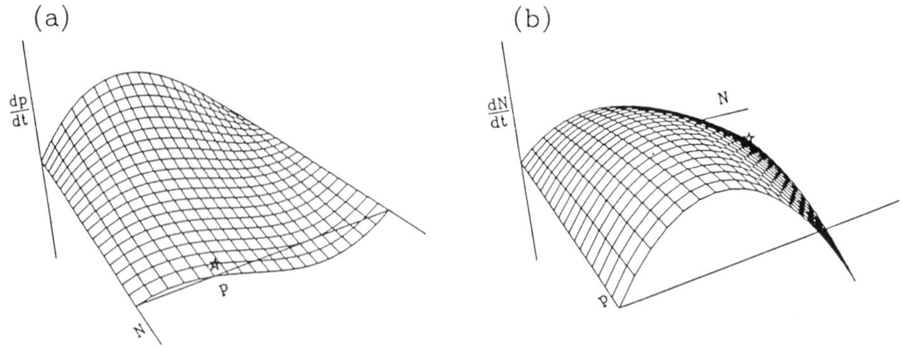

Fig. 6.5. Selection on both recruitment and mortality. (a) The change in allele frequency, dp/dt, and (b) the change in population size, dN/dt, as functions of allele frequency, p, and population size, N, for the model given by (6.7) and (6.8). The density-dependent functions are $r(N) = \exp(-cN)$ and $m(N) = \exp(aN)$ and the following parameter values were used: $b_{11} = 2$, $b_{12} = 1.5$, $b_{22} = 1$, $\mu_{11} = 0.5$, $\mu_{12} = \mu_{22} = 0.2$, $c = 0.02$, and $a = 0.003$. The stable equilibria are shown by stars. [From R. A. Desharnais and R. F. Costantino (1983), Natural selection and density-dependent population growth. *Genetics* **105**: 1029–1040.]

6.3 Experimental Results on the Corn Oil Sensitive Mutant

The corn oil sensitive mutant was serendipitously identified in the Spring of 1964. Unknown to us at the time, a single stray adult female had somehow been swept into a culture of *T. castaneum* that had undergone 30 generations of family selection for small larval weight. Several of the progeny from this particular culture were *nine times* heavier than the normal larval weight for the selected line. The occurrence of these incredibly large larvae was completely unexpected. Single pair matings among these "giant" individuals resulted in a progeny distribution with a 3 to 1 phenotypic ratio of large to small larvae. Our hypothesis was that we were dealing with the inheritance pattern of an autosomal recessive allele. To complete the story, the wandering adult female was homozygous for the wild type allele and the "giant" larvae were heterozygous which, when mated, gave the 3 to 1 ratio of large to small larvae.

Further study of the selected line revealed that it was a component of the culture medium, corn oil, that inhibited larval growth. The genetic basis of the inhibition was an autosomal recessive gene symbolized *cos* for corn oil sensitive and that the degree of larval growth inhibition of the *cos* homozygote was a function of the concentration of unsaturated fatty acids in the diet and the degree of saturation of the fatty acids involved (Costantino et al. 1966, 1967, 1968, 1970, 1972). While far from complete, we had some understanding of the biology of the mutant.

Following these experiments our research program focused on the utilization of the mutant in experimental population dynamic studies. The new questions addressed the manner in which populations initially segregating at the *cos* locus approached both genetic and demographic equilibria and, more generally, how these equilibria were to be characterized: fixed points, cycles, stationary distributions. In the following pages we report on the results of two long term experiments. The first is referred to as the *genetic polymorphism experiment* (68 weeks) and the second as, the *genetic perturbation experiment* (128 weeks).

6.3.1 Predictions

What are the long term projections for a population of beetles initially segregating at the *cos* locus? A stable genetic polymorphism? Maximization of population size? In the tradition of population genetics, we summarized a great deal of biology in the assignment of numerical values to the three genotypes: the numbers of eggs surviving to pupation, the number of days to sexual maturity, the number of eggs produced in a 24 hour interval, etc. These and other measures indicated that the relationship among the genotypes is $b_{11} < b_{12} > b_{22}$ (Scully and Costantino 1975, Moffa and Costantino 1977) where a 1 refers to the *cos* allele and 2 denotes the wild type allele. The mutant does not affect adult mortality nor the *per capita* rate of cannibalism, consequently, these values were assumed to be the same for all individuals. Selection is on recruitment.

The long term prediction is that a population initially segregating at this locus will reach a frequency of the *cos* allele given by

$$p^* = \frac{(b_{22} - b_{12})}{(b_{11} - 2b_{12} + b_{22})} \tag{6.22}$$

and attain the fixed point equilibrium population size

$$N^* = c^{-1} \log\left(\frac{b^*}{\mu}\right). \tag{6.23}$$

This joint allele frequency-population size equilibrium is globally stable. The population size N^* is expected to be maximized by natural selection maximizing the average rate of recruitment.

6.3.2 Genetic Polymorphism Experiment

The hypothesis of a genetic polymorphism at the *cos* locus was tested by Moffa and Costantino (1977). Eleven experimental treatments in which the initial *cos* allele frequency ranged from 0 to 1 in increments of 0.1 were established. Five replicates were maintained for each of the initial allele frequencies 1.0, 0.9, 0.8, 0.7, 0.6, 0.5; three replicates were maintained for each of the allele frequencies 0.4, 0.3, 0.2, 0.1, 0.0. The initial genotypic arrays were constructed from combinations of the +/+ and *cos/cos* homozygotes; the initial demographic array consisted of 10 newly emerged adult females and 10 newly emerged adult males. Each population was grown on 20 g of corn oil media

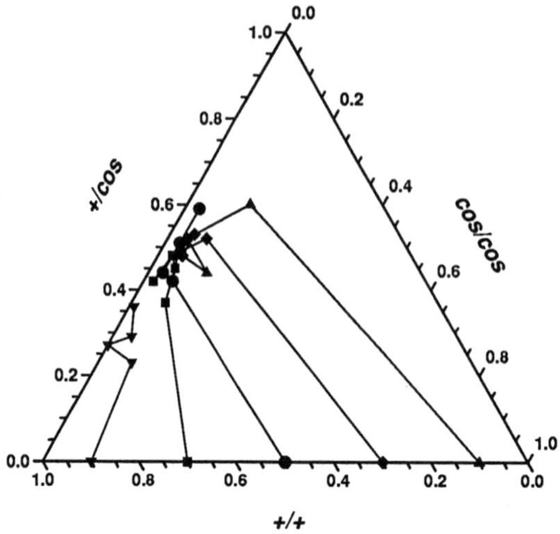

Fig. 6.6. Triangular coordinate plot of the genotypic arrays in the genetic polymorphism study. [Data from A. M. Moffa and R. F. Costantino (1977), Genetic analysis of a population of *Tribolium*. VI. Polymorphism and demographic equilibrium. *Genetics* **87**: 785–805, Table 3].

(percentage composition: 90% wheat flour, 5% dried brewers yeast, and 5% liquid corn oil) and kept in an unlighted incubator at 33±1°C and 42±6% relative humidity. The population was censused and the media was changed every two weeks. The cultures were maintained for 68 weeks.

At weeks 52 and 68, the *cos* allele frequency was estimated in those cultures with an initial allele frequency of 0.9, 0.7, 0.5, 0.3, and 0.1. Of the 10 cultures examined, the estimated mean allele frequency was 0.25±0.03 and the genotypic array composed of approximately equal numbers of +/+ and +/*cos* individuals. No *cos*/*cos* individuals were observed. With this information, we established the working hypothesis that cultures initially segregating for the *cos* allele (or *cos* chromosome region) have converged to a stable polymorphic genetic equilibrium.

In a separate experiment, conducted identically to the above "68-week" study, allele frequency and the genotypic array were estimated in 12 replicates for each initial allele frequency of 0.9, 0.7, 0.5, 0.3, and 0.1. To assure independence of samples, 3 replicates from each initial frequency were sampled at 2, 6, 12 and 16 weeks.

The data (Fig. 6.6) support the hypothesis of a polymorphic genetic equilibrium. Of the 60 cultures examined only one culture, where $p(0)=0.1$, lacked traces of the *cos* allele. The largest magnitude of change in the allele frequencies and the genotypic arrays occurred in the first two weeks. This is consistent with the life history observation that differences among genotypes for total number of offspring were due to larval viability. The genotypic arrays showed an immediate reduction in the frequency of the *cos*/*cos* genotype and an increase in the frequency of +/*cos* individuals. At week 16 the +/+ and +/*cos* genotypes were nearly equal with only a small fraction of *cos*/*cos* individuals. The *cos* allele frequency declined in the test populations started at 0.9, 0.7, 0.5, 0.3 and increased in the population with an initial frequency of 0.1 which suggested that the equilibrium was globally stable. At week 16 the allele frequencies ranged from 0.29 to 0.19.

The population size data in this experiment consisted of the numbers of small and large larvae, pupae, and adults recorded at biweekly intervals for 68 weeks. We will address these data later in this chapter and again in Chapter 7.

6.3.3 Genetic Perturbation Experiment

The design of experiment of Desharnais and Costantino (1980) was to establish populations at or near both demographic and genetic equilibrium and then to examine the local stability properties of the equilibrium by deliberate perturbation. The question of genetic stability was addressed by observing both the genetic and the age structure changes following the introduction of genetically different (+/+) individuals into the equilibrium populations. An equilibrium was said to be genetically stable if it was locally stable to changes in allele frequency. A similar experiment using the antennapedia allele was conducted recently by Delgado and Englert (1986).

Thirteen cultures of *T. castaneum*, homozygous for the corn oil sensitive allele (*cos/cos*), were initiated with the equilibrium age distribution of 64 adults, 16 pupae, 20 large larvae, and 70 small larvae. Each population was contained in a one-half pint milk bottle with 20 grams of corn oil media and kept in an unlighted incubator at 33 ± 1°C and $56 \pm 11\%$ relative humidity. All age classes were censused and placed in fresh medium at biweekly intervals for 128 weeks.

After ten weeks of culture, each population was randomly assigned to one of four treatments. Three of the treatments, each with three replicates, involved genetic perturbations and one treatment, with four replicates, served as a control. The genetic perturbations were (i) one +/+ adult female added, (ii) three +/+ adult females added, and (iii) three +/+ male and three +/+ female adults added. The control cultures were not disturbed.

Examination of the estimated genotypic frequencies (Fig. 6.7) show that the homozygous corn oil sensitive population was *genetically unstable* to the introduction of the wild type gene. From week 10 to 22, the *cos* allele frequency decreased in each of the treatments and then leveled off for the remainder of the experiment at an overall mean value of 0.42. The populations also seemed to converge to a common genotypic array. At equilibrium, i.e. from weeks 22 to 34, the pupae were composed of approximately 35% wild-type homozygotes, 43% heterozygotes, and 22% sensitive homozygotes. We could not detect any statistically significant differences among the three treatments.

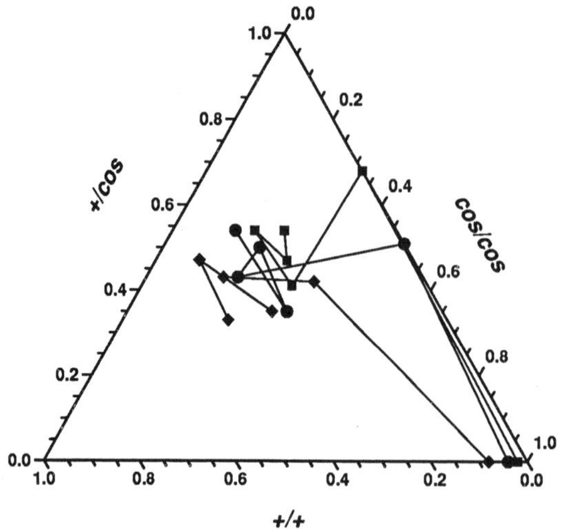

Fig. 6.7. Triangular coordinate plot of the genotypic arrays in the genetic perturbation study. The lower right vertex represents the unstable genetic equilibrium $p^*=1$. [Data from R. A. Desharnais and R. F. Costantino (1980), Genetic analysis of a population of *Tribolium*. VII. Response to genetic and demographic perturbations. *Can. J. Genet. Cytol.* **22**: 577–589, Table 2].

6.3.4 Stationary Distribution of Population Numbers

The fixed point equilibrium number of adults predicted by the model was not realized in the experiments with the corn oil sensitive mutant. Instead, these data support the concept of a *stochastic equilibrium* (Chapters 4 and 5). The stationary probability distribution of population numbers $f^*(N)$ is a more realistic prediction for the size of a beetle population.

We followed the statistical protocol outlined in Chapter 4 for the data obtained from the polymorphism experiment of Moffa and Costantino (1977). Maximum likelihood estimates for the parameters α and β in the gamma model were computed using the adult census data for each of the nine polymorphic experimental treatments (Desharnais et al. 1989). In every case, the null hypothesis that the gamma model fits these data was accepted (Fig. 6.8).

Did the test cultures initially segregating at the *cos* locus converge to a common gamma (see Fig. 6.9)? We used a likelihood ratio test of H_0: one gamma versus H_1: more than one gamma for the nine data sets ($p(0)=0.1, 0.2, ..., 0.9$). For the null hypothesis, one gamma was fitted simultaneously to the nine data sets (not pooled). We computed

$$G_0^2 = 2\sum_{i=1}^{9}\sum_{j=1}^{17} y_{ij}\log\left(\frac{y_{ij}}{k\,q_j(\alpha,\beta)}\right), \tag{6.24}$$

where y_{ij} is the frequency count in the jth abundance interval for the ith population ($p(0)=i/10$), k is the total sample size, and

$$q_j(\alpha,\beta) = \int_{s_{j-1}}^{s_j}\left(\frac{\beta^\alpha}{\Gamma(\alpha)}\right)x^{\alpha-1}\exp(-\beta x)\,dx \tag{6.25}$$

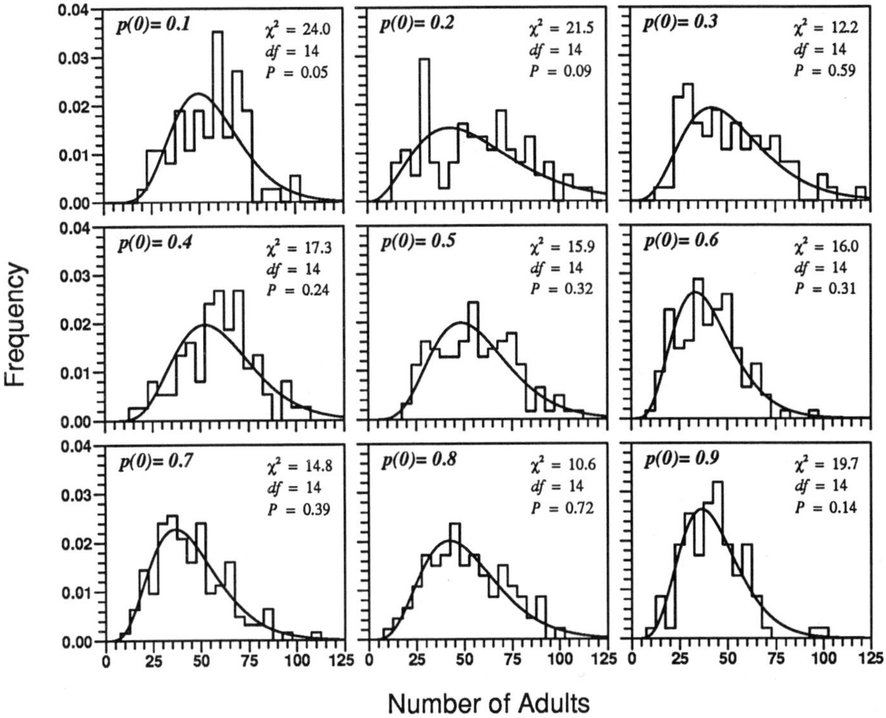

Fig. 6.8. The fitted gamma density functions (smooth curves) and the observed frequency distributions (histograms) of adult numbers for the nine experimental cultures initially segregating at the *cos* locus. The test statistics (χ^2), degrees of freedom (*df*), and probability levels (*P*) are given for the goodness-of-fit tests. [From R. A. Desharnais, B. Dennis and R. F. Costantino (1990), Genetic analysis of a population of *Tribolium*. IX. Maximization of population size and the concept of a stochastic equilibrium. *Genome* (in press).]

for the class frequency intervals $0 = s_0 < s_1 < \ldots < s_9 = \infty$. The maximum likelihood estimates of $\alpha = 5.4338$ and $\beta = 0.1079$ minimize (6.24) (solid curve in Fig. 6.9). Under the alternate hypothesis of one or more of the α's, β's not equal, the nine gammas were fitted separately to the nine data sets (dotted curves in Fig. 6.9). We computed

$$G_1^2 = 2 \sum_{i=1}^{9} \sum_{j=1}^{17} y_{ij} \log\left(\frac{y_{ij}}{k_i q_j(\alpha_i, \beta_i)}\right), \qquad (6.26)$$

where y_{ij} is the frequency count, k_i is the sample size for the *ith* population, and α_i, β_i are the maximum likelihood estimates for the *ith* population. The likelihood ratio test statistic of H_0 vs. H_1 is

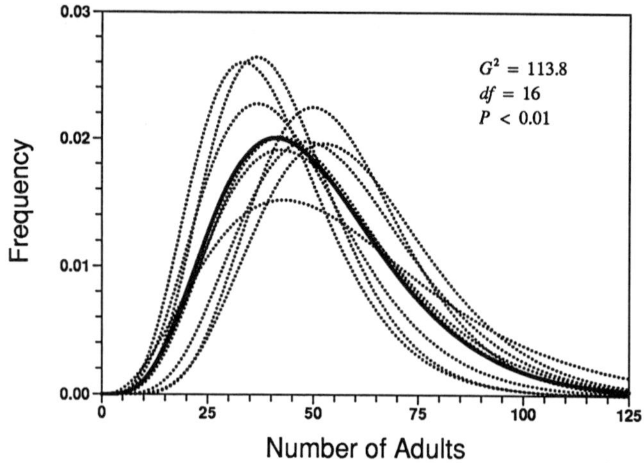

Fig. 6.9. A comparison of the nine gamma distributions fitted independently to each of the polymorphic groups (dotted curves) versus the single best fit gamma distribution for all the populations combined (but not pooled). The test statistic (G^2), degrees of freedom (*df*), and probability level (*P*) are given for a test of the null hypothesis that all these distributions are the same. [From R. A. Desharnais, B. Dennis and R. F. Costantino (1990), Genetic analysis of a population of *Tribolium*. IX. Maximization of population size and the concept of a stochastic equilibrium. *Genome* (in press).]

$$G^2 = G_0^2 - G_1^2 = 275.1707 - 161.3843 = 113.7864 \qquad (6.27)$$

with 16 degrees of freedom (18 parameters estimated in the alternate hypothesis minus the 2 parameters estimated in the null hypothesis). The null hypothesis of one common gamma is rejected.

In the deterministic genetic theory of natural selection if $p(0) = p^*$ then the population is in genetic equilibrium; if $p(0) \neq p^*$ then eventually p is expected to reach p^*. In these data $p^* \approx 0.3$, so the test cultures with that allele frequency were at or near genetic equilibrium at the beginning of the experiment and all nonequilibrium cultures should, eventually, be similar to these cultures. With respect to the dynamics of adult numbers, the 0.3 test cultures act as a "reference population" (Desharnais and Costantino 1982b, Desharnais 1986). Did any of the nonequilibrium cultures converge to the stationary distribution of adult numbers attained by the $p(0) = 0.3$ cultures? To answer this equation we used the D^2 test statistic (4.34) for the pairwise comparison of the gamma parameters for the 0.3 treatment with the other eight treatments (Fig. 6.10). We accepted the null hypothesis at the 0.05 level of significance that the gammas for $p(0) = 0.1$, 0.2, 0.5, 0.7 and 0.8 are equal to the $p(0) = 0.3$ cultures. On the other hand, the null hypotheses for the $p(0) = 0.4$, 0.6 and 0.9 cultures were rejected.

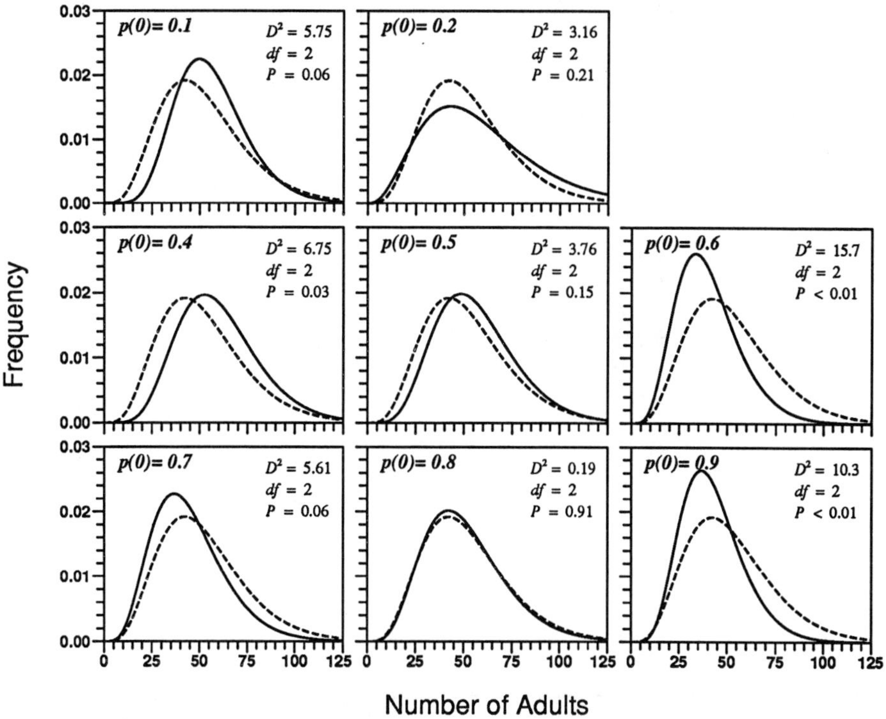

Fig. 6.10. Pairwise comparisons of the fitted gamma distributions for the nonequilibrium polymorphic populations (solid curves) versus the reference population with $p(0) = 0.3$ (dashed curve). The test statistics (D^2), degrees of freedom (*df*), and probability levels (*P*) are given for a test of the null hypothesis that the two distributions are the same. [From R. A. Desharnais, B. Dennis and R. F. Costantino (1990), Genetic analysis of a population of *Tribolium*. IX. Maximization of population size and the concept of a stochastic equilibrium. *Genome* (in press).]

While the idea of a shared or common gamma for all nine cultures was not accepted using the G^2 test statistic (6.27), the individual pairwise comparisons suggests that *some* convergence to a common stochastic equilibrium did occur (Fig. 6.10). Overall the polymorphic populations seem more alike than different.

The adult census data for the wild type (+/+) and sensitive (*cos*/*cos*) populations, in the region of the steady-state, were also fitted to the gamma abundance model (Fig. 6.11). The gamma was accepted in both cases, however, these strains have statistically different parameter values according to the D^2 test.

In the perturbation experiment of Desharnais and Costantino (1980, 1985), the adult census data for the homozygous *cos* and polymorphic cultures were fitted to the gamma density. The observed and expected distributions are presented in Fig. 6.12. The hypothesis that the observed distributions were gamma was accepted at the 0.01 level of probability. However, the null hypothesis that these gammas have identical parameters was rejected using the D^2 test statistic.

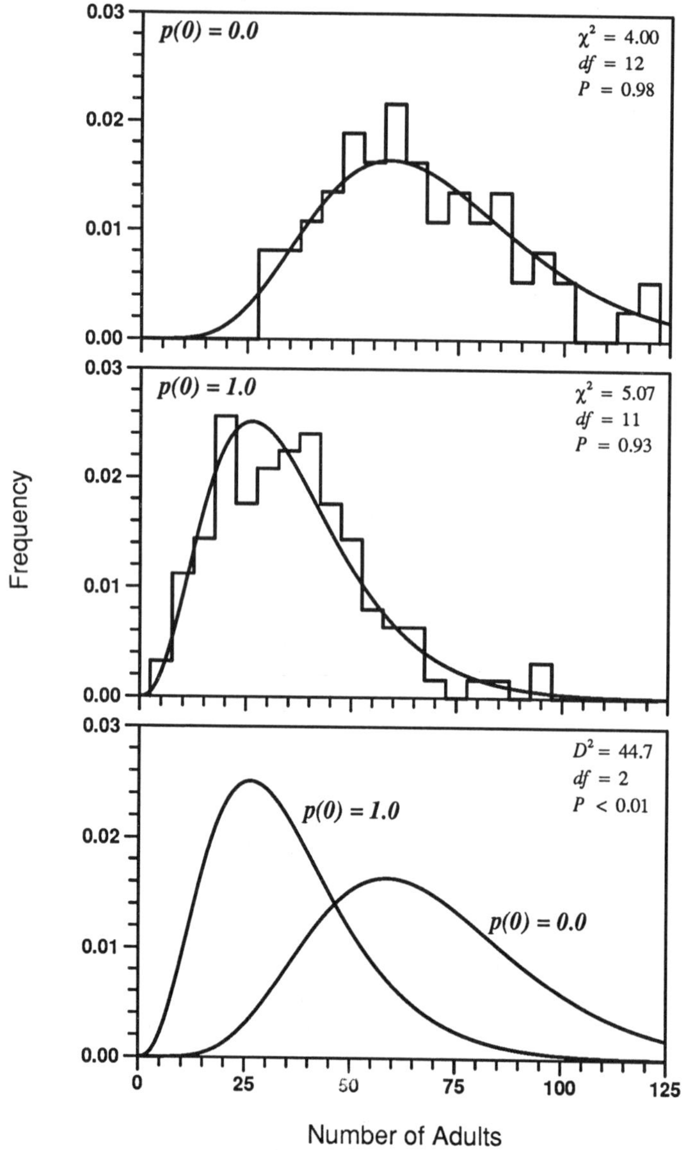

Fig. 6.11. Stationary distributions of adult numbers for the sensitive (*cos/cos*) and wild type (+/+) populations. The histograms represent the observed distributions and the smooth curves are the fitted gamma density functions. The test statistics (χ^2), degrees of freedom (*df*), and probability levels (*P*) are given for the goodness-of-fit tests. In the bottom panel the test statistic (D^2) is given for a test of the null hypothesis that these two distributions are the same. [From R. A. Desharnais, B. Dennis and R. F. Costantino (1990), Genetic analysis of a population of *Tribolium*. IX. Maximization of population size and the concept of a stochastic equilibrium. *Genome* (in press).]

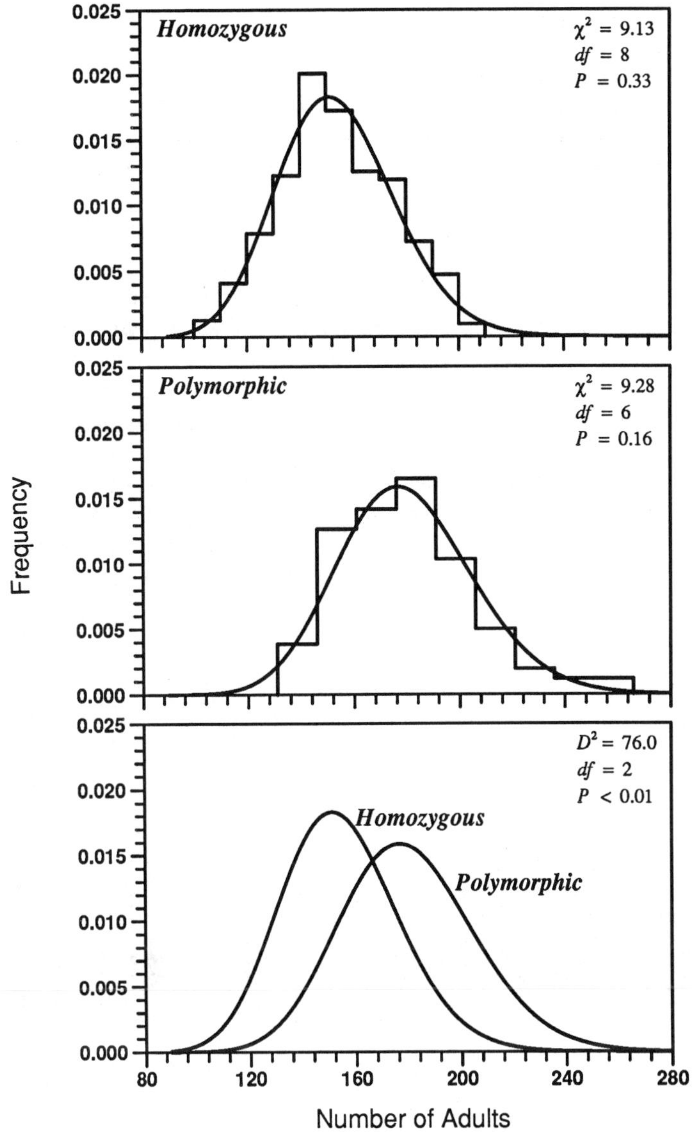

Fig. 6.12. Stationary distributions of adult numbers for the *cos* homozygous and polymorphic populations. The histograms represent the observed distributions and the smooth curves are the fitted gamma density functions. The test statistics (χ^2), degrees of freedom (df), and probability levels (P) are given for the goodness-of-fit tests. In the bottom panel the test statistic (D^2) is given for a test of the null hypothesis that these two distributions are the same. [From R. A. Desharnais and R. F. Costantino (1985), Genetic analysis of a population of *Tribolium*. VIII. The stationary stochastic dynamics of adult numbers. *Can. J. Genet. Cytol.* **27**: 341–350.]

6.3.5 Autocorrelation of Population Numbers

The mean and standard error of adult numbers over time from the experiment of Desharnais and Costantino (1980, 1985) are plotted in Fig. 6.13. These data were used to compute the autocorrelations of adult numbers as discussed in §4.5.2. We estimated the autocorrelation of order τ by pairing the adult census number at week τ with the adult number at week $t+\tau$ for each replicate. These paired observations were pooled within the homozygous and polymorphic groups, and an overall correlation was calculated in the usual way for both groups. This procedure was repeated for values of τ ranging from two to twenty weeks at intervals of two weeks, which is the average period between censuses. For both groups, this yielded ten serial correlations of adult numbers. We then corrected the estimated correlation coefficients of order τ by dividing the values by $[1-(\tau/k)]$, where k is the total number of paired data points used in the computation. This procedure removes the bias in an autocorrelation estimate. These corrected estimates are plotted in Fig. 6.14.

The estimated autocorrelations were used to evaluate the theoretical predictions based upon the linear stochastic differential equation (4.35). Although these predictions use a linear model, this approximation is valid if the fluctuations in population size are

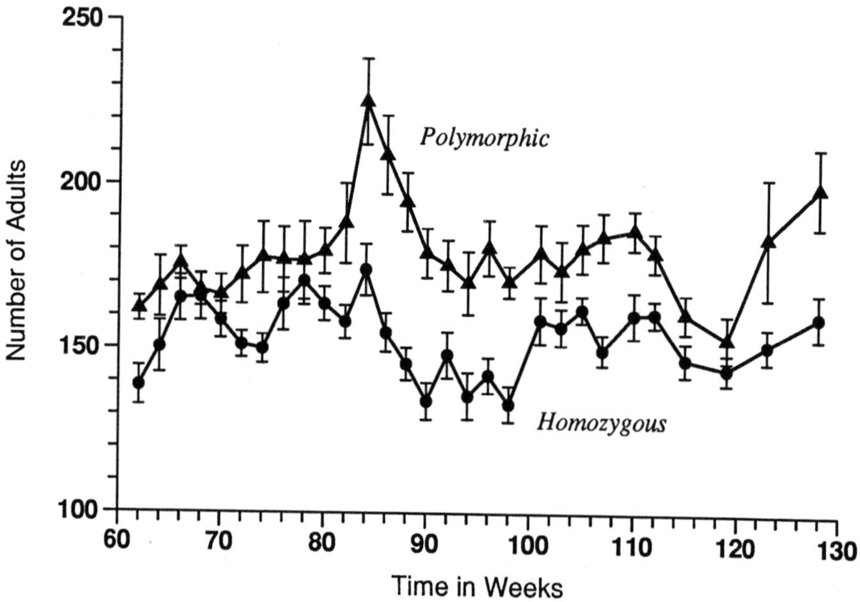

Fig. 6.13. Mean ± standard error of the number of adults plotted against census week for the *cos* homozygous populations (circles) and the polymorphic populations (triangles). [From R. A. Desharnais and R. F. Costantino (1985), Genetic analysis of a population of *Tribolium*. VIII. The stationary stochastic dynamics of adult numbers. *Can. J. Genet. Cytol.* **27**: 341–350.]

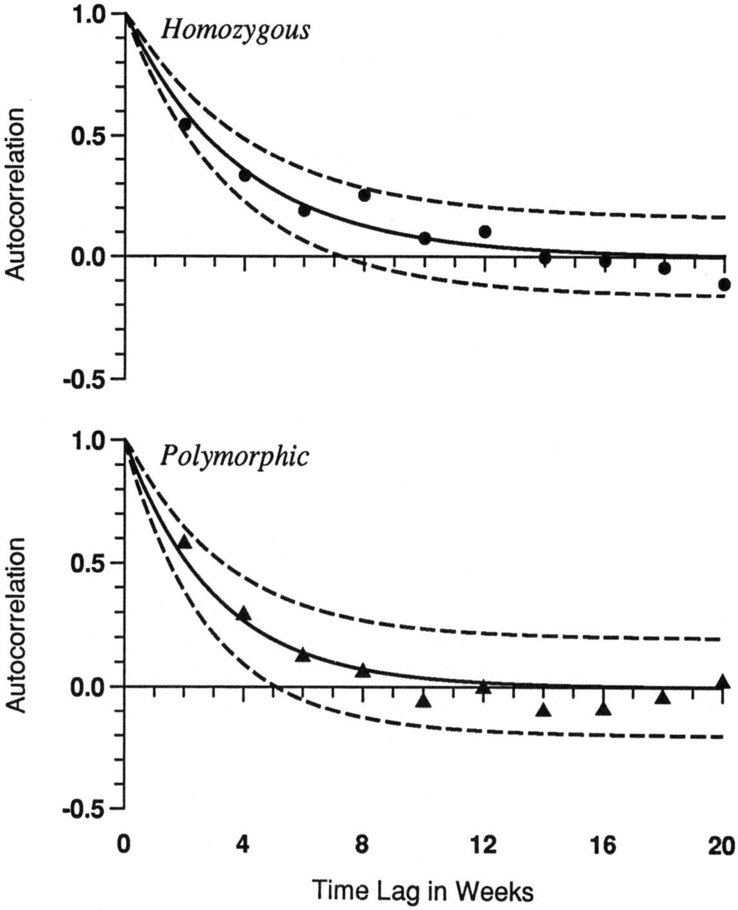

Fig. 6.14. Stationary autocorrelations of adult numbers versus time lag for the *cos* homozygous and polymorphic populations. The circles and triangles are the observed autocorrelations. The solid curves are the fitted theoretical autocorrelation functions. The dashed curves form bands of ±2 standard errors. [From R. A. Desharnais and R. F. Costantino (1985), Genetic analysis of a population of *Tribolium*. VIII. The stationary stochastic dynamics of adult numbers. *Can. J. Genet. Cytol.* **27**: 341–350.]

not very large. For both distributions (Fig. 6.12), the coefficient of variation is 0.14, suggesting that our use of the linear model is reasonable. Using a nonlinear least squares procedure, we fit the equation (4.39) to the serial correlations. The fitted functions appear in Fig. 6.14 together with the confidence interval. For both groups, the regression accounted for 92% of the variation in the autocorrelations. The least-squares estimates for the eigenvalues are $\lambda = 0.258 \pm 0.025$ for the homozygous populations and $\lambda = 0.330 \pm 0.035$ for the polymorphic group. Since $\lambda = c^{-1}\log(b/\mu]$, the larger estimated value for the polymorphic populations is consistent with the assumption that the parameter b is larger for this group. We also used the least-square procedure to evaluate the alternative prediction (4.41) which is based on the assumption of random pink noise. For the homozygous autocorrelations, the least-squares fit continued to improve as $\phi \to \infty$, yielding our original estimate of λ from the white noise model. For the polymorphic autocorrelations, the pink noise model did provide a slightly better fit, with a coefficient of determination of 0.94. However, the estimated values of $\lambda = 0.456 \pm 0.173$ and $\phi = 1.144 \pm 0.939$ also suggest that the time scale for the fluctuations in pink noise is shorter than the time scale for the fluctuations in adult number.

6.3.6 Maximization of Population Numbers

Several evolutionary models, including the one presented in this chapter, predict that natural selection will maximize the equilibrium size of a population. Was that the situation with respect to the corn oil sensitive mutant?

In the genetic polymorphism experiment (Moffa and Costantino 1977) the stationary distributions for the wild type ($p(0) = 0.0$) and sensitive ($p(0) = 1.0$) populations have statistically different parameter values (Fig. 6.11). Nevertheless, the distributions do overlap. Deterministic ecological-genetic theory predicts that the wild type populations will attain a larger equilibrium density than the sensitive populations. The gamma distribution provides a way to refine this prediction. Consider two populations, N_1 and N_2, both subject to independent random fluctuations. Let $f_1(x)$ and $f_2(x)$ represent the stationary probability density functions for N_1 and N_2. The probability that N_1 exceeds N_2 is given by

$$\text{Prob}\,\{N_1 > N_2\} = \int_0^{\infty} f_1(x)\,\text{Prob}\,\{N_2 < x\}\,\mathrm{d}x = \int_0^{\infty} f_1(x)\,F_2(x)\,dx\,, \tag{6.28}$$

where $F_2(x)$ is the cumulative distribution function for N_2. Substituting the gamma density function (4.15) into (6.28) gives

$$\text{Prob}\,\{N_1 > N_2\} = \left(\frac{\beta_1^{\alpha_1}\beta_2^{\alpha_2}}{\Gamma(\alpha_1)\Gamma(\alpha_2)}\right)\int_0^{\infty} x^{\alpha_1 - 1}\exp(-\beta_1 x) \times \left(\int_0^{x} z^{\alpha_2 - 1}\exp(-\beta_2 z)\,\mathrm{d}z\right)\mathrm{d}x\,. \tag{6.29}$$

The integrals in (6.29) can be computed numerically. Using the maximum likelihood estimates for the wild type populations as α_1 and β_1 and the sensitive populations as α_2 and β_2, the probability that adult numbers in a wild type culture exceed the adult numbers

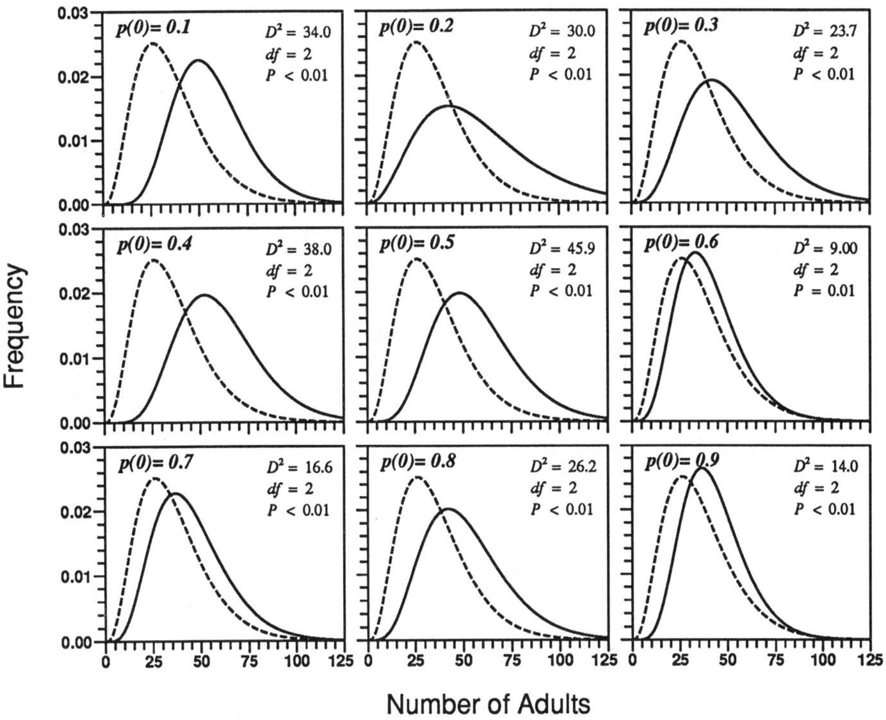

Fig. 6.15. Pairwise comparisons of the fitted gamma distributions for the polymorphic populations (solid curves) versus the homozygous *cos*/*cos* population (dashed curve). The test statistics (D^2), degrees of freedom (*df*), and probability levels (*P*) are given for a test of the null hypothesis that the two distributions are the same. [From R. A. Desharnais, B. Dennis and R. F. Costantino (1990), Genetic analysis of a population of *Tribolium*. IX. Maximization of population size and the concept of a stochastic equilibrium. *Genome* (in press).]

in a sensitive culture was computed as 0.8666. Although it is clear that the wild type populations have higher densities of adults (Fig. 6.11), it is interesting to note that 13.34% of the time one would observe *cos*/*cos* cultures with larger numbers of adults.

The fitted gamma density functions for each of the populations initially segregating at the *cos* locus were compared with the fitted gammas for the sensitive (*cos*/*cos*) and wild type (+/+) populations. In Fig. 6.15 we show the pairwise comparisons of the gamma parameters for the $p(0)=1$ (*cos*/*cos*) cultures with the other nine treatments ($p(0)=0.1$, 0.2, 0.3, ..., 0.9). Using the D^2 test statistic, we concluded that each polymorphic treatment is different than the sensitive population. In Fig. 6.16 we present similar pairwise comparisons with the $p(0)=0$ (+/+) distribution. In this case, we rejected the null hypothesis at the 0.05 level of probability for every treatment except for $p(0)=0.4$, In general, the polymorphic cultures have gamma distributions that are different from the wild type and sensitive populations.

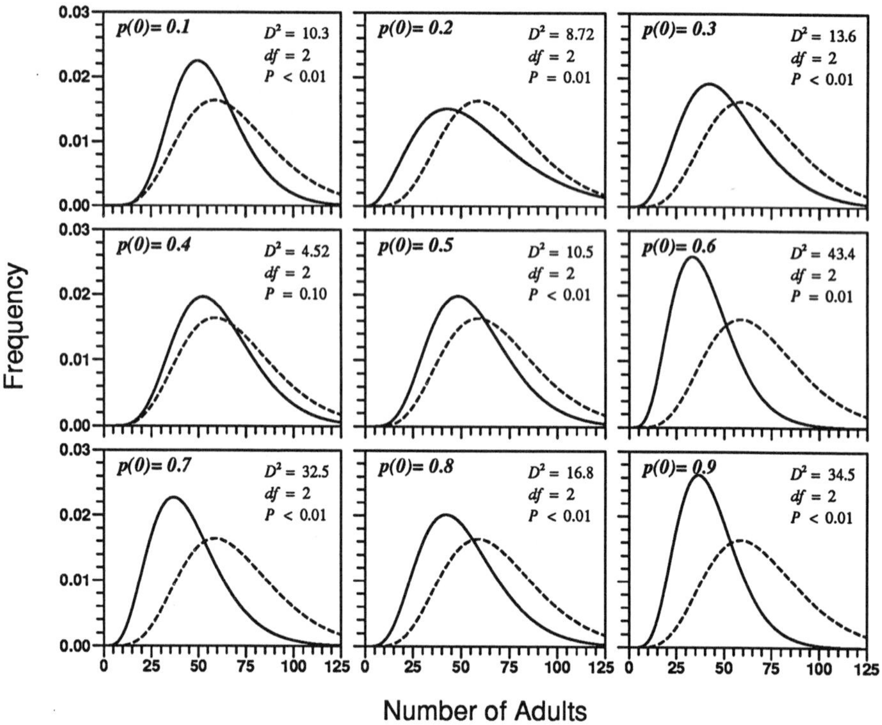

Fig. 6.16. Pairwise comparisons of the fitted gamma distributions for the polymorphic populations (solid curves) versus the homozygous +/+ population (dashed curve). The test statistics (D^2), degrees of freedom (*df*), and probability levels (*P*) are given for a test of the null hypothesis that the two distributions are the same. [From R. A. Desharnais, B. Dennis and R. F. Costantino (1990), Genetic analysis of a population of *Tribolium*. IX. Maximization of population size and the concept of a stochastic equilibrium. *Genome* (in press).]

To emphasize the stochastic nature of these equilibria, we used (6.29) to make pairwise comparisons of the *random variables* for population size. For each treatment segregating at the *cos* locus, we computed the probability that the population size will exceed that of the homozygous and reference populations (Table 6.1). Values near 0.5 indicate no tendency for a higher or lower population density. In comparison to the wild type homozygotes, the probabilities for the polymorphic populations are consistently below 0.5; they range from 0.17 for $p(0)=0.6$ to 0.41 for $p(0)=0.4$ In comparison to the sensitive homozygotes, the probabilities for the polymorphic populations are consistently above 0.5; they range from 0.59 for $p(0)=0.6$ to 0.82 for $p(0)=0.4$. The reference population is intermediate; the probabilities range from 0.34 for $p(0)=0.6$ to 0.61 for $p(0)=0.4$. It is worth noting that none of these probabilities fall outside of the range 0.15 to 0.85. In general, there is a significant overlap among the stationary distributions.

Table 6.1. Probabilities that the population size, N, associated with a particular initial frequency of the *cos* allele will be larger than the wild-type, sensitive, and reference populations.

Initial *cos* frequency	Prob $\{N > \text{wild type}\}$	Prob $\{N > \text{sensitive}\}$	Prob $\{N > \text{reference}\}$
0.1	0.3544	0.8013	0.5710
0.2	0.3744	0.7465	0.5483
0.3	0.3094	0.7268	0.5000
0.4	0.4062	0.8243	0.6141
0.5	0.3616	0.7880	0.5663
0.6	0.1696	0.5947	0.3411
0.7	0.2235	0.6522	0.4065
0.8	0.2979	0.7246	0.4919
0.9	0.1917	0.6384	0.3777

6.4 Summary and Conclusions

According to the single locus genetic theory of natural selection with density-dependent population growth, the (deterministic) equilibrium density of the polymorphic populations should be larger than either homozygote equilibrium. Furthermore, since the genetic equilibrium is globally stable, all populations initially segregating at the genetic locus should converge to the same equilibrium population size. Neither of these predictions were supported by the data. Overall, *Tribolium* populations segregating at the *cos* locus were intermediate in population density to the two homozygous groups. This result is true whether one uses the estimates of the deterministic equilibria or the probability comparisons for population size. Among the polymorphic populations, there was a significant amount of heterogeneity for the estimated gamma density functions (Fig. 6.9), although, on the whole, the polymorphic populations are more alike than different when compared to the two homozygous distributions (Fig. 6.10 vs. Figs 6.15, 6.16). Although there is little doubt that natural selection has affected the dynamics of population size, the maximization principle was not supported.

What could account for these differences? Clearly, the theory is an oversimplification of any real biological system. The model ignores age structure, age class interactions such as larvae eating eggs, and many of the other details that influence population dynamics in *Tribolium* (see Chapters 2 and 3). It is possible that the theory, while sufficiently general and robust enough to support the prediction of a gamma stationary density function, lacks sufficient detail to account for the influence of natural selection. On the other hand, we must point out that our stochastic model is for population size alone; it does not explicitly include any genetic dynamics. The maximization prediction is based on a model that includes genetic and population size dynamics (Roughgarden 1971; Charlesworth 1971), but this model is *deterministic*. It may be the case that a stochastic model which combines both genetic and population size dynamics may yield new predictions which are in agreement with the experimental observations.

The need for a combined *stochastic* theory of natural selection and population growth can be illustrated with a simple example. Consider two populations, N_1 and N_2, which are subject to independent stochastic fluctuations. Assume that the first population is polymorphic and has gamma parameters $\alpha_1 = 3$ and $\beta_1 = 0.01$, and that the second population is homozygous and has gamma parameters $\alpha_2 = 34$ and $\beta_2 = 0.1$. Using (4.16) and Ito calculus, the deterministic equilibria are given by $N_1^* = (\alpha_1 + 1)/\beta_1 = 400$ and $N_2^* = (\alpha_2 + 1)/\beta_2 = 350$, and, in accordance with the theory, $N_1^* > N_2^*$. However, from (6.29) we compute Prob$\{N_1 > N_2\} = 0.3526$. Even though the homozygous population has the smaller deterministic equilibrium, N_2 will exceed N_1 about 65% of the time! Although this is a contrived example, it points out the need to reformulate the genetic maximization principle in the context of a stochastic model.

One should not overlook a major success of the stochastic model – the prediction of a gamma stationary probability distribution for population size. For each of the eleven experimental groups, the gamma distribution provided an adequate description of the variation in adult numbers.

In the next chapter, we continue with an analysis of the dynamics of natural selection and population growth but with an emphasis on the approach to equilibrium.

Chapter 7 Genetic Disequilibrium

In the "ecological-genetic" models of the previous chapter, equilibria are defined by two entities: the vector of equilibrium allele frequencies, $\mathbf{p}^*$, and the equilibrium population size, N^*. If the fixed point $(\mathbf{p}^*, N^*)$ is locally stable, then trajectories in the neighborhood of the equilibrium will converge to this point. If the genetic equilibrium is polymorphic $(p_i^* \neq 0)$, then it will be globally stable and all populations with nonzero initial conditions will converge to $(\mathbf{p}^*, N^*)$. After a long span of time, the equilibria can be used to characterize the population. The identification and analysis of equilibria is a main theme of population research.

But what are the dynamics of the approach to equilibrium? This is a difficult question, because, except for a few special cases, even the simplest models have no known closed-form solutions. The equilibria are determined by the parameters of the model, but the trajectories themselves also depend on the initial conditions. For example, we know that the equilibrium population density N^* is maximized when allele frequencies are fixed at a globally stable $\mathbf{p}^*$, but what is the effect of nonequilibrium initial allele frequencies on the trajectory $N(t)$?

In this chapter we look at the consequences of initial conditions. Specifically, we study the situation in which the initial allele frequencies, $\mathbf{p}(0)$, are not equal to the globally stable equilibrium allele frequency, $\mathbf{p}^*$. We refer to this condition of $\mathbf{p}(0) \neq \mathbf{p}^*$ as "genetic disequilibrium." Armed with a theory relating the dynamics of natural selection and population growth, we show how the *Tribolium* experimental model can be used to test theoretical predictions.

7.1 An Entropy Measure of Genetic Disequilibrium

7.1.1 Ginzburg's "Fitness Entropy"

Consider the single locus model of natural selection (6.1) introduced in Chapter 6. As $t \to \infty$, natural selection will move an arbitrary vector of allele frequencies, $\mathbf{p}(t)$, towards a fixed point equilibrium $\mathbf{p}^*$. Without loss of generality, we can label alleles so that $p_i^* > 0$ for $i = 1, 2, \ldots, k$ and $p_i^* = 0$ for $i = k+1, k+2, \ldots, n$. In the context of this model, Ginzburg (1972, 1974, 1977a) introduced an "entropy" function to measure the deviation between $\mathbf{p}(t)$ and $\mathbf{p}^*$:

$$H(t) = H(\mathbf{p}(t), \mathbf{p}^*) = -\sum_{i=1}^{k} p_i^* \log\left[\frac{p_i(t)}{p_i^*}\right]. \tag{7.1}$$

Ginzburg called this function "fitness entropy" because of its resemblance to other entropy functions used in physics and information theory.

The entropy function (7.1) can be used as a measure of genetic disequilibrium. Consider a diallelic population and, dropping subscripts, let $p(0)$ represent the initial frequency of one of the two alleles. In Fig. 7.1 we have plotted $H(0)$ as a function of $p(0)$ for three different cases: $p^* = 0.1$, 0.3, and 0.5. The entropy function is U-shaped; i.e. $H(0) = 0$ when $p(0) = p^*$, $H(0) > 0$ when $p(0) \neq p^*$, and $H(0) \to \infty$ as $p(0) \to 0$ or 1. In the more general case of n alleles, $H(\mathbf{p}, \mathbf{p}^*)$ is a convex function of all the allele frequencies p_i for $i = 1, 2, \ldots, k$ with a global minimum at $\mathbf{p} = \mathbf{p}^*$. In other words, the further one moves in any direction away from $\mathbf{p}^*$, the larger the value of H, and if $\mathbf{p}^*$ is unattainable because $p_i = 0$ for $i \leq k$, then H is infinite. Given these properties, $H(\mathbf{p}, \mathbf{p}^*)$ seems reasonable as a measure of genetic disequilibrium.

Another important property of fitness entropy is that it is a Lyapunov function for the genetic model (6.1) when the fitness values are constant. Assume $\mathbf{p}^*$ is a stable polymorphic equilibrium involving all the alleles ($k = n$). Taking the time-derivative of $H(t)$ gives

$$\frac{\mathrm{d}H}{\mathrm{d}t} = -\sum_{i=1}^{n}\left(\frac{p_i^*}{p_i}\right)\left(\frac{\mathrm{d}p_i}{\mathrm{d}t}\right) = -\sum_{i=1}^{n} p_i^*[W_i - W]$$

$$= W - \sum_{i=1}^{n} p_i^* W_i = W - \sum_{i=1}^{n}\sum_{j=1}^{n} p_i^* p_j W_{ij}, \qquad (7.2)$$

where use has been made of (6.3) and (6.4). Since fitnesses are symmetric, $W_{ij} = W_{ji}$ for all ij. Furthermore, with full polymorphism, $\mathrm{d}p_i/\mathrm{d}t = 0$ implies $W_i^* = W^*$ for $i = 1, 2, \ldots, n$. Therefore,

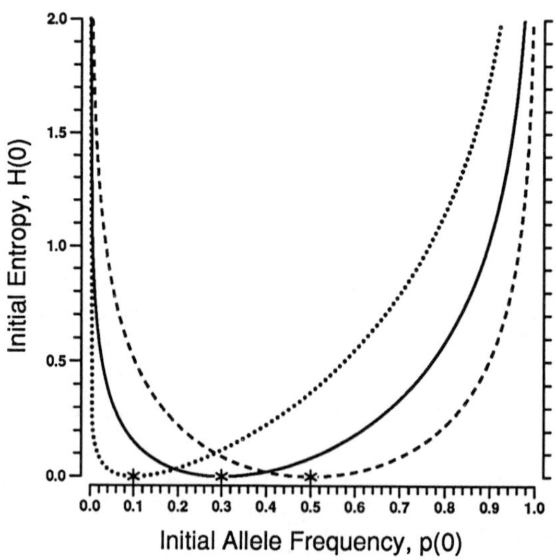

Fig. 7.1. Fitness entropy (7.1) as a function of initial allele frequency when the equilibrium allele frequency equals 0.1 (dotted line), 0.3 (solid line), and 0.5 (dashed line). The asterisks represent points where the entropy equals zero.

$$\frac{dH}{dt} = W - \sum_{j=1}^{n} \sum_{i=1}^{n} p_j p_i^* W_{ji} = W - \sum_{j=1}^{n} p_j W_j^* = W - W^* . \tag{7.3}$$

From Fisher's fundamental theorem of natural selection we know that W^* is a global maximum at a stable polymorphism. Hence $W(t) \leq W^*$ with the equality holding only when $\mathbf{p} = \mathbf{p}^*$. Thus (7.3) implies $dH/dt \leq 0$ and $H(t)$ is a Lyapunov function for the single locus model of natural selection. Using (7.3) we can say that the "velocity" of the genetic disequilibrium is equal to the difference between the mean fitness and its global maximum.

We can also take Fisher's theorem one step further and relate the entropy measure of disequilibrium to the additive genetic variance in fitness. For the single locus model, the additive genetic variance is defined as

$$V_A = 2 \sum_{i=1}^{n} p_i (W - W_i)^2 . \tag{7.4}$$

With this definition one can easily show that

$$\frac{dW}{dt} = V_A , \tag{7.5}$$

or, in Fisher's (1958, p. 37) words, "The rate of increase in fitness of any organism at any time is equal to its genetic variance in fitness at that time." Using (7.3) we have the following corollary:

$$\frac{d^2 H}{dt^2} = V_A \tag{7.6}$$

which shows that the "deceleration" in the genetic disequilibrium is equal to the additive genetic variance in fitness.

7.1.2 Haldane's "Cost of Natural Selection"

The ideas discussed in this Chapter have a predecessor in population genetics theory: Haldane's (1957) "cost of natural selection." Haldane tried to place a limit on the rate of evolution by showing that if selection is too intense, a species will become extinct. For example, if selection on a trait culls half the population in a single generation and selection acts independently on ten different traits, then only $(1/2)^{10}$, or one in 1024, of the original genotypes will survive. Haldane (1957, p. 514) defined the cost of selection as the total "number of deaths or their equivalents in lowered fertility."

The cost of selection can be illustrated with a simple numerical example. Consider a haploid population with two genotypes, the first one being very rare and the second being common. Let W_1 and W_2 denote the fitnesses of the genotypes. Assume that the environment undergoes a rapid change so that suddenly the rare genotype is favored by selection, that is, $W_1 > W_2$. Let $p(t)$ be the frequency of the rare genotype. Selection will increase the frequency of the rare genotype at a rate given by

$$\frac{\mathrm{d}p}{\mathrm{d}t} = p\,(1-p)(W_1-W_2)\,. \tag{7.7}$$

Eventually, the initially rare genotype will be selected to fixation. At any given time, the proportion of "selective deaths" is given by the frequency of the inferior genotype times the fitness differential for that genotype: $(1-p)(W_1-W_2)$. Integrating this quantity from $t=0$ to ∞ gives the total cost of selection:

$$\text{cost} = \int_0^\infty (1-p)(W_1-W_2)\,\mathrm{d}t = \int_0^\infty \left(\frac{1}{p}\right)\left(\frac{\mathrm{d}p}{\mathrm{d}t}\right)\mathrm{d}t = -\log[p(0)]\,. \tag{7.8}$$

The total number of selective deaths will be the population number multiplied by the cost. If $p(0)=0.001$ and the population size is 1000, then, from (7.8), 6908 selective deaths must occur before the rare allele can become fixed in the population. If the population stays roughly constant, this cannot occur in fewer than 7 generations, and it would probably take much longer. If the environmental change affects ten different traits with roughly the same initial frequencies, then 70 generations would be needed. This, Haldane argued, puts a limit on the rate of evolution.

The concept of a cost of natural selection generated much attention and controversy (Van Valen 1963; Brues 1964, 1969; Feller 1966, 1967; Crow 1968; Maynard Smith 1968b; Sved 1968; O'Donald 1969; Moran 1970). A detailed account of this controversy is beyond the scope of this Chapter. However, in our opinion, one of the main problems with the cost concept is that it was formulated without a consideration of population dynamics. Hence, the "number of selective deaths" really referred to a measure of fitness and not to actual numbers of individuals in the population. In Roughgarden's (1979, pp. 212–213) interpretation of Haldane's concept, cost is defined as the cumulative difference between the asymptotic mean fitness, which is maximized by selection, and the actual mean fitness as it evolves over time:

$$\text{cost} = \int_0^\infty [W^* - W(t)]\,\mathrm{d}t\,. \tag{7.9}$$

In the example described above, the mean fitness is $W(t)=p(t)W_1+[1-p(t)]W_2$ and the equilibrium fitness is $W^*=W_1$, which makes (7.8) consistent with (7.9). Using this definition of cost, we show that the entropy function (7.1) can be used to tie together the cost concept, genetic disequilibrium, and the dynamics of population growth.

7.2 Rate of Genetic Adaptation

In this section we show how the entropy measure of genetic disequilibrium can be used to compute the cost of natural selection, as defined by (7.9), and the rate at which mean fitness converges to its equilibrium. Both predictions are evaluated using genetic data from laboratory populations of *Tribolium castaneum*. These results are based on the work of Ginzburg and Costantino (1979).

7.2.1 Entropy and the Cost of Selection

Assume $\mathbf{p}(t) \to \mathbf{p}^*$, where $\mathbf{p}^*$ is a stable polymorphism with all alleles present ($k=n$). In our discussion of fitness entropy as a Lyapunov function, we have already shown that $dH/dt=W-W^*$. Integrating from $t=0$ to ∞ gives

$$H(\infty)-H(0) = \int_0^\infty [W(t)-W^*]\,dt\,. \tag{7.10}$$

However, $H(\infty)=H(\mathbf{p}^*,\mathbf{p}^*)=0$. Thus (7.10) gives

$$H(0) = \int_0^\infty [W^*-W(t)]\,dt\,, \tag{7.11}$$

or, from (7.9), $H(0)=$cost. Haldane's cost of natural selection is equal to the initial genetic disequilibrium when disequilibrium is measured using (7.1).

What happens when some alleles are lost in the selection process? Ginzburg and Costantino (1979) show that when $k<n$,

$$\int_0^\infty [W^*-W(t)]\,dt = H(0)+\sum_{j=k+1}^{n}(W^*-W_j)\int_0^\infty p_j(t)\,dt\,. \tag{7.12}$$

If the equilibrium is stable, then alleles k+1 through n cannot increase from low frequencies when the remaining alleles are near their equilibrium values. This implies $W_j^*<W^*$ for $j=k+1, k+2, \ldots, n$, so that the second term on the right side of (7.12) is always positive. Therefore, in general,

$$H(0) \leq \int_0^\infty [W^* - W(t)]\,\mathrm{d}t\,. \tag{7.13}$$

In the special (but biologically important) case when a single dominant allele is selected to fixation, $k=1$, $W^*=W_{1j}=W_{j1}$, and the second term on the right side of (7.12) is zero. Hence (7.11) applies despite $k<n$.

7.2.2 Entropy and the Adaptation Rate

As a consequence of Fisher's (1930) theorem, we know that the mean fitness of a population, $W(t)$, will increase monotonically with time until it reaches an asymptotic maximum value W^*. Intuitively, we might think that the rate of approach of $W(t)$ to W^* depends on the magnitude of the differences between the fitnesses of the genotypes, i.e., the selection intensity s where

$$s = \max_{ijkm} |W_{ij} - W_{km}|\,. \tag{7.14}$$

In fact, using entropy, Ginzburg and Costantino (1979) discovered a bound for the rate of adaptation which is independent of s.

Assume that $\mathbf{p}^*$ is fully polymorphic ($k=n$) and stable. We can now show that

$$W^* \geq W(t) \geq W^* - \frac{H(0)}{t}. \tag{7.15}$$

The left inequality is a direct consequence of Fisher's theorem. To prove the right inequality, assume $W(t_0)<W^*-H(0)/t_0$ for some time t_0 where $0<t_0<\infty$. Then, by Fisher's theorem, it must be true that $W(t)<W^*-H(0)/t_0$ for $0\leq t\leq t_0$. This implies

$$W^* - W(t) > \frac{H(0)}{t_0}, \qquad 0 \leq t \leq t_0\,,$$

$$\Rightarrow \int_0^{t_0} [W^* - W(t)]\,\mathrm{d}t > H(0) = \int_0^\infty [W^* - W(t)]\,\mathrm{d}t\,, \tag{7.16}$$

where use has been made of (7.11). Since $W^*\geq W(t)$, (7.16) is impossible. Therefore t_0 cannot exist and (7.15) is true for all $0\leq t\leq\infty$. The inequalities in (7.15) place a bound on the adaptation rate which depends only on the initial conditions, $\mathbf{p}(0)$, and not on the strength of selection, s.

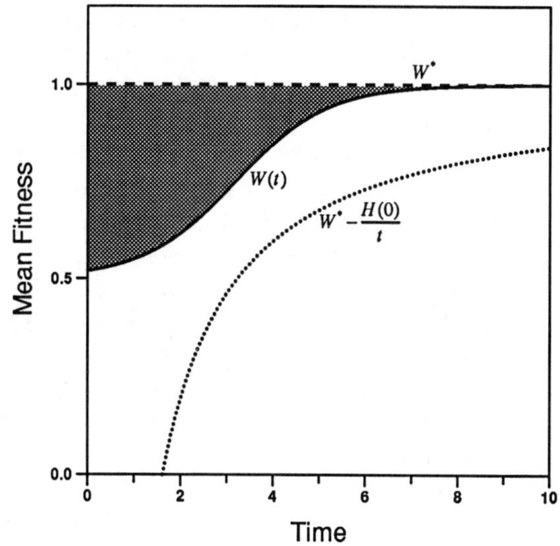

Fig. 7.2. Mean fitness (solid line) as a function of time in the single locus genetic model of natural selection. The dashed line is the equilibrium fitness. The dotted line is a lower bound on the mean fitness which is independent of the strength of selection. The shaded area is Haldane's cost of natural selection (7.9). [Modified from L. R. Ginzburg and R. F. Costantino (1979), On the rate of genetic adaptation under natural selection. *J. Theor. Biol.* **77**: 307–316.]

Figure 7.2 illustrates the relationships among the mean fitness, the cost of natural selection, and the entropy measure of disequilibrium. In this example, $W_{11} = W_{22} = 0.5$, $W_{12} = 1.5$, and $p(0) = 0.01$. At equilibrium $p^* = 0.5$, $W^* = 1$, and $H(0) = 1.6145$. The solid line is the trajectory of $W(t)$ which was determined by numerically integrating the differential equations (6.1) using a Runge-Kutta fifth order algorithm. The dashed line represents the equilibrium fitness W^*. The shaded area is the cost as defined by (7.9). From (7.11), it equals the initial entropy $H(0)$. The dotted line represents the lower bound on the adaptation rate in (7.15). No matter how small the selection intensity, s, the mean fitness will never cross below the dotted line and the shaded area will remain constant.

7.2.3 An Experimental Test

The data of Moffa and Costantino (1977) on the corn oil sensitive (*cos*) mutant strain of *Tribolium castaneum* were used by Ginzburg and Costantino (1979) to test the predictions (7.11) and (7.15). Using the same data, we repeated the analysis with some modifications. The experimental design was described earlier (§6.3.2). We begin with a brief summary of the relevant experimental data.

Cultures were established in which the frequencies of the *cos* allele ranged from 0.0 to 1.0 in increments of 0.1. The initial populations consisted of ten males and ten females of the +/+ and *cos/cos* genotypes. The allele and genotypic frequencies were estimated in twelve replicates for each initial allele frequency of 0.1, 0.3, 0.5, 0.7, and 0.9. To assure independence of samples, three replicates from each initial frequency were sampled at 2, 6, 12, and 16 weeks. By week six, the allele frequencies and genotypic arrays had converged. These data were used to test the theory.

Using life history data, Moffa and Costantino (1977) estimated intrinsic rates of increase for the three genotypes. In the continuous model of natural selection, these can be used to approximate fitnesses (Charlesworth 1980). For the *cos/cos*, *+/cos*, and *+/+* genotypes, respectively, we have $W_{11}=0.18$, $W_{12}=0.70$, and $W_{22}=0.63$. These values were combined with the observed genotypic frequencies to obtain estimates of the mean fitnesses for each treatment over time. The observed *cos* frequencies converged to a value between 0.25 and 0.30; to be commensurate with the analyses in the later sections of this Chapter, we will choose the upper range of $p^*=0.3$. This value differs from the predicted polymorphic equilibrium frequency of $p^*=0.12$ which one obtains from the fitness values. This discrepancy was discussed by Moffa and Costantino (1977) and it appears that the predicted value represents a lower boundary of p^*; we used $p^*=0.3$ in all our computations. With this value of p^*, we can use (7.1) to compute the initial entropy, $H(0)$, for each treatment. Finally, the equilibrium fitness, $W^*=0.644\pm0.003$, was computed by averaging over the mean fitnesses for each treatment for weeks 6, 12, and 16.

The data are presented in Fig. 7.3. The circles are the computed mean fitness values for each treatment. The mean fitnesses quickly approach an asymptotic value which approximates the computed value of W^* (dashed line). The shaded area represents the cost (7.9) which was computed by integrating the data using the trapezoid rule. The dotted line is the predicted lower bound on the adaptation rate from (7.15). Except for $p(0)=p^*=0.3$, the observed fitnesses lie above this theoretical lower bound. In the case of $p(0)=0.3$, the deviation of the initial mean fitness from the equilibrium value is due to the lack of heterozygotes in the initial genotypic array.

A test of the prediction (7.11) is given in Fig. 7.4. In Fig. 7.4a, the observed costs (shaded areas in Fig. 7.3) are plotted as a function of the initial entropy measures of disequilibrium. The filled circles are the observations. The theoretical prediction $\text{cost}=H(0)$ is represented by the dashed line which has a slope of one and goes through the origin. Using least squares linear regression, we computed a slope of 0.526 ± 0.056 and an intercept of 0.148 ± 0.028 which differ significantly from the predicted values of one and zero, respectively ($P<0.05$). However, there is a significant linear relationship between the cost and the disequilibrium ($P<0.01$).

One possible explanation for the discrepancy between the computed costs of selection and the genetic disequilibria is that the genotypic fitnesses were not estimated accurately. Fitness is an elusive concept and estimates of absolute fitness are difficult to obtain. It is easier to obtain estimates of relative fitness which, for the purposes of making genetic predictions, are equivalent to the absolute fitnesses. Let ω_{ij} be an estimate of the relative fitness of genotype ij. The relative fitness is linearly related to the absolute fitness, i.e.,

$$\omega_{ij} = \alpha + \beta W_{ij}\,, \tag{7.17}$$

where α and β are constants. Using (7.17), it is easy to show that, for any time $T<\infty$,

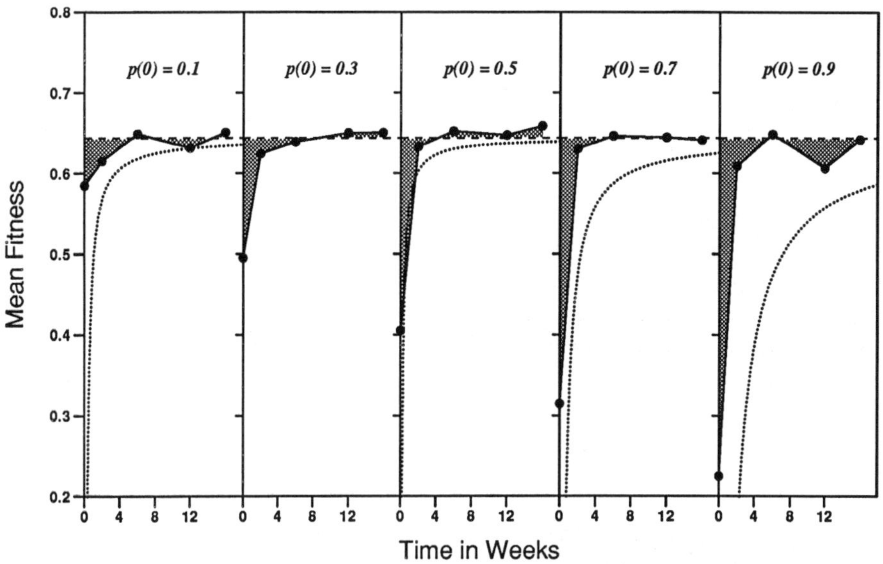

Fig. 7.3. Experimentally observed values of mean fitness (circles) versus time for five populations of *T. castaneum* with different initial frequencies of the *cos* allele. The dashed line is the average of the observed equilibrium fitnesses. The dotted line is a theoretical lower bound on the mean fitness based on the entropy function. The shaded area is an estimate for the cost of natural selection. [Modified from L. R. Ginzburg and R. F. Costantino (1979), On the rate of genetic adaptation under natural selection. *J. Theor. Biol.* **77**: 307–316.]

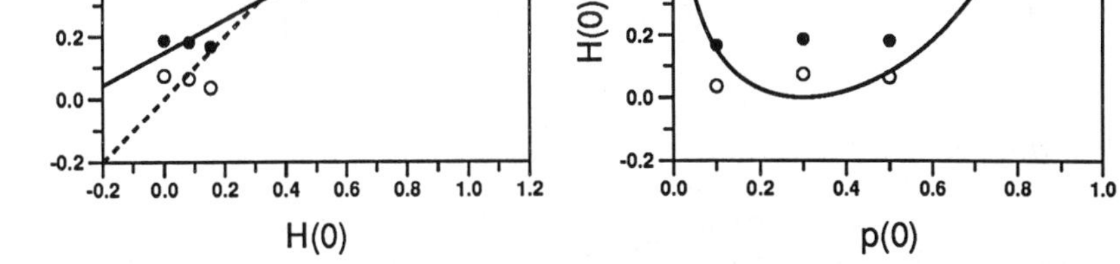

Fig. 7.4. An experimental evaluation of the relationship between the cost of natural selection and the entropy measure of genetic disequilibrium. The filled circles are the estimates of the cost based on the assumption that the data represent absolute fitnesses. The open circles are the estimates of the cost based on the assumption that the data represent relative fitnesses. In **a**, the solid line is a least squares regression for the filled circles and the dashed line is the theoretical prediction. In **b**, the solid curve is the theoretical entropy function.

$$\int_0^T \omega(t)\,\mathrm{d}t = \omega^* T - \beta\,[H(0) - H(T)]\,. \tag{7.18}$$

If T is sufficiently large so that $p(t) \approx p^*$, then $H(T) \approx 0$ and (7.18) predicts a linear relationship between the integral of the relative fitness and the initial entropy. We integrated the observed mean fitness curves using the trapezoid rule with $T = 16$ weeks and used least squares linear regression to obtain estimates of $\omega^* T = 10.15 \pm 0.028$ and $\beta = 0.526 \pm 0.056$. Dividing the intercept by $T = 16$ weeks gives $\omega^* = 0.635$ which differs slightly from the estimate $W^* = 0.644$ obtained above by averaging the observed mean fitnesses. Rearranging (7.18) with $H(T) = 0$ gives the prediction

$$H(0) = \beta^{-1} \int_0^T [\omega^* - \omega(t)]\,\mathrm{d}t\,. \tag{7.19}$$

These values are represented in Fig. 7.4 by the open circles. These open circles can be viewed as the linear transformation of the closed circles which provides the best fit to the values of $H(0)$. In Fig. 7.4b we have plotted the untransformed and transformed estimates of the cost, together with the predicted disequilibrium curve, as a function of initial allele frequency. Both sets of estimates display the trend of increasing cost with distance from p^*; the transformed data fit the entropy function quite well.

7.3 Density-Independent Population Growth

We now explore the connection between the entropy measure of disequilibrium and population growth. This connection comes from the fact that, in (6.2), $W(t)$ is the instantaneous rate of population growth. From (7.3), $\mathrm{d}H/\mathrm{d}t = W(t) - W^*$. This relationship between entropy and the population growth rate allows us to make predictions about the effects of genetic disequilibrium on the population density. In this section, we begin with the simplest case of population growth – the Malthusian model.

7.3.1 The Reference Population

Haldane's concept of the cost of natural selection involved the comparison of the average population fitness with its maximum equilibrium value. The equilibrium fitness served as a "reference" value which could be compared to the mean fitness to determine the effects of genetic disequilibrium. In dealing with population growth, we also need a point of reference. However, a single population density will not suffice, since even a genetically homogeneous population can experience changes in density. For this reason, we introduce the concept of a "reference population."

Consider an arbitrary population in the context of the model (6.1)–(6.2). This population is defined by a set of genotypic fitnesses W_{ij}, an initial allele frequency vector $\mathbf{p}(0)$, and an initial density $N(0)$. This information is sufficient to determine the trajectory of the population $[\mathbf{p}(t), N(t)]$ through time. Eventually, $\mathbf{p}(t) \to \mathbf{p}^*$. This population will be referred to as an *evolving population.* Given an evolving population, its *reference population* is defined using

$$\begin{aligned} \tilde{W}_{ij} &= W_{ij}, & i,j &= 1, 2, 3, \ldots, n, \\ \tilde{p}_i(0) &= p_i^*, & i &= 1, 2, 3, \ldots, n, \\ \tilde{N}(0) &= N(0). \end{aligned} \tag{7.20}$$

In other words, the reference population has the same fitness values and initial population density as the evolving population, but its initial genetic structure is equal to the evolving population's equilibrium structure. A comparison of the trajectories $N(t)$ and $\tilde{N}(t)$ allows an assessment of the effects of genetic disequilibrium.

7.3.2 Entropy and Selective Delay

When the fitness values, W_{ij}, are positive constants, $W^* > 0$ and the population density will continue to grow exponentially. For this case, Ginzburg (1977) derived an expression which relates the initial entropy measure of genetic disequilibrium to the dynamics of population growth. Assume $\mathbf{p}^*$ is fully polymorphic ($k = n$) and stable. The densities of the evolving and reference populations are given by

$$\begin{aligned} N(t) &= N(0) \exp\left[\int_0^t W(\xi)\, \mathrm{d}\xi\right], \\ \tilde{N}(t) &= N(0) \exp\,[W^* t]\,. \end{aligned} \tag{7.21}$$

Taking the logarithmic ratio of $\tilde{N}(t)$ to $N(t)$ as $t \to \infty$ gives

$$\lim_{t\to\infty} \log\left[\frac{\tilde{N}(t)}{N(t)}\right] = \int_0^\infty [W^* - W(t)]\, \mathrm{d}t = H(0)\,. \tag{7.22}$$

Since the ratio of the reference and evolving population numbers approaches a constant asymptotic value, $\tilde{N}(t)$ and $N(t)$ eventually grow at the same asymptotic rate and become separated in time by a constant asymptotic time lag, τ. Ginzburg called τ the *selective delay*. Formally, τ is the solution to

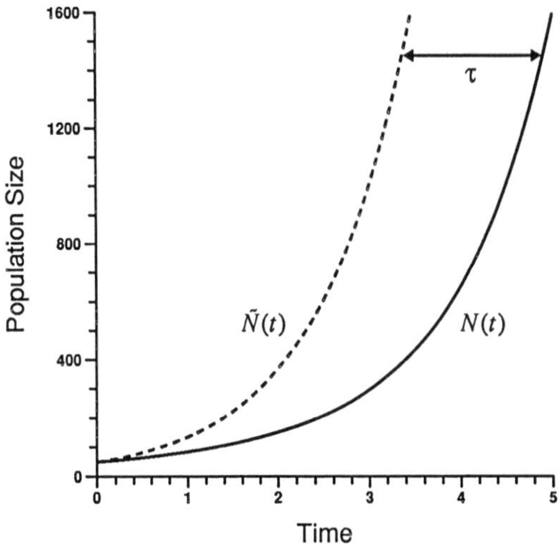

Fig. 7.5. Population size versus time for natural selection with unconstrained population growth. The dashed line is the trajectory of the reference population with allele frequencies in equilibrium. The solid line is the trajectory of the evolving population with nonequilibrium initial allele frequencies. The evolving population lags behind the reference population by an asymptotic time value, τ, which equals the selective delay.

$$\lim_{t \to \infty} \left[\frac{\tilde{N}(t-\tau)}{N(t)} \right] = 1 \; ; \tag{7.23}$$

i.e., τ is the extra amount of time the evolving population needs in order to achieve the same density as the reference population. Using (7.22) in (7.23), we have

$$H(0) = W^* \tau \, . \tag{7.24}$$

This simple expression relates the time lost due to selection to the entropy measure of genetic disequilibrium. The abstract concept of a cost of natural selection can now be measured in concrete terms as the amount of time the population loses because of a nonequilibrium initial genetic state.

The selective delay is illustrated in Fig. 7.5. In this example $W_{11} = W_{22} = 0.5$, $W_{12} = 1.5$, and $p(0) = 0.01$. At equilibrium $p^* = 0.5$, $W^* = 1$, and $H(0) = 1.6145$. The solid line is the trajectory of $N(t)$ which was determined by numerically integrating the differential equations (6.1)–(6.2) using a Runge-Kutta fifth order algorithm. The dashed line represents the reference population $\tilde{N}(t) = N(0)\exp(W^* t)$. Since $W^* \geq W(t)$ by Fisher's theorem, $\tilde{N}(t) \geq N(t)$. As $p(t) \to p^*$, $W(t) \to W^*$ and the asymptotic growth rates of $\tilde{N}(t)$ and $N(t)$ become equal. However, the evolving population, $N(t)$, has been delayed by $\tau = H(0)/W^* = 1.6145$ time units due to its initial genetic disequilibrium.

7.3.3 An Experimental Test

In the experiment of Moffa and Costantino (1977) described in §7.2.3, the populations experienced an initial exponential phase of growth. These data were used by Costantino et al. (1977) to test the prediction (7.24). We now present their analysis.

Exponential growth rates were estimated for the adult numbers in each treatment with an initial *cos* allele frequency of $p(0) = 0.1, 0.2, 0.3, \ldots, 0.9$. Since the populations were started with adults only, there was no recruitment into the adult life stage during the first two weeks of the experiment; consequently, time was shifted so that $t=0$ corresponds to week 2. A least squares linear regression was computed for the logarithm of adult numbers *versus* time for $t=0$, 2, and 4 weeks when population growth appeared exponential. The form of the linear regression equation was $\log[N(t)] = \log[N(0)] + Wt$, where W is the growth rate. The coefficients of determination, which measure the proportion of the sums of squares that are accounted for by the linear model, ranged from 0.964 to 0.995. However, due to the single degree of freedom, the probability levels for the regressions fell into the range 0.046 to 0.121.

The average fitness of the population at equilibrium was determined from data on eighteen new cultures started with twenty randomly chosen sexed adults taken from the populations at week 68. Using linear regression on the exponential growth phase, the estimated value of W^* was 0.713 ± 0.044. As in §7.2.3, we used $p^* = 0.3$ for the equilibrium frequency of the *cos* allele.

It is impossible to find an exact value of the selective delay for it is the asymptotic concept as time tends to infinity. To estimate the selective delay, τ, we used the estimated growth rates to compute the extra time required for each evolving population to reach the same adult density as the reference population at $t=4$. This gives $\tau = 4(W^* - W)/W$ for each treatment. The experimental results are summarized in Table 7.1.

In Fig. 7.6a, we have plotted $W^*\tau$ versus the initial entropy $H(0)$. The theoretical prediction (7.24) is given by the dashed line which has a slope of one and goes through the origin. The solid line represents a linear least squares regression of the data. The

Table 7.1. Experimental results for selective delay (Costantino et al. 1977)

Initial *cos* frequency, $p(0)$	Estimated exponential growth rate, $W \pm$ s.e. (per week)	Selective delay, τ (weeks)	Equilibrium growth rate times selective delay, $W^*\tau$	Initial genetic disequilibrium, $H(0)$
0.1	0.634 ± .104	0.498	0.355	0.154
0.2	0.676 ± .053	0.217	0.155	0.028
0.3	0.702 ± .065	0.061	0.044	0.000
0.4	0.712 ± .115	0.005	0.004	0.022
0.5	0.674 ± .049	0.234	0.167	0.082
0.6	0.634 ± .113	0.496	0.353	0.184
0.7	0.624 ± .071	0.567	0.404	0.339
0.8	0.610 ± .117	0.673	0.480	0.583
0.9	0.489 ± .061	1.828	1.303	1.033

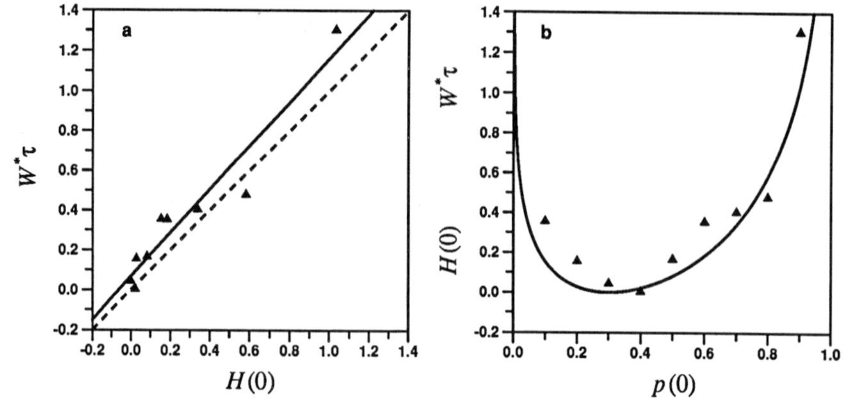

Fig. 7.6. An experimental evaluation of the relationship between the selective delay and the entropy measure of genetic disequilibrium. The triangles are estimates of the product of the asymptotic population growth rate and the selective delay. In **a**, the solid line is a least squares regression for the data and the dashed line is the theoretical prediction. In **b**, the solid curve is the theoretical entropy function.

estimated slope is 1.094 ± 0.121 and the estimate for the intercept is 0.068 ± 0.051. Neither the slope nor the intercept differ significantly from the theoretical values of one and zero, respectively, at the 5% probability level.

In Fig. 7.6b we have plotted the estimates of $W^*\tau$ and the curve $H(p(0), p^* = 0.3)$ versus the initial allele frequency. In general, the agreement between the observations and predictions is quite good.

7.4 Density-Dependent Population Growth

In §6.2 we looked at extensions of the classical model of natural selection where absolute fitness was a function of population size. In these models, population growth is density-dependent and $N(t) \to N^*$ as $t \to \infty$. We now look at the consequences of genetic disequilibrium when population is regulated by density.

7.4.1 Logistic Population Growth

The simplest assumption of density-dependent selection is the case where genotypic fitnesses decrease linearly with density:

$$W_{ij}(N) = r_{ij} - aN . \tag{7.25}$$

For each genotype, r_{ij} represents the intrinsic or Malthusian rate of increase and a is a *density-sensitivity* parameter which describes the rate of decrease in fitness with increases in population size. Substituting (7.25) into the equations (6.1)–(6.2) for natural selection and population growth we have

$$\begin{aligned} \frac{dp_i}{dt} &= p_i(r_i - r), \qquad i = 1, 2, 3, \ldots, n, \\ \frac{dN}{dt} &= N(r - aN), \end{aligned} \tag{7.26}$$

where r_i and r are given by

$$\begin{aligned} r_i &= \sum_{j=1}^{n} p_i r_{ij}, \qquad i = 1, 2, 3, \ldots, n, \\ r &= \sum_{i=1}^{n} \sum_{j=1}^{n} p_i p_j r_{ij}. \end{aligned} \tag{7.27}$$

Note that the rate of population growth is simply the *logistic* equation. However, the parameter r is a time-varying average which depends on allele frequencies. Hence the dynamics of population growth are tied to natural selection.

The r_{ij}'s in (7.26) play the same role as the genotypic fitnesses in (6.1). This means that $r(t) \rightarrow r^*$ as $t \rightarrow \infty$ and r^* is a local maximum of r with respect to allele frequencies. For the reference population defined by (7.20), $\tilde{p}(t) = p^*$ for $t \geq 0$. Hence the reference population grows according to the logistic equation

$$\tilde{N}(t) = \frac{N^*}{1 + [N^*/N(0) - 1] \exp(-r^* t)}, \tag{7.28}$$

where $N^* = r^*/a$. Since $r^* \geq r(t)$, $\tilde{N}(t) \geq N(t)$ for $t \geq 0$.

7.4.2 Entropy and Life Not Lived

Desharnais and Costantino (1982b) investigated the impact of genetic disequilibrium on population growth in density-regulated populations. They considered fitness functions in which the genotypic differences were located in a single parameter that enters the fitness value as a linear coefficient. The W_{ij}'s defined in (7.25) are a special case of this class of models. We shall consider only the dynamics of this logistic fitness function. See Desharnais and Costantino (1982b) for the more general case.

Consider the case where all n alleles are maintained by selection ($k = n$). From (7.26) and (7.28) we know that

$$\frac{\mathrm{d}\log N(t)}{\mathrm{d}t} = W(t) = r(t) - a\,N(t)\,,$$
$$\frac{\mathrm{d}\log \tilde{N}(t)}{\mathrm{d}t} = \tilde{W}(t) = r^* - a\,\tilde{N}(t)\,. \tag{7.29}$$

Since the r_{ij}'s play the role of fitnesses in (7.26), (7.3) implies

$$\begin{aligned}\frac{\mathrm{d}H(t)}{\mathrm{d}t} &= r(t) - r^* \\ &= \frac{\mathrm{d}\log N(t)}{\mathrm{d}t} + a\,N(t) - \frac{\mathrm{d}\log \tilde{N}(t)}{\mathrm{d}t} - a\,\tilde{N}(t) \\ &= \frac{\mathrm{d}\log [N(t)/\tilde{N}(t)]}{\mathrm{d}t} + a\,[N(t) - \tilde{N}(t)]\,. \end{aligned} \tag{7.30}$$

Integrating (7.30) with respect to time, one obtains

$$H(t) - H(0) = \log\left[\frac{N(t)}{\tilde{N}(t)}\right] - \log\left[\frac{N(0)}{\tilde{N}(0)}\right] + a\int_0^t [N(\xi) - \tilde{N}(\xi)]\,\mathrm{d}\xi\,. \tag{7.31}$$

By definition, $\tilde{N}(0) = N(0)$. In the limit as $t \to \infty$, $\tilde{N}(t) \to N^*$, $N(t) \to N^*$, and $H(t) \to 0$. Therefore,

$$H(0) = a\int_0^\infty [\tilde{N}(t) - N(t)]\,\mathrm{d}t\,. \tag{7.32}$$

We shall denote the integral in (7.31) as $L(t)$ and its limit in (7.32) as L^*. It represents the amount of *life not lived* due to genetic disequilibrium. The value of L^* can be predicted from genetic data using

$$L^* = \frac{H(0)}{a}\,. \tag{7.33}$$

The units of L^* are "individual-time," for example, animal-hours or beetle-days. It is the extra amount of life the evolving population would have realized if its initial allele frequencies were in genetic equilibrium. Desharnais and Costantino (1982a) also referred to L^* as the "selective difference area."

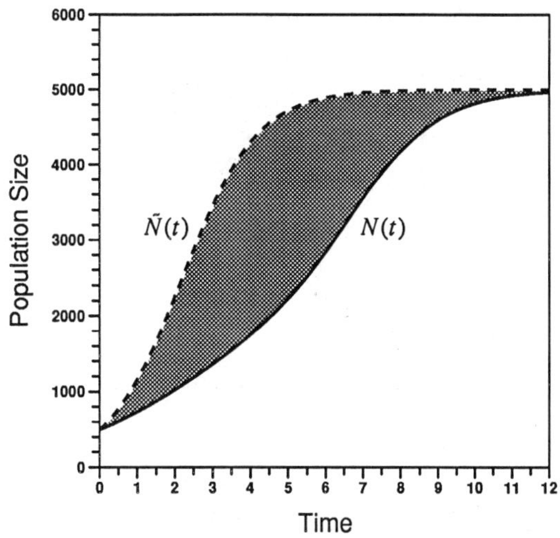

Fig. 7.7. Population size versus time for natural selection with logistic population growth. The dashed line is the trajectory of the reference population with allele frequencies in equilibrium. The solid line is the trajectory of the evolving population with nonequilibrium initial allele frequencies. The total difference between the trajectories of the reference and evolving populations (shaded area) equals the life not lived due to genetic disequilibrium.

The concept of life not lived is illustrated in Fig. 7.7. In this numerical example $r_{11}=r_{22}=0.5$, $r_{12}=1.5$, $a=0.0002$, $p(0)=0.001$, and $N(0)=500$. At equilibrium $p^*=0.5$, $r^*=1$, $N^*=r^*/a=5000$ and $H(0)=2.7612$. The solid line is the trajectory of $N(t)$ which was determined by numerically integrating the differential equations (7.26) using a Runge-Kutta fifth order algorithm. The dashed line represents the reference population, $\tilde{N}(t)$, which was computed using (7.28). Since $r^* \geq r(t)$ by Fisher's theorem, we have $\tilde{N}(t) \geq N(t)$. As $p(t) \rightarrow p^*$, $r(t) \rightarrow r^*$ and the trajectories of $\tilde{N}(t)$ and $N(t)$ both converge to N^*. However, the evolving population, $N(t)$, has failed to realize $L^*=H(0)/a=13{,}806$ individual-time units of life (shaded area) due to its initial genetic disequilibrium.

7.4.3 Properties of Life Not Lived

The life not lived has some interesting, and perhaps counterintuitive, properties. Consider the expression (7.33). The life not lived, L^*, depends upon the density-sensitivity parameter a and the initial entropy $H(0)$. In turn, $H(0)$, as given by (7.1), depends only upon the initial and equilibrium allele frequencies. The initial population size, $N(0)$, the strength of selection, s, and the specific values of the parameters r_{ij} have no influence on the value of L^*. Of course, all of these factors will influence the trajectory of $N(t)$, but when this trajectory is subtracted from $\tilde{N}(t)$ and integrated, the influence of the factors disappears. For example, consider the selection strength s as defined in (7.14) using r_{ij}'s in place of W_{ij}'s. If s is large, there will be a large difference in $N(t)$ compared to $\tilde{N}(t)$, but selection will be rapid and this difference will quickly converge to zero. If s is small, there will be a small difference in $N(t)$ *versus* $\tilde{N}(t)$, but selection will proceed slowly, and this difference will last for a long time. In either case, the integrated difference will be the same.

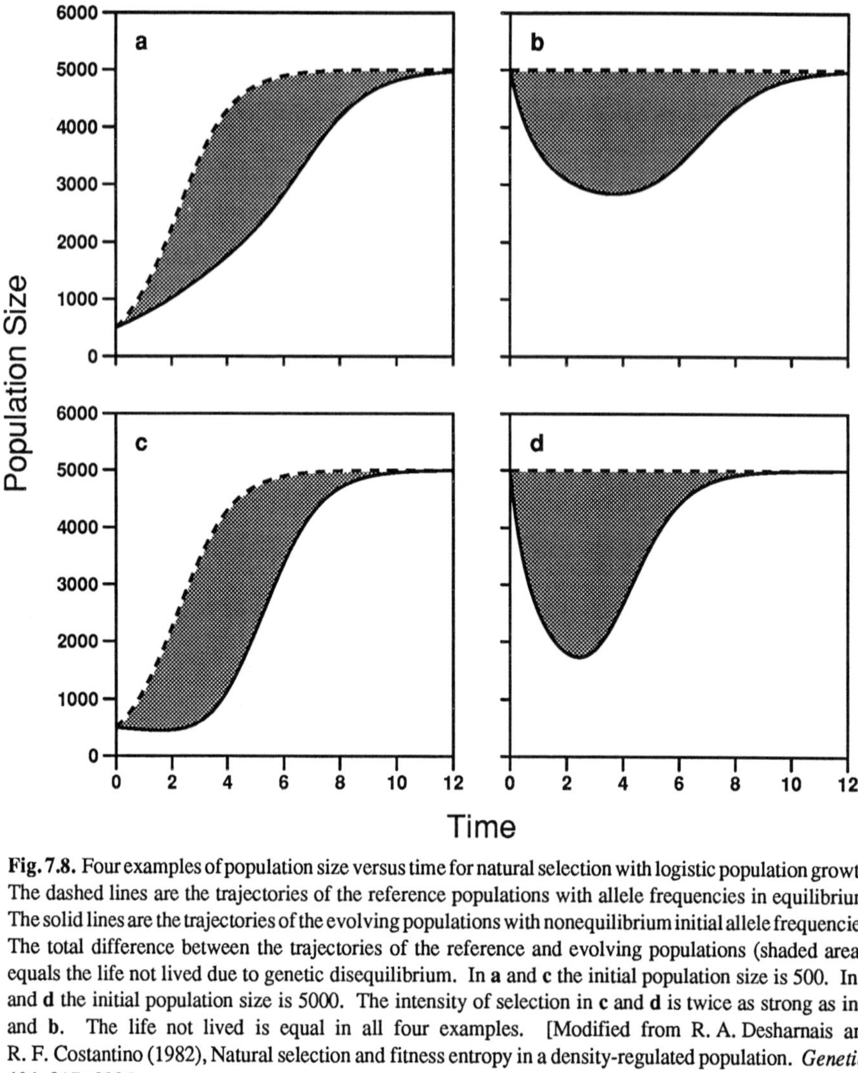

Fig. 7.8. Four examples of population size versus time for natural selection with logistic population growth. The dashed lines are the trajectories of the reference populations with allele frequencies in equilibrium. The solid lines are the trajectories of the evolving populations with nonequilibrium initial allele frequencies. The total difference between the trajectories of the reference and evolving populations (shaded areas) equals the life not lived due to genetic disequilibrium. In **a** and **c** the initial population size is 500. In **b** and **d** the initial population size is 5000. The intensity of selection in **c** and **d** is twice as strong as in **a** and **b**. The life not lived is equal in all four examples. [Modified from R. A. Desharnais and R. F. Costantino (1982), Natural selection and fitness entropy in a density-regulated population. *Genetics* **101**: 317–329.]

Figure 7.8 illustrates the lack of dependence of life not lived on initial population size and selection strength. In Fig. 7.8a, the initial conditions and parameter values are identical to those given in Fig. 7.7. In this case, $L^* = 13{,}806$. The example in Fig. 7.8b is identical to the one in Fig. 7.8a, except for the fact that $N(0) = N^* = 5000$. Note that the reference population remains in both genetic and demographic equilibrium, while the genetic disequilibrium of the evolving population causes a temporary drop in the population density. The examples in Figs 7.8c and 7.8d are identical to those in Figs 7.8a and 7.8b, respectively, except for the genotypic values, which have been changed to

$r_{11}=r_{22}=0$, and $r_{12}=2$. Although the selection intensity has changed from $s=1$ to $s=2$, we still have $p^*=0.5$ and $H(0)=2.7612$, and therefore the L^*'s (shaded areas) are equal in all four examples. For $s=2$, the early difference between $\tilde{N}(t)$ and $N(t)$ is greater, but convergence occurs more quickly (Fig. 7.8).

The definition (7.20) of a reference population requires the initial densities $N(0)$ and $\tilde{N}(0)$ to be equal. This restriction was made for the sake of convenience. If we allow $N(0)\neq\tilde{N}(0)$, then, from (7.31), in the limit as $t\rightarrow\infty$, it is obvious that

$$H(0) = \log\left[\frac{N(0)}{\tilde{N}(0)}\right] + a\int_0^\infty [\tilde{N}(t) - N(t)]\,\mathrm{d}t\,. \tag{7.34}$$

To test the theory, one need only add the logarithm of the ratio of the initial population densities to the life not lived L^*. Since population size data are needed to compute L^*, this represents no additional burden for the experimentalist.

The independence of the life not lived on the selection intensity raises the interesting possibility of using population size data to distinguish weak selection from neutrality. Consider two populations with different frequencies of the same alleles at some genetic locus. Although it is known that these populations differ genetically, it is not possible or is too costly to monitor changes in allele frequencies. Let ω_{ij} represent the relative fitness of the genotypes and set $r_{ij}=r-s\omega_{ij}$. Assume this organism can be cultured under replicated conditions in the laboratory or field. Let $N(t)$ and $N'(t)$ be the experimental densities of two populations and let $H(0)$ and $H'(0)$ represent their initial entropies. If $N(0)=N'(0)$, then

$$\int_0^\infty [N(t) - N'(t)]\,\mathrm{d}t = \begin{cases} \dfrac{H'(0) - H(0)}{a} & \text{if } s\neq 0\,, \\ 0 & \text{if } s=0\,. \end{cases} \tag{7.35}$$

The important point is that the total difference in the two population trajectories depends only on the dissimilarity in their genetic compositions, $H(0)\neq H'(0)$, and not upon the magnitude of selection, s. Hence, weak selection ($s\neq 0$) could be distinguished from neutrality ($s=0$). In practice, one must still contend with random effects when comparing $N(t)$ with $N'(t)$, especially when selection is weak. Nevertheless, the prospect of using population size data to detect selection is intriguing.

7.4.4 An Experimental Test

Once again we turn to the data of Moffa and Costantino (1977) in order to test the relationship between genetic disequilibrium and the life not lived. Although our analysis is based on the logistic equation and not the familiar "adult-immature" model (4.1), we shall assume that the logistic equation is a crude, but sufficient, approximation to density dependence in *Tribolium*. In §7.5.1, we show that the life not lived is a valid concept

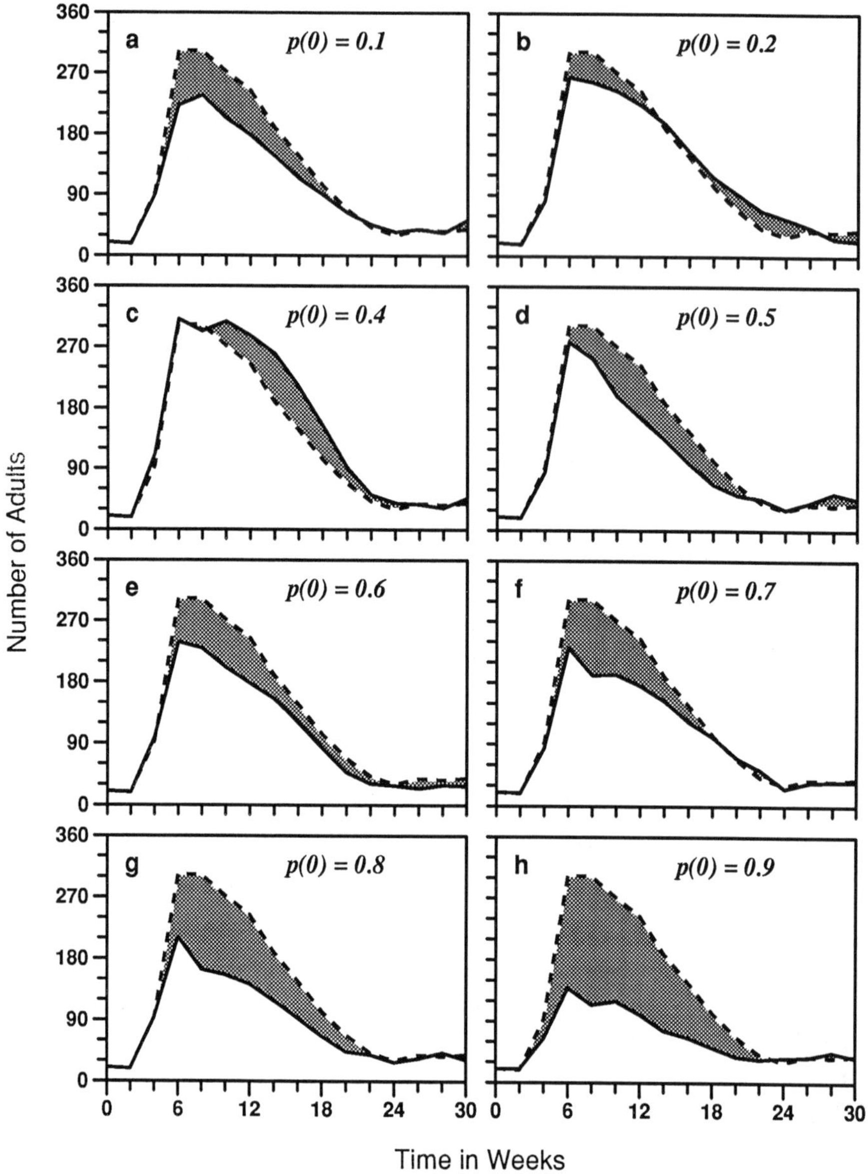

Fig. 7.9. Experimentally observed numbers of adults versus time for nine populations of *T. castaneum* with different initial frequencies of the *cos* allele. In each case, the dashed line is the population with an initial *cos* frequency of $p(0)=0.3$ which serves as a reference population. The solid lines are for the remaining treatments where $p(0)\neq 0.3$. The shaded area is the estimate for the life not lived due to genetic disequilibrium.

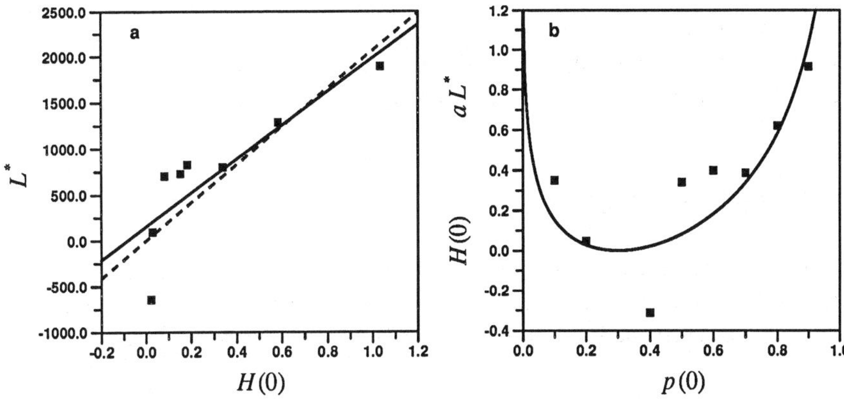

Fig. 7.10. An experimental evaluation of the relationship between the life not lived and the entropy measure of genetic disequilibrium. In **a**, the squares are estimates of the life not lived. The solid line is a least squares regression for the data and the dashed line is a regression based on the theoretical prediction of a zero intercept. In **b**, the squares are estimates of the life not lived divided by the slope of the dashed line in **a**. The solid curve is the theoretical entropy function.

for any general density-dependent model of population growth. Desharnais and Costantino (1982a) conducted a similar analysis of these data using fitness expressions based on the adult-immature model.

Recall that populations of *T. castaneum* were initiated with *cos* allele frequencies from $p(0)=0$ to $p(0)=1$ in increments of 0.1 and that the frequency $p(t)$ converged to an equilibrium in the range of 0.25 to 0.30 after six weeks. After an initial exponential growth phase, adult numbers decreased and then began to fluctuate around an equilibrium level. Our analysis is based on the census number of adults from weeks 0 through 30 for those populations with $p(0)$ in the range from 0.1 to 0.9.

Our first step was to calculate the life not lived. For the reference population, we set $p^*=0.3$ and used the census data for the treatment where $p(0)=0.3$ as an approximation to the trajectory $\tilde{N}(t)$. For the remaining values of $p(0)$, we subtracted the number of adults from $\tilde{N}(t)$ and integrated the difference using the trapezoid rule. The adult census data for each treatment appears in Fig. 7.9. In each graph, the dashed line is the reference population ($p(0)=0.3$) and the solid line is the evolving population ($p(0)\neq 0.3$). The shaded area represents the estimate of the life not lived. Note that any shaded areas falling below the dashed line were subtracted from the shaded areas falling above the dashed line. Hence negative values of L^* are possible (eg. $p(0)=0.4$).

Our test of the prediction (7.33) appears in Fig. 7.10. The squares represent the observed values of L^*. From (7.33) we expect a linear relationship between the initial entropy, $H(0)$, and L^* which goes through the origin. A linear least squares regression (solid line in Fig. 7.10a) gave a slope of 1834.2 ± 464.1 and an intercept of 154.0 ± 206.6. The regression is statistically significant ($P<0.01$), but the intercept is not significantly different from zero ($P=0.48$). We performed a second regression with no intercept (dashed line in Fig. 7.10a). In this case we obtained a slope of 2069.5. The inverse of this slope gave an estimate of the parameter $a=0.0004832$. In Fig. 7.10b, the products

aL^* and the curve $H(p(0), p^* = 0.3)$ are plotted as a function the initial allele frequency. Although there are some large deviations from the curve, especially in the range of $p(0) = 0.4$ to 0.6, the overall trend in the data supports the theory.

7.5 Extensions to the Theory

Research involving the use of Ginzburg's entropy function continues. Several theoretical extensions have recently been made. In this section we briefly summarize some of these results. The original publications should be consulted for details.

7.5.1 A Generalization for Weak Selection

The results reported above for density-dependent population growth are based on the logistic population model. In work which extends these results to the Lotka-Volterra model of interacting species, Desharnais (1986) outlines a generalization for arbitrary fitness functions and weak selection. We now sketch this argument for a single population undergoing selection.

Let $W_{ij}(N)$ be any arbitrary fitness function and suppose selection is weak. As a first approximation, assume $dN/dt \approx 0$ before allele frequencies change very much. In other words, $N(t)$ quickly approaches a "quasi-equilibrium" value $\theta(\mathbf{p})$, where, for any given allele frequency vector $\mathbf{p}$, θ is the solution to

$$W(\theta) = \sum_{i=1}^{n} \sum_{j=1}^{n} p_i p_j W_{ij}(\theta) = 0 . \tag{7.36}$$

As allele frequencies move slowly towards their equilibrium values, $\theta(\mathbf{p}) \to N^*$ and $N(t)$ tracks these changes in θ. Since selection is weak, the genotypes do not differ greatly in their fitness values. This means that for any arbitrary vector $\mathbf{p}$, $\theta(\mathbf{p})$ is relatively close to the value N^*. Expanding each fitness function in a Taylor series around N^* gives

$$W_{ij}(N) \approx W_{ij}(N^*) - a_{ij}(N - N^*) , \tag{7.37}$$

where

$$a_{ij} = -\left(\frac{dW_{ij}}{dN}\right)_{N=N^*} . \tag{7.38}$$

Combining this approximation with the entropy function gives

$$H(0) \approx \log\left[\frac{N(0)}{N^*}\right] + a\int_0^\infty [N^* - N(t)]\,[1-\varepsilon(t)]\,\mathrm{d}t\,, \tag{7.39}$$

where

$$a = \sum_{i=1}^{n}\sum_{j=1}^{n} p_i^* p_j^* a_{ij}\,, \tag{7.40}$$

$$\varepsilon(t) = 1 - a^{-1}\sum_{i=1}^{n}\sum_{j=1}^{n} p_i^* p_j(t) a_{ij}\,. \tag{7.41}$$

Obviously, $\varepsilon(t) \to 0$ as $t \to \infty$. Since selection is weak, it is reasonable to assume that $a_{ij}/a \approx 1$ for all genotypes. This implies $|\varepsilon(t)| \ll 1$ for $t \geq 0$. Hence,

$$H(0) \approx \log\left[\frac{N(0)}{N^*}\right] + a\int_0^\infty [N^* - N(t)]\,\mathrm{d}t\,. \tag{7.42}$$

Now consider a reference population as defined by (7.20). We can repeat all the steps from (7.36) through (7.42) for $\tilde{p}(0) = p^*$. In this case, $\tilde{H}(0) = 0$ and (7.42) gives

$$\log\left[\frac{\tilde{N}(0)}{\tilde{N}^*}\right] \approx a\int_0^\infty [\tilde{N}(t) - \tilde{N}^*]\,\mathrm{d}t\,. \tag{7.43}$$

Since $\tilde{N}(0) = N(0)$ and $\tilde{N}^* = N^*$, we can substitute (7.43) into (7.42) to obtain

$$H(0) \approx a\int_0^\infty [\tilde{N}(t) - N(t)]\,\mathrm{d}t\,. \tag{7.44}$$

This result suggests that the relationship between entropy and life not lived may be fairly robust. Some preliminary numerical work supports this idea for reasonably strong levels of selection. These problems are the subjects of current investigations.

7.5.2 Several Interacting Species

Desharnais (1986) modeled natural selection in a *community* composed of several interacting populations. Assume that there are m species with densities $N_1(t), N_2(t), \ldots, N_m(t)$ and within species k there are n_k alleles segregating at a single genetic locus. The density-dependent fitness functions are given by

$$W_{k,ij}(N_1, N_2, \ldots, N_m) = r_{k,ij} - \sum_{l=1}^{m} a_{kl} N_l(t)\,, \qquad \begin{array}{l} k = 1, 2, \ldots, m, \\ i,j = 1, 2, \ldots, n_k, \end{array} \tag{7.45}$$

where $r_{k,ij}$ is the Malthusian growth rate for genotype ij within species k and a_{kl} is the *per capita* effect of species l on species k. Using (7.45) in the genetic model (6.1)–(6.2), one obtains equations for the population growth rates of the form

$$\frac{\mathrm{d}N_k}{\mathrm{d}t} = N_k(t)\left[r_k - \sum_{l=1}^{m} a_{kl} N_l(t) \right], \qquad k = 1, 2, \ldots, m. \tag{7.46}$$

This represents a *Lotka-Volterra* model of species interactions. However, the parameter r_k is a time-varying average which depends on the allele frequencies within species k. Since genetic changes within each species affect their population growth rates, and the population density of each species affects the growth rate of other species in the community, the growth rate of each species will depend on genetic changes in the community as a whole. Thus we have a model of density-dependent *coevolution.*

The effects of genetic disequilibrium can be studied in the context of this simple coevolutionary model. Let $p_{k,i}(t)$ represent the frequency of allele i within species k. If any $p_{k,i}(0) \neq p_{k,i}^*$, we will refer to this system as an *evolving community*. Define a *reference community* as one where

$$\begin{array}{lcll} \tilde{W}_{k,ij} & = & W_{k,ij}\,, & i,j = 1, 2, 3, \ldots, n_k, \\ \tilde{p}_{k,i}(0) & = & p_{k,i}^*\,, & i = 1, 2, 3, \ldots, n_k, \\ \tilde{N}_k(0) & = & N_k(0)\,, & k = 1, 2, 3, \ldots, m. \end{array} \tag{7.47}$$

The reference community is identical to the evolving community except that *all* species are initially in genetic equilibrium. If use (7.1) to define $H_k(0)$ as the initial entropy disequilibrium of each species k, then a comparison of the evolving and reference communities yields the following result:

$$H_k(0) = \sum_{l=1}^{m} a_{kl} \int_0^{\infty} [\tilde{N}_l(t) - N_l(t)]\,\mathrm{d}t\,, \qquad k = 1, 2, \ldots, m. \tag{7.48}$$

This result, which was derived by Desharnais (1986), relates the genetic disequilibrium to a linear combination of the life not lived of all the species in the community.

The result (7.48) can be expressed in matrix form. Let L_k^* represent the life not lived for species k and let $\mathbf{L}^*$ denote a catenation of the L_k^*'s into a vector. Similarly, let $\mathbf{H}(0)$ denote the vector of initial entropies. The species interaction parameters can be used to form the $m \times m$ matrix $\mathbf{A} = \{a_{kl}\}$. The main result can be expressed as

$$\mathbf{H}(0) = \mathbf{A}\,\mathbf{L}^*. \tag{7.49}$$

If the matrix $\mathbf{A}$ is nonsingular, then

$$\mathbf{L}^* = \mathbf{A}^{-1}\mathbf{H}(0). \tag{7.50}$$

With (7.50) one can predict the life not lived for each species using a linear combination of the genetic disequilibria in the community as a whole. Note that (7.49) and (7.50) are the multidimensional extensions of (7.32) and (7.33) for the logistic model. In fact, (7.32) and (7.33) are recovered when $m=1$.

One new feature of this multi-species analysis is that the life not lived can take on *negative* values. When $L_k^*<0$, one can view L_k^* as the "extra life lived" due to genetic disequilibrium in the community as a whole. For example, genetic disequilibrium in a predator can result in extra life lived by its prey. Desharnais (1986) discusses several specific examples.

7.5.3 Asexual Age-Structured Populations

Desharnais and Cohen (1986) considered natural selection in asexual age-structured populations. Some of their models and results are presented below. The original paper should be consulted for a more complete account, including technical details and proofs.

Consider a large age-structured population composed of n demographically distinct asexual clones. Let the age structure of the ith genotype be described by a column vector $\mathbf{K}_i(t)$ of length m_i. In the absence of density dependence, the $n_i \times n_i$ Leslie matrix $\mathbf{M}_i$ will be used to project the vector $\mathbf{K}_i(t)$ forward one unit in time. (See §2.1.2 for a discussion of the Leslie matrix model.) As a first approximation to density dependence, assume that every individual, regardless of its genotype and age, contributes equally and independently to the mortality of other individuals, regardless of their genotype and age, and this extra mortality is independent of other causes of death. Since the fecundity terms in the first row of the Leslie matrix also include the survivorship of the parents and offspring within each time interval (Keyfitz, 1968, p. 30), all the elements of the projection matrix will experience the same proportional change. This change will be modeled using a negative exponential function:

$$\mathbf{K}_i(t+1) = \mathbf{M}_i\,\mathbf{K}_i(t)\exp[-aN(t)]\,, \qquad i=1,2,\ldots,n. \tag{7.51}$$

$N(t)$ is the total number of individuals of all ages and genotypes. As in the logistic model (7.25), the parameter a is a measure of density sensitivity.

The dynamics of the model (7.51) depend on the matrices $\mathbf{M}_i$ and the density-sensitivity parameter a. When $a=0$, each genotypic subpopulation grows at the exponential rate $r_i=\log(\lambda_i)$, where λ_i is the dominant real root of the matrix $\mathbf{M}_i$. Eventually, the genotype(s) with the largest growth rate, $r=\max(r_1,r_2,\ldots,r_n)$, dominates the population as a whole. When $a>0$, the population size remains bounded. If $r<2$, then the population approaches a finite equilibrium density, $N^*=r/a$ (Desharnais and Cohen 1986). At equilibrium, only those genotypes with $r_i=r$ survive.

We introduce some additional variables. Let $N_i(t)$ denote the total number of individuals with the ith genotype. Let $p_i(t)=N_i(t)/N(t)$ represent the frequency of each genotype when age is ignored. Finally, let $\mathbf{q}_i(t)=\mathbf{K}_i(t)/N_i(t)$ represent an *age distribution vector* for the ith genotype.

Genetic disequilibrium will be measured using Ginzburg's entropy function. Without any loss of generality, we can label genotypes so that $r_i=r$ for $i=1,2,\ldots,k$ and $r_i<r$ for $i=k+1,k+2,\ldots,n$. Hence $p_i(t)\rightarrow p_i^*>0$ for $i\leq k$ and $p_i(t)\rightarrow 0$ for $i\geq k$. For the initial genetic disequilibrium,

$$H_p(0) = -\sum_{i=1}^{k} p_i^* \log\left[\frac{p_i(0)}{p_i^*}\right]. \tag{7.52}$$

An explicit formula for the p_i^*'s can be found in Desharnais and Cohen (1986).

Demographic disequilibrium will be measured using a similar function. Let $\mathbf{u}_i$ and $\mathbf{v}_i$ represent the right and left eigenvectors of $\mathbf{M}_i$ which correspond to λ_i. We can scale these nonnegative vectors so that the elements of $\mathbf{u}_i$ sum to one and the first element of $\mathbf{v}_i$ equals one. For each genotype, $\mathbf{u}_i$ represents the *equilibrium age distribution*, i.e. $\mathbf{q}_i(t)\rightarrow\mathbf{q}_i^*=\mathbf{u}_i$, and $\mathbf{v}_i$ is called the vector of *reproductive values* (Fisher 1930). Let the average reproductive value for the ith genotype be given by the inner product $V(t)=\mathbf{v}_i^{\mathrm{T}}\mathbf{q}_i(t)$ and let $V_i^*=\mathbf{v}_i^{\mathrm{T}}\mathbf{q}_i^*$. The initial demographic disequilibrium will be measured using

$$H_V(0) = -\sum_{i=1}^{k} p_i^* \log\left[\frac{V_i(0)}{V_i^*}\right]. \tag{7.53}$$

Unlike $H_p(0)$, this function can take on both positive and negative values. However, when $\mathbf{q}_i(0)=\mathbf{q}_i^*$ for $i\leq k$, $H_V(0)=0$.

The effects of disequilibrium will be evaluated, once again, by comparing $N(t)$ for an evolving population with $\tilde{N}(t)$ for a reference population. The reference population is assumed to be identical to the evolving population with the following exceptions:

$$\begin{aligned} \tilde{\mathbf{q}}_i(0) &= \mathbf{q}_i^*, \qquad i=1,2,3,\ldots,n, \\ \tilde{p}_i(0) &= p_i^*, \qquad i=1,2,3,\ldots,n. \end{aligned} \tag{7.54}$$

In other words, the reference population is initially in both genetic *and* demographic equilibrium. In the density-independent case ($a=0$), the trajectories of $\tilde{N}(t)$ and $N(t)$ become separated in time by an asymptotic time lag τ, where

$$H_p(0)+H_V(0) = r\tau. \tag{7.55}$$

In the density-dependent case ($a>0$), the total life not lived can be related to disequilibrium using

$$H_p(0) + H_V(0) = a \sum_{t=0}^{\infty} [\tilde{N}(t) - N(t)] \ . \tag{7.56}$$

These results are extensions of (7.24) and (7.32) to an age-structured model with asexual reproduction. The functions $H_p(0)$ and $H_V(0)$ partition the effects of disequilibrium into genetic and demographic components.

7.6 Summary and Conclusions

In a strictly genetic model, disequilibrium results in Haldane's (1957) cost of natural selection, which is defined in terms of the total difference between the evolving mean fitness and its maximum equilibrium value. Using Ginzburg's (1972, 1974, 1977) entropy function, one can relate the cost of selection to the initial disequilibrium in allele frequencies. Entropy can also be used to place a bound on the adaptation rate, i.e., the rate at which mean fitness moves towards its equilibrium value. While these predictions are conceptually pleasing, they are difficult for the experimentalist, because he/she must use genetic data and fitness estimates in order to verify the result. Furthermore, it is not quite clear what the implications of these predictions are for the population as biological entity. Nevertheless, an attribute of the entropy function is that it makes precise the relationship between genetic disequilibrium and cost.

A better approach includes a consideration of population size. Using the concept of a reference population which is initially in genetic equilibrium, Ginzburg (1977) was able to show that the effect of disequilibrium is to cause a lag in the Malthusian growth curve, which he called the selective delay. The advantage of selective delay is that it is a concrete concept, measured in units of time, and observable from a knowledge of the population numbers. Fitness estimates are not necessary and only the initial and equilibrium allele frequencies are required. Unfortunately, selective delay is based on the assumption of Malthusian growth, and few populations grow unconstrained for more than a generation or two. But in principle, selective delay is easily measured, and its biological meaning is simple and clear.

The best formulation uses a density-dependent model. Desharnais and Costantino (1982a) considered fitness functions which depend on population density. When population growth is logistic, the effect of disequilibrium is to temporarily suppress population growth, so that the total life not lived is proportional to the entropy function. As with selective delay, predictions are based on population numbers; fitness estimates are not necessary and only initial and equilibrium allele frequencies are needed. However, from an experimentalist's point of view, this prediction is the easiest to test and does not require a truncation of the data so that only the exponential growth phase is used. Furthermore, as we argued in §7.5.1, the concept of life not lived seems valid for any form of density-dependence, not just the logistic model. Desharnais (1986) showed that life not lived can be studied in multi-species communities. Desharnais and Cohen (1986) extended the concepts of life not lived and selective delay to age structured populations. But much theoretical work remains to be done.

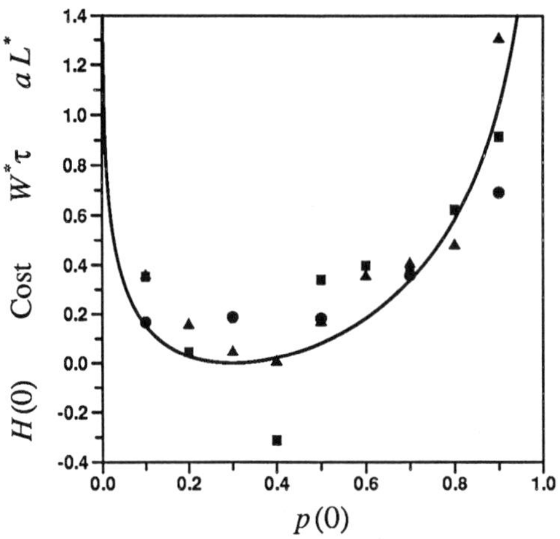

Fig. 7.11. A composite of the experimental results obtained from *Tribolium* data which were used to test predictions based on the entropy function. The circles are the estimates of the cost of natural selection. The triangles are estimates of the product of the asymptotic population growth rate and the selective delay. The squares are estimates of the product of the density sensitivity parameter and the life not lived. The solid curve is the theoretical entropy function.

In three separate instances, we have used the data of Moffa and Costantino (1977) on the *cos* strain of *Tribolium castaneum* to test predictions based on the entropy function. As a summary, we have plotted those experimental results together in Fig. 7.11. The solid line is the theoretical prediction, $H(p(0), p^*)$, with $p^* = 0.3$. The circles are from Fig. 7.4 and are based on estimates of the cost of natural selection using fitness values and genetic data. The triangles are from Fig. 7.6 and are based on estimates of the asymptotic growth rate and the selective delay using census data for the exponential growth phase. Lastly, the squares are from Fig. 7.10 and are based on estimates of the life not lived using census data for the approach to equilibrium. These results, when considered together, support the idea that entropy is the proper "scale" for measuring the impact of genetic disequilibrium on the dynamics of population growth. Secondarily, they point to the utility of the *Tribolium* laboratory model for testing predictions on the interactions between natural selection and population growth.

Chapter 8 Two Interacting Species

Flour beetles of the genus *Tribolium* can be routinely maintained in the laboratory for long periods of time. Monthly census records on populations continuously cultured for up to six years are available (Park and Frank 1950). In single species cultures extinction is rare. But in cultures with two species present one or the other *nearly always* goes extinct. The rule is reversed: extinction is common.

It is tempting to invoke Gause's (1934) *competitive exclusion principle* to explain the outcome of two-species *Tribolium* experiments. Loosely stated, this principle says that two species with the same resource requirements cannot coexist. While the usefulness of this principle has been the subject of some debate (eg. Hardin 1960, Cole 1960), Thomas Park's experiments are usually cited in support of this principle. While it is certainly true that in laboratory cultures two *Tribolium* species are utilizing the same resource, it would be a mistake to oversimplify the problem to one of resource utilization. One must examine the nature of the species interaction to fully appreciate the results of *Tribolium* competition experiments.

In this chapter we address the problem of extinction and coexistence by focusing on the biological basis of the species interaction. Cannibalism immediately comes to mind. But is the eating of eggs and pupae by larvae and adults indiscriminate or discriminate? Do the adults of one species preferentially cannibalize the immatures of their own species or the other species? Initially, it was assumed that eggs and pupa were eaten indiscriminately (Landahl 1955b, Neyman et al. 1956), but new data altered that assumption (Leslie 1962, Park et al. 1968, Ryan et al. 1970). Furthermore, cannibalism is not the only mechanism of species interaction in this genus. Adults, present in large numbers, can inhibit the metamorphosis of larvae into pupae (Jillson and Costantino 1980). This gives us three qualitatively different types of interactions in two species *Tribolium* communities: (1) two cannibalistic species, (2) a cannibalistic species with an inhibiting species, and (3) two inhibiting species. Our goal is to place the experimental data in the context of a simple and general theory of species interaction.

8.1 Competition as an Adult-Immature Interaction

8.1.1 Two-Species Equations

Our quantitative analysis is based on the demographically reduced adult recruitment model introduced in §4.1. In single species cultures the change in adult number is given by

$$\frac{dN}{dt} = N\,[b\,\exp(-cN) - \mu]\;. \tag{8.1}$$

This equation reflects the balance between density-dependent recruitment and mortality of adults. Our analysis of the two species problem is a straightforward extension of the single species model:

$$\begin{aligned} \frac{dN_1}{dt} &= f_1(N_1,N_2) = N_1\,[b_1 \exp(-c_{11}N_1 - c_{12}N_2) - \mu_1] \\ \frac{dN_2}{dt} &= f_2(N_1,N_2) = N_2\,[b_2 \exp(-c_{21}N_1 - c_{22}N_2) - \mu_2] \end{aligned} \tag{8.2}$$

where, for species i, N_i is the number of adults, b_i is the rate at which potential recruits are produced per adult, and μ_i is the rate of adult mortality. The parameter c_{ij} has at least two possible biological interpretations: (i) the cannibalism rate at which an adult of species j consumes eggs, pupae or callow adults of species i or (ii) the rate at which an adult of species j inhibits larval metamorphosis of species i. But other mechanisms are feasible. In fact, reductions in fecundity and/or juvenile survival, age-specific predation, toxin production, pheromone secretion and other factors can also be modeled using the same or similar equations.

Notice that if the adults of one species do not interact in any way with the immature life stages of the other species, i.e., $c_{12} = c_{21} = 0$, then we recover the single species model (8.1). Our analysis is on the "slow" time scale which is dominated by the adult population dynamics (see Chapters 3 and 4).

8.1.2 Analysis of the Model

Setting $f_1 = 0$ and $f_2 = 0$ in (8.2), we can solve for the equilibrium number of adults for each species:

$$\begin{aligned} N_1^* &= \frac{\left(c_{22}\log\left(\frac{b_1}{\mu_1}\right) - c_{12}\log\left(\frac{b_2}{\mu_2}\right)\right)}{(c_{11}c_{22} - c_{12}c_{21})}, \\ N_2^* &= \frac{\left(c_{11}\log\left(\frac{b_2}{\mu_2}\right) - c_{21}\log\left(\frac{b_1}{\mu_1}\right)\right)}{(c_{11}c_{22} - c_{12}c_{21})}. \end{aligned} \tag{8.3}$$

This equilibrium is only defined for $c_{11}c_{22} \neq c_{12}c_{21}$. A biologically feasible coexistence equilibrium, $N_1^* > 0$ and $N_2^* > 0$, implies

$$\left(\frac{b_2}{\mu_2}\right)^{c_{12}/c_{22}} < \left(\frac{b_1}{\mu_1}\right) < \left(\frac{b_2}{\mu_2}\right)^{c_{11}/c_{21}} \quad \text{for} \quad c_{11}c_{22} > c_{12}c_{21}\,,$$

$$\left(\frac{b_2}{\mu_2}\right)^{c_{12}/c_{22}} > \left(\frac{b_1}{\mu_1}\right) > \left(\frac{b_2}{\mu_2}\right)^{c_{11}/c_{21}} \quad \text{for} \quad c_{11}c_{22} < c_{12}c_{21}\,. \tag{8.4}$$

If neither of these conditions are satisfied, then $N_1^* > 0$ and $N_2^* = 0$, or $N_1^* = 0$ and $N_2^* > 0$, or, in the trivial case where $b_1 < \mu_1$ and $b_2 < \mu_2$, then $N_1^* = N_2^* = 0$.

The stability of the equilibrium is determined in the usual way by linearizing the model (8.2) around the equilibrium point. The coefficients of the linear model are the elements of the following stability matrix:

$$\mathbf{Z} = \{z_{ij}\} = \left\{\frac{\partial f_i}{\partial N_j}\right\}_{\substack{N_1 = N_1^* \\ N_2 = N_2^*}} = \begin{pmatrix} -\mu_1 c_{11} N_1^* & -\mu_1 c_{12} N_1^* \\ -\mu_2 c_{21} N_2^* & -\mu_2 c_{22} N_2^* \end{pmatrix}. \tag{8.5}$$

The eigenvalues λ_1 and λ_2 of this matrix are the roots of the characteristic equation $\det(\mathbf{Z} - \lambda\mathbf{I}) = 0$, where $\mathbf{I}$ is the identity matrix. This gives the quadratic polynomial

$$\lambda^2 + (\mu_1 c_{11} N_1^* + \mu_2 c_{22} N_2^*)\lambda + \mu_1\mu_2 N_1^* N_2^* (c_{11}c_{22} - c_{12}c_{21}) = 0\,. \tag{8.6}$$

with roots

$$\lambda_1, \lambda_2 = \frac{-(\mu_1 c_{11} N_1^* + \mu_2 c_{22} N_2^*) \pm \sqrt{(\mu_1 c_{11} N_1^* - \mu_2 c_{22} N_2^*)^2 + 4\mu_1\mu_2 N_1^* N_2^* c_{12}c_{21}}}{2}\,. \tag{8.7}$$

Note that the term under the radical is positive. Thus both eigenvalues are real and $\lambda_1 > \lambda_2$.

Consider the coexistence equilibrium $N_1^* > 0$ and $N_2^* > 0$. For the equilibrium to be stable, the largest eigenvalue must be negative, $\lambda_1 < 0$. From (8.7) one can show that this is true if and only if

$$c_{11}c_{22} > c_{12}c_{21}\,. \tag{8.8}$$

Combining (8.8) with (8.4) we see that stable coexistence is possible if and only if

$$\left(\frac{b_2}{\mu_2}\right)^{c_{12}/c_{22}} < \left(\frac{b_1}{\mu_1}\right) < \left(\frac{b_2}{\mu_2}\right)^{c_{11}/c_{21}}\,. \tag{8.9}$$

Reversing this inequality gives the necessary and sufficient conditions for an unstable equilibrium. In this latter case, the identity of the surviving species will depend on the initial numbers. Species 1 will prevail if $N_1(0) > kN_2(0)$ where

$$k = \frac{\log\left[\left(\frac{b_1}{\mu_1}\right)^{c_{22}}\left(\frac{b_2}{\mu_2}\right)^{-c_{12}}\right]}{\log\left[\left(\frac{b_1}{\mu_1}\right)^{-c_{21}}\left(\frac{b_2}{\mu_2}\right)^{c_{11}}\right]}. \tag{8.10}$$

If $N_1(0) < kN_2(0)$, species 2 will survive. If $N_1(0) = kN_2(0)$, the unstable equilibrium (8.3) will be approached asymptotically along the trajectory $N_1(t) = kN_2(t)$, but any perturbation off this trajectory will cause one of the species to go extinct.

In cases where a coexistence equilibrium does not exist (i.e. (8.4) is violated), species 1 will displace species 2 for *all* positive initial conditions whenever

$$\left(\frac{b_1}{\mu_1}\right) \geq \left(\frac{b_2}{\mu_2}\right)^{c_{11}/c_{21}} \quad \text{and} \quad \left(\frac{b_1}{\mu_1}\right) \geq \left(\frac{b_2}{\mu_2}\right)^{c_{12}/c_{22}}, \tag{8.11}$$

provided one of the two conditions is a strict inequality. With the direction of the inequalities in (8.11) reversed, species 2 will displace species 1. Finally, if both conditions in (8.11) are equalities, then $c_{11}c_{22} = c_{12}c_{21}$ and there exists a continuous set of "neutral equilibria" which satisfy the relationship

$$N_1^* = \left(\frac{\log\left(\frac{b_1}{\mu_1}\right)}{c_{11}}\right) - \left(\frac{c_{22}\log\left(\frac{b_1}{\mu_1}\right)}{c_{11}\log\left(\frac{b_2}{\mu_2}\right)}\right)N_2^*. \tag{8.12}$$

The specific values of N_1^* and N_2^* that the species attain will depend on the initial conditions.

It should be pointed out that these results parallel the dynamics of the classical Lotka-Volterra model (Volterra 1926, Lotka 1932) of two competing species:

$$\begin{aligned} \frac{dN_1}{dt} &= N_1 r_1\left(\frac{K_1 - N_1 - \alpha N_2}{K_1}\right), \\ \frac{dN_2}{dt} &= N_2 r_2\left(\frac{K_2 - N_2 - \beta N_1}{K_2}\right). \end{aligned} \tag{8.13}$$

Here, r_i is the growth rate of species i, K_i is the carrying capacity of species i, α is the effect of species 2 on species 1 relative to the intraspecific effect, and β is the effect of species 1 on species 2 relative to the intraspecific effect. If we identify K_i with the single-species equilibrium $N_i^\circ = \log(b_i/\mu_i)/c_{ii}$, α with c_{12}/c_{11}, and β with c_{21}/c_{22}, then conclusions concerning the dynamics of the two models are the same. The advantage of the *Tribolium* model (8.2) is that it allows us to consider the specific biological mechanisms of cannibalism or developmental inhibition that mediate the species interactions.

8.2 Two Cannibalistic Species

8.2.1 *T. castaneum* versus *T. confusum*

In his 1954 paper, Thomas Park reported that *T. castaneum* and *T. confusum* did not coexist when they shared a common habitat in the laboratory. Competitive exclusion was the rule.

Park (1954) cultured these species in six different climatic conditions defined by three temperatures (24, 29, 34°C) and two relative humidities (30, 60% rh). When *T. castaneum* and *T. confusum* were cultured together in the hot-wet climate (HW = 34°C, 70% rh) *T. castaneum* was the sole survivor. In the cold-dry climate (CD = 24°C, 30% rh) *T. castaneum* was unable to grow even when cultured alone. In the hot-dry (HD = 34°C, 30% rh), temperate-wet (TW = 29°C, 70% rh), temperate-dry (TD = 29°C, 30% rh), and cold-wet (CW = 24°C, 70% rh) climates the results were mixed: in some replicates *T. castaneum* prevailed and in other cultures *T. confusum* was the winner. The percentage of the 20-30 replicates in which one or the other species won reflected the conditions of temperature and humidity: *T. castaneum* won in 100% of the HW, 86% of the TW, and 29% of the CW climates; *T. confusum* won in 90% of the HD, 87% of the TD, and 100% of the CD climates. As one moves from warm temperatures and high humidities to cold temperatures and low humidities, *T. confusum* gains the advantage.

Park's experiments were significant in two respects. First, he demonstrated that even along an environmental gradient from a point where the first species is favored to a point where the second species is favored, coexistence is not a possibility. Second, Park introduced a new concept into ecological theory: the idea of *indeterminate* competition.

8.2.2 The Fan-Shaped Outcome Pattern

Two years after Park's 1954 study, his data were presented (Neyman et al. 1956) in a graph that had a powerful impact on the interpretation of his results. In fact, the graph itself has come to characterize Park's (1948, 1954, 1957) competition experiments. It is the *pattern* revealed in the graph that motivated the mathematical analysis of Neyman et al. (1956).

The adult census data for each replicate were plotted on the (N_1, N_2)–plane. The time sequence of points traced a path that eventually led to the survival of either species 1 or species 2 (Fig. 8.1). After all of the replicates were plotted, two solid lines were drawn on the figure. One line identified the region where the trajectories of the replicates

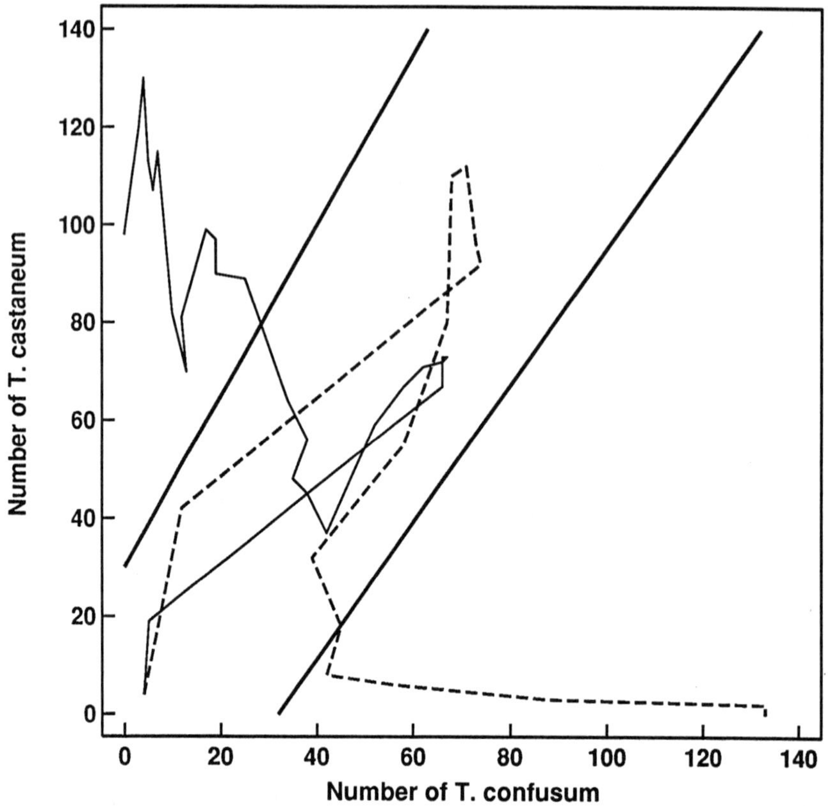

Fig. 8.1. An empirically established fan-shaped outcome pattern. The solid and dashed trajectories represent two independent replicates with identical initial conditions. The solid lines bound the area where the outcome of competition is uncertain. [From J. Neyman, T. Park and E. L. Scott (1956), Struggle for existence, the *Tribolium* model: Biological and statistical aspects. Third Berkeley Symposium for Mathematics, Statistics, and Probability, University of California Press, Berkeley.]

led irreversibly to the survival of species 2. The other line located a similar region for species 1. In the middle zone the numbers of both species changed in an erratic manner. This observation is the *fan-shaped outcome pattern.*

Two fundamentally different biological hypotheses were proposed to explain these data: indiscriminate and discriminate cannibalism. We will consider each in turn.

8.2.3 Indiscriminate Cannibalism

In two mathematical papers that were published immediately following Park's (1948, 1954) experimental reports, a key assumption with regards to the population dynamics was made: the two species when grown together were assumed to eat each other's eggs and pupae indiscriminately. In the first paper, H. D. Landahl (1955b, p. 131) wrote, "... there is no differential resistance to cannibalism, that is, a given stage of either type cannibalizes another developmental stage to the same extent whether the latter type is the same or different from its own type (c.f. Birch, Park and Frank 1951)." In the second paper, Neyman et al. (1956, p. 68) made the following statement: "The theory given here is based on the assumption that the two species of *Tribolium* eat each other's eggs, as well as their own, indiscriminately." Cannibalism was recognized and acknowledged as operating in the beetle cultures. However, with the assumption of indiscriminate egg-eating the outcome of competition is independent of cannibalism.

Following Landhal and Neyman, let us assume that the adults do not selectively eat the immature life stages of one species over another. With respect to the parameters c_{ij}, we can write $c_{11}=c_{21}=c_1$ and $c_{22}=c_{12}=c_2$. Substituting into the original equations (8.2) we have

$$\frac{d\log N_1}{dt} = \frac{dN_1}{N_1\,dt} = b_1\exp(-c_1N_1-c_2N_2)-\mu_1\,,$$
$$\frac{d\log N_2}{dt} = \frac{dN_2}{N_2\,dt} = b_2\exp(-c_1N_1-c_2N_2)-\mu_2\,. \tag{8.14}$$

Multiplying the first equation by b_2 and the second by b_1 and then subtracting the second from the first gives

$$\frac{d\log\left(N_1^{b_2}N_2^{-b_1}\right)}{dt} = b_1\mu_2-b_2\mu_1\,. \tag{8.15}$$

Integrating (8.15) from 0 to t and exponentiating we find

$$\frac{N_1^{b_2}(t)}{N_2^{b_1}(t)} = \left(\frac{N_1^{b_2}(0)}{N_2^{b_1}(0)}\right)\exp[(b_1\mu_2-b_2\mu_1)t]\,. \tag{8.16}$$

Hence the ratio on the left hand side of (8.16) decays to zero when $b_1\mu_2<b_2\mu_1$ or grows to infinity when $b_1\mu_2>b_2\mu_1$.

With indiscriminate cannibalism species coexistence is impossible: species 1 will survive or species 2 will survive but never both. Furthermore, it is the ratio b_i/μ_i that determines the outcome. As an example, let $b_1/\mu_1>b_2/\mu_2$. From (8.16) it is clear that species 1 will be the sole long term survivor. If $b_1/\mu_1<b_2/\mu_2$ species 2 will persist. Note that the proposed biological mechanism of the interaction between these two species – cannibalism – is of no consequence in determining the outcome. The model places c_1 and c_2 in the background; the critical entity is the ratio b_i/μ_i. The conclusions for this special case are consistent with (8.11).

A similar argument was used by Leslie (1957) with a pair of logistic differential equations in his study of Gause's experiments with the protozoa *Paramecium aurelia* and *P. caudatum.* Leslie also pointed out that this special case was discussed by Volterra (1926) in his original memoir (see Scudo and Ziegler 1978, p. 73). The model was also considered in detail by Lotka (1932).

We can interpret the model geometrically by plotting the set of points which solves $f_1(N_1,N_2)=0$ and the set which solves $f_2(N_1,N_2)=0$ in (8.2). These zero isoclines are given by

$$L_1(N_1,N_2) = c_1N_1 + c_2N_2 = \log\left(\frac{b_1}{\mu_1}\right),$$
$$L_2(N_1,N_2) = c_1N_1 + c_2N_2 = \log\left(\frac{b_2}{\mu_2}\right), \tag{8.17}$$

on the N_1,N_2-plane (Fig. 8.2). The lines are always parallel and the species with the isocline closest to the origin always goes extinct.

We now consider the experimental results of Park (1954). Under hot-wet (HW) culture conditions *T. castaneum* (species #1) prevailed over *T. confusum* (species #2) and under cold-dry (CD) conditions *T. castaneum* failed to grow in single species cultures. Under the assumption of indiscriminate cannibalism, we would conclude from the model

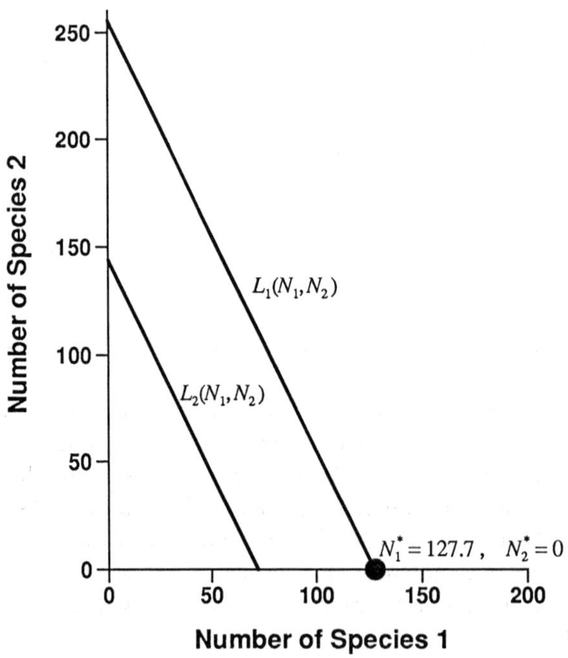

Fig. 8.2. Zero growth isoclines for indiscriminate cannibalism. The isoclines for species 1, $L_1(N_1,N_2)$, and species 2, $L_2(N_1,N_2)$, were computed using (8.17) with $c_1=0.004$, $b_1=0.5$, $\mu_1=0.3$, $c_2=0.002$, $b_2=0.4$ and $\mu_2=0.3$. The filled circle marks the globally stable equilibrium.

that $b_1/\mu_1 > b_2/\mu_2$ under HW conditions; i.e., the isocline for *T. castaneum* lies above the isocline for *T. confusum*. In the CD climate, $b_1 < \mu_1$ and $b_2 > \mu_2$; the rate of reproduction falls below the mortality rate for *T. castaneum* but not for *T. confusum*. Of course, *T. confusum* wins in 100% of the mixed species cultures.

The outcome of competition was indeterminate for the four intermediate culture conditions. Clearly, a deterministic mathematical model is incapable of explaining these observations. How, then, are these data to be explained? The answer is obvious: stochasticity is present. Recognizing this fact, Neyman et al. (1956, p. 69) wrote, "...the criterion [deciding the outcome] will vary from one replicate of the experiment to the next and that, therefore, depending on the chance variation, the winning of the competition of one group of *T. confusum*, opposed by a particular group of *T. castaneum*, may well be observed along with the opposite result of the competition of two other groups of beetles." However, determining the biological basis of this "chance variation" is not a simple task. We address this question in §8.2.5. First we must consider another possible relationship among c_{ij}'s: the case of discriminate cannibalism.

8.2.4 Discriminate Cannibalism

The association of the empirical fan-shaped outcome pattern (Fig. 8.1) with a mathematical model based on discriminate cannibalism was hinted at by Bartlett (1957) and was stated explicitly by Leslie (1958), Leslie and Gower (1958), Bartlett (1960a,b), Barnett (1962), Leslie (1962), Park et al. (1964), Leslie et al. (1968), Mertz (1972), Mertz et al. (1976), and Mangel and Ludwig (1977). These authors all considered models which have two features in common. First, a coexistence equilibrium exists but it is globally unstable. Second, all of the models are stochastic.

It is of historical interest to note that the seminal Neyman et al. (1956) paper on indiscriminate cannibalism was read at a meeting in July of 1955. In a section added to the paper in September of that year these authors pointed out that a coexistence equilibrium is possible and that it may be stable or unstable. Nevertheless, their explanation of the data did not invoke the hypothesis of an unstable coexistence equilibrium (see Neyman et al. 1956, p. 69-71; also see Chiang 1954, p. 200).

Empirical evidence for discriminate cannibalism also exists. Assays of cannibalism rates where immatures of each species are exposed to adults of one or the other species often reveals that interspecific cannibalism rates exceed intraspecific cannibalism rates (Park et al. 1965, Park et al. 1968, Ryan et al. 1970). However, the mechanism of species recognition is not known.

If we accept the fact that adults and larvae of one or both species selectively eat the eggs and pupae of their competitor, then $c_{11}c_{22} > c_{12}c_{21}$. Reversing the inequalities in (8.9) gives the necessary and sufficient conditions for an unstable coexistence equilibrium. In this case, the isoclines are given by

$$L_1(N_1,N_2) = c_{11}N_1 + c_{12}N_2 = \log\left(\frac{b_1}{\mu_1}\right),$$
$$L_2(N_1,N_2) = c_{21}N_1 + c_{22}N_2 = \log\left(\frac{b_2}{\mu_2}\right). \tag{8.18}$$

These isoclines are drawn in Fig. 8.3. The unstable equilibrium lies at the intersection of the isoclines. The dashed line is given by $N_1 = kN_2$, where k is computed using (8.10). This line separates the domains of attraction for the two locally stable equilibria. A mixed species culture is expected to be drawn into one of the two locally stable equilibria depending on the initial number of beetles. Initial numbers above the dashed line result in species 2 excluding species 1; the opposite outcome is expected below the dashed line. Replicates with different initial numbers of animals may or may not give rise to the same outcome; however, those replicates with the same number of animals always have the same outcome. That is the quandary we face with Park's data: replicates with the same initial conditions did not give the same outcome.

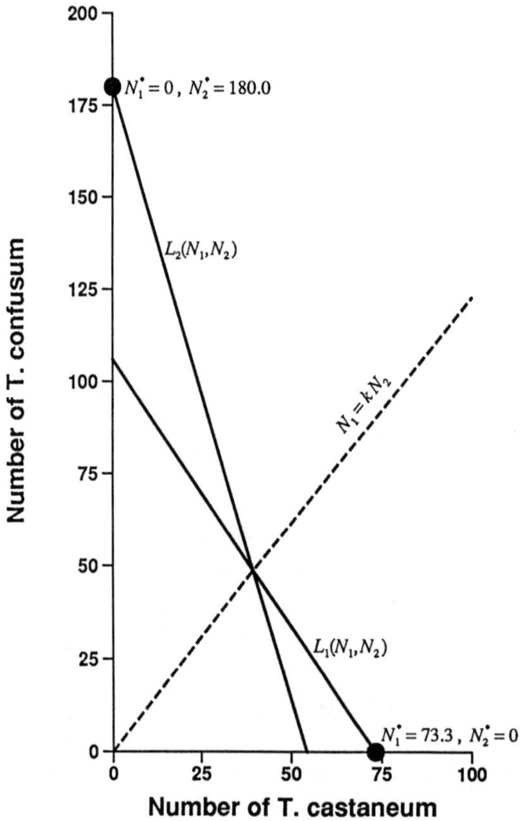

Fig. 8.3. Zero growth isoclines for discriminate cannibalism. The isoclines for species 1, $L_1(N_1,N_2)$, and species 2, $L_2(N_1,N_2)$, were computed using (8.18) with the parameter estimates from Table 8.1 The filled circles mark the two locally stable equilibria. The intersection of the isoclines marks an unstable equilibrium. The dashed line given by $N_1 = kN_2$ was computed using (8.10). This line connects the unstable equilibrium with the origin and separates the domains of attraction for the two stable equilibria.

As an example, let us consider competition between two specific genetic strains: *T. castaneum* cIV and *T. confusum* bI. A great deal of information is available on these two strains: the geographic origins, fecundity, fertility, rates of development and adult longevity (Park et al. 1961); the long term growth of single and mixed species cultures (Park et al. 1964); cannibalism by adults and larvae (Park et al. 1965); the mortality patterns (Mertz et al. 1965); the cannibalism of pupae by adults (Park et al. 1968); a stochastic model of adult cannibalism of pupae (Mertz and Davies 1968); age distribution and population size (Mertz 1969); larval cannibalism on eggs (Mertz and Robertson 1970); discriminate predation (Ryan et al. 1970); computer simulation of single and mixed species cultures (Taylor 1971); predation by adults on pupae and predation by larvae on eggs (Young 1970); sensory receptors of larvae (Ryan and Behan 1973); cannibalism of eggs by larvae (Park et al. 1974); factors affecting pupal cannibalism by adults (Park et al. 1974); the emigration of beetles and the influence of age (Ziegler 1976); density related mortality (Daly and Ryan 1983); and stimuli governing intraspecific egg predation (Craig 1986). We use only a small fraction of the available biological information for these two strains.

Park et al. (1964, p. 146) reported that in one replicate containing strains cIV and bI, "a temporary balance, or pseudo-stationary state, had been set up." Later, in a 1968 study, strains cIV-a and bI-a were maintained as a mixed species culture for over 2.5 years (Leslie et al. 1968, Table 7). Could these be cases of stable coexistence? The rates of cannibalism of adults on pupae provided by Park et al. (1965, Table 6) do not support this idea. These rates (Table 8.1) satisfy the condition $c_{12}c_{21} > c_{11}c_{22}$ for competitive exclusion (also Park et al. 1968). The product of the intraspecific cannibalism rates are less than the interspecific rates.

Is the coexistence equilibrium given by equation (8.3) unstable? The necessary and sufficient conditions are

$$\left(\frac{b_2}{\mu_2}\right)^{c_{12}/c_{22}} > \left(\frac{b_1}{\mu_1}\right) > \left(\frac{b_2}{\mu_2}\right)^{c_{11}/c_{21}}. \tag{8.19}$$

To obtain an estimate of the ratio b_i/μ_i, recall that for species i, the single species equilibrium density is given by

Table 8.1. Cannibalism rates, population densities, and recruitment-mortality ratios for the genetic strains bI-a and cIV-a.

Genetic strain	Label, i	Cannibalism rates, c_{ij}		Single-species equilibrium, N_i^o	Recruitment-mortality ratio, b_i/μ_i
cIV-a	1	$c_{11}=0.00391$	$c_{21}=0.00603$	73.3	1.33
bI-a	2	$c_{22}=0.00182$	$c_{12}=0.00270$	180.0	1.39

Cannibalism rates from Park et al. (1965, Table 6); single-species equilibrium densities from gamma distributions (Figs 4.11, 4.12).

$$N_i^o = \frac{\log\left(\frac{b_i}{\mu_i}\right)}{c_{ii}}, \tag{8.20}$$

so that

$$\left(\frac{b_i}{\mu_i}\right) = \exp(c_{ii}N_i^o)\,. \tag{8.21}$$

To solve (8.21), we set N_i^o equal to the mean of the observed 1968 gamma densities (Figs 4.11b, 4.12b) and used the c_{ii} values from Park et al. (1965). These means and the estimated ratios are given in Table 8.1. We find that condition (8.19) is satisfied. The coexistence equilibrium of $N_1^*=39.61$ adults for strain cIV-a and $N_2^*=48.74$ adults for bI-a is globally unstable. In fact, the isoclines in Fig. 8.3 are based on these estimates. The two species will not, in the long run, coexist. Individual replicates will finally be drawn into one of the two locally stable equilibria at ($N_1^o=0$, $N_2^o=180.0$) or ($N_1^o=73.3$, $N_2^o=0$).

The hypothesis of an unstable coexistence equilibrium was tested by Leslie et al. (1968). In their experiment, initiated with 25 different combinations of bI-a and cIV-a adult densities, 22 of the cultures, as predicted, converged to the locally stable equilibrium which included just strain cIV-a. The remaining 3 cultures, which were the only combinations that gave probabilities of greater than 50% of strain bI-a surviving, did not converge to this equilibrium (Leslie et al. 1968, Table 7). In the replicates with the initial numbers of adults of (4 cIV-a, 64 bI-a) and (8 cIV-a, 64bI-a) the numbers of both increased at the outset of the experiment and then, consistent with the unstable coexistence equilibrium, the cultures were drawn into the locally stable equilibrium in which bI-a was the sole survivor. The culture started with 4 cIV-a and 16 bI-a adults followed a nearly identical pattern with the latter two replicates but was not drawn into this local attractor. The culture after one year fell away from this equilibrium and lost large numbers of bI-a adults. When the experiment was ended after 960 days of continuous culture it appeared headed toward the equilibrium in which cIV-a was the sole survivor.

8.2.5 Stochastic Effects: Demographic or Genetic?

Speculation on the role of a stochastic effect was made because of one key feature of the 1954 experiments: the initial population size consisted of just 4 animals (2 males and 2 females) of each species. There has arisen two ideas concerning the biology of this small population size effect. On the one hand, the effects were attributed to the fact that births and/or deaths are, themselves, probabilistic events. When a population is started with just two females this so called "demographic stochasticity" (see Chapter 5) could result in variation among the replicate cultures. Individuals with perhaps the same genotype simply laid more eggs due to happenstance. Genetic differences, per se, were not specifically invoked because it was not deemed necessary to do so. This was the interpretation of P. H. Leslie, M. S. Bartlett, J. C. Gower, V. D. Barnett, D. B. Mertz and T. Park. On the other hand, evidence on genetic variation in *Tribolium* populations was accumulating so that Lerner and Ho (1961), Lerner and Dempster (1962), Dawson and Lerner (1962, 1966), and Dawson (1970) argued that indeterminate competition was a small population size effect due to sampling differences among the genotypes represented in the initial cultures.

Let us first consider demographic stochasticity in the context of indiscriminate cannibalism (Fig. 8.2). Random variation in the number of eggs laid, developmental rates, cannibalism, mortality, etc. could result in one or the other species eventually "drifting" to extinction. This is most likely to happen when the ratio b_1/μ_1 is close in magnitude to the ratio b_2/μ_2. In this case, the isoclines in Fig. 8.2 are almost superimposed and the numbers of each species undergo a "random walk" along the "common" isocline. The actual probability of one or the other species going extinct depends on which isocline is higher and on the separation between them. In the HW climate, *T. castaneum*'s isocline lies well above *T. confusum*'s isocline and *T. castaneum* wins in 100% of the replicates. In temperate-wet (TW) conditions, *T. castaneum* still has the advantage, but the isoclines are sufficiently close to allow *T. confusum* to win in 14% of the replicates. In the cold-wet (CW), hot-dry (HD), and temperate-dry (TD) climates, *T. confusum*'s isocline lies above *T. castaneum*'s, and, consequently, *T. confusum* wins in 71%, 90% and 87% of the replicates, respectively. This interpretation implicates variation in population growth rates as the major cause of indiscriminate competition.

Another possibility is genetic variation. Since each population was initiated with four adults from each species, random genetic sampling may have resulted in different replicates having different ratio's of b_i/μ_i. Under this hypothesis, *T. castaneum* still does better in the TW climate, but in 14% of the replicates $b_1/\mu_1 < b_2/\mu_2$ due to random sampling of genotypes. A similar argument holds for the remaining climates. For each climate, one can envision a genetically-based probability distribution for b_1/μ_1 and b_2/μ_2 with different *population mean values*. Random samples from these stock populations may have *sample mean values* which differ from the population means. Furthermore, although the *T. castaneum* population mean may be slightly larger than the *T. confusum* population mean, the *T. confusum* sample mean could exceed the *T. castaneum* due to statistical variation. In this scenario, the outcome of competition is determined at the *outset* of the experiment, and the probability of one or the other species winning is determined by the difference in the population mean values of b_i/μ_i for any given climate. Random population growth is not the major cause of indeterminate competition.

Now let us consider discriminate cannibalism (Fig. 8.3). The first insights into the solution of a stochastic model with an unstable stationary state were obtained by the numerical work of Leslie and Gower (1958). Their discrete time model is

$$N_i(t+1) = f_i(N_1, N_2) + Z_i(t+1), \qquad i = 1, 2, \tag{8.22}$$

where the first term is given by the logistic growth equation (see Chapter 1) and $Z_i(t+1)$ is a random normal variable with zero mean and specified variance. Bartlett (1957, 1960a,b) considered a continuous time version of (8.19) which was also studied by Barnett

(1962). The role of the stochasticity in the mathematical scheme is to account for Park's observation that all mixed species cultures with the same initial conditions do not yield the same final outcome. The biological nature of the randomness is not specified.

With the unstable coexistence equilibrium hypothesis, if cultures are started too close to one or the other of the two locally stable equilibria, then the outcome, even in the stochastic model, is unequivocal. The outcome is most uncertain in the region near the dashed line in Fig. 8.3, especially where the number of beetles is very small (Barnett 1962). This forecast is consistent with the observations. In Park's (1948) report on competition, the only condition that yielded what he then called "reversals" was when the experiment was started with 4 males and 4 females (2 of each species) in 8 grams of media. Cultures started with 10 pairs of each species (40 total animals in 40 grams of medium) or 20 pairs of each species (80 total animals in 80 grams of medium) displayed no reversals (Park 1948, Table 14). Perhaps this was one reason why Park chose to begin all of his 1954 experiments with 8 beetles?

Unfortunately, the observation of indeterminate competition with small numbers of adults can be used to support either the hypothesis of demographic stochasticity or the hypothesis of genetic founder effect. For example, the dashed line in Fig. 8.3, which is based on real data, is given by the equation $N_1 = kN_2$, with $k = 1.23$. Hence when there are four adults of both *T. castaneum* and *T. confusum*, the model predicts *T. castaneum* winning in competition. But if one of the *T. castaneum* adults accidently dies after being introduced into the culture, then four *T. confusum* adults are sufficient to let it win in competition. Since adults come in integral units, the effects of demographic stochasticity are more pronounced for low initial numbers. But this is exactly the same condition that leads to a genetic founder effect. In the latter case, random sampling of individuals leads to different ratios of b_i/μ_i in different cultures. From (8.18) we see that the isoclines themselves move closer to or further from the origin (provided the c_{ij}'s are unchanged). This causes the value of k to change. If $k > 1$, then *T. castaneum* is predicted to win in Park's experiments; if $k < 1$, *T. confusum* wins. Variation in the c_{ij}'s only adds to the uncertainty.

Using a clever experimental design, Mertz et al. (1976) tried to determine experimentally the relative importance of demographic stochasticity versus genetic effects in indeterminate competition. To control for a genetic effect, virgin adults of *T. castaneum* were mated in cultures with 1, 2, 4, or 125 pairs. The offspring of each these cultures were used to initiate one mixed species culture. All the *T. confusum* cultures were husbanded with 125 pairs of beetles. The demographic treatments consisted of three sets of initial population densities. The numbers of *T. confusum* to *T. castaneum* were either 40:10, 10:10, or 10:40. The four genetic treatments and three demographic treatments formed a two-way factorial design with 20 replicates for each treatment combination. After 780 days of culture, the outcome of the competition was ascertained for all but two of the populations. Consistent with the hypothesis of a unstable coexistence equilibrium, the initial population numbers were the main determinant of the competitive outcome. *T. confusum* prevailed in the majority of cases where it had an initial advantage; the same was true of *T. castaneum*. In these cases, the probability of winning was not greatly affected by the genetic treatments. However, consistent with the genetic hypothesis, the probabilities were most variable across genetic treatments when the initial numbers of both species were equal. Mertz et al. (1976, p. 1370) "conclude that the competitive indeterminancy exhibited between [*T. confusum*] and [*T. castaneum*] in this experiment is compatible with levels of demographic stochasticity predicted by stochastic competition theory. Classical genetic founder effect seems to have played an undetectably small part in determining the experimental outcomes. However, in confirmation of earlier reports, the genetic and/or ecological properties of the founder beetles can, and do, affect the identity of the winning species."

In our minds, the practical problem of experimentally distinguishing between "demographic stochasticity" and a "genetic founder effect" as the biological basis of the randomness is not completely resolved. Our interpretation is to simply extend the discriminate cannibalism hypothesis of Leslie (1962), which at the time it was put forth did not have a genetic argument, to include the genetic hypothesis of Lerner and Dempster (1962). The recognition of genetic heterogeneity does not negate discriminate cannibalism nor vice-versa. Together these ideas form a plausible explanation of the experimental observations.

8.3 An Inhibiting Species with a Cannibalistic Species

8.3.1 *T. castaneum* versus *T. brevicornis*

Not all species in the genus *Tribolium* regulate their numbers *solely* by cannibalism. For *T. castaneum* and *T. confusum*, egg-eating and pupa-eating by larvae and adults are the primary means of maintaining population size, although "larval arrest" or delays in larval development have been reported in *T. castaneum* (Botella and Mensua 1986). But *T. brevicornis*, in addition to the usual egg-eating behavior of larvae and adults, has another density regulating mechanism. The adults of this species are able to inhibit larval metamorphosis. Recruitment into the adult class is prevented by adults stopping the larval to pupal transition (Jillson and Costantino 1980).

Inhibition of pupation due to crowding has been demonstrated in several tenebrionid beetles including the large Central American beetle *Zophobas rugipies* Kirsch, *Tenebrio molitor* L., *Argoporis alutacea* Csy., *Neatus tenebrioides* Beauv., *Eleodesarmata* LeC. and in *Amphidora littoralis* Esch. (Tschinkel and Willson 1976). The primary factor causing the inhibition is the mechanical stimulation resulting from larval-larval contacts. In *T. brevicornis*, the inhibition also appears to be the consequence of mechanical stimulation but not so much by larvae as by adult-larva contacts (Jillson and Costantino 1980). What is the physiological bases of this response? Tschinkel and Willson (1976, p. 143) state, "We now propose that in tenebrionid species whose larvae are affected by mechanical stimulation, a heightened internal secretion of juvenile hormone (JH) is induced by the nervous impulses arising from external irritation. Isolation (cessation of stimulation) would then result in a reduction of the JH level, eventually allowing pupation." The Tschinkel-Willson proposal is consistent with the data on *T. brevicornis*.

What is the role of this adult-larval interaction in mixed species cultures? Jillson and Costantino (1980) examined competition between *T. brevicornis* and *T. castaneum*. Once again, competitive exclusion was the rule. In the presence of *T. castaneum* adults, *T. brevicornis* larvae were unable to pupate. *T. brevicornis* was eliminated in every mixed species culture (Jillson and Costantino 1980, Table 1).

8.3.2 Parameter Estimates and Model Predictions

The *brevicornis-castaneum* data can be explained in terms of the adult-immature interaction equations (8.2). The c_{ij}'s, formerly viewed as rates of cannibalism, must be interpreted more generally as the rates at which adults prevent the recruitment of immatures. This can occur through cannibalism, developmental inhibition, or both.

The data of Jillson and Costantino (1980) can be used to obtain estimates of the c_{ij}'s. To assay intra- and interspecific effects, they established vials in which 26 larvae were confined with various numbers of adults. All four possible combinations of larvae and adults were used. For example, if we label *T. castaneum* as species 1 and *T. brevicornis* as species 2, then c_{12} was estimated from the vials containing *T. castaneum* larvae and *T. brevicornis* adults. The rates and their standard errors are given in Table 8.2. We find $c_{21} \gg c_{11} > c_{22} > c_{22}$. *T. castaneum* adults exploit the adult-larval interaction employed by *T. brevicornis* for the regulation of its population size to such an extent that the larval to pupal transition is virtually stopped. On the other hand, *T. brevicornis* adults cannot reciprocate and inhibit larval metamorphosis in *T. castaneum* because this adult-larval interaction is not present.

We can also predict the outcome of competition. The interaction rates fail to satisfy the necessary condition $c_{11}c_{22} > c_{12}c_{21}$ for coexistence (Table 8.2). The product of the intraspecific rates is less than the product of the interspecific rates. Furthermore, using single species adult densities and the relationship (8.21) we can estimate the ratios b_1/μ_1 and b_2/μ_2 (Table 8.2). We find

Table 8.2. Recruitment inhibition rates, population densities, and recruitment-mortality ratios for the species *T. castaneum* and *T. brevicornis*

Species	Label, i	Recruitment Inhibition Rates, c_{ij} ($\pm$s.e.)		Single-species equilibrium, N_i^o	Recruitment-mortality ratio, b_i/μ_i
T. castaneum	1	$c_{11}=0.01288\pm0.00059$	$c_{21}=0.04848\pm0.00098$	154.2	7.287
T. brevicornis	2	$c_{22}=0.00828\pm0.00049$	$c_{12}=0.00455\pm0.00014$	207.4	5.569

Inhibition rates computed from Jillson and Costantino (1980, Table 2); single-species equilibrium densities from Desharnais and Costantino (1985, Fig. 3) for *T. castaneum* and Desharnais and Costantino (1982, Fig. 2) for *T. brevicornis*.

$$\left(\frac{b_1}{\mu_1}\right) > \left(\frac{b_2}{\mu_2}\right)^{c_{12}/c_{22}} > \left(\frac{b_2}{\mu_2}\right)^{c_{11}/c_{21}} . \tag{8.23}$$

In every case we expect *T. castaneum* to be the surviving species.

The relatively small magnitude of c_{12} in Table 8.2 has some interesting implications. If we let $c_{12}=0$ and hold constant the number of *T. brevicornis* adults, then from (8.3) and (8.20) we obtain $N_1^*=N_1^{\circ}$; the single and "mixed" species equilibria are the same. Jillson (1977, Table 15) has shown that *T. castaneum* adults reached the same equilibrium size in the presence of large numbers of *T. brevicornis* as when they were alone. Using the approximations $c_{12}=0$ and $N_1=N_1^*$ in (8.2), the rate of change in the numbers of *T. brevicornis* is given by

$$\frac{\mathrm{d}N_2}{\mathrm{d}t} = N_2\left[b_2\left(\frac{\mu_1}{b_1}\right)^{c_{21}/c_{11}}\right]\exp(-c_{22}N_2) - \mu_2 N_2 . \tag{8.24}$$

The term in square brackets can be considered the "effective" recruitment rate. Using the estimates of c_{11}, c_{21}, and b_1/μ_1 from Table 8.2, this rate equals $0.000567\,b_2$. Recruitment is effectively zero and as a first approximation we have

$$\frac{\mathrm{d}N_2}{\mathrm{d}t} \approx -\mu_2 N_2 \quad \Rightarrow \quad N_2(t) \approx N_2(0)\exp(-\mu_2 t) . \tag{8.25}$$

The number of *T. brevicornis* adults decays to zero at the adult mortality rate. This is exactly what Jillson and Costantino (1980) observed when they gave *T. brevicornis* a large advantage in initial numbers. Recruitment was zero and μ_2 was small. *T. brevicornis* moved slowly toward extinction.

The effects of inhibitory phenomena on the population dynamics of competing species have been discussed for several years (Crombie 1947, Slobodkin 1961, Miller 1967, Gill 1974). More recently, Rothstein (1979) introduced and discussed the evolution of inhibition competition as distinct from exploitation and interference competition. While we have not addressed the evolution of inhibitory traits, we can characterize the *T. castaneum*-*T. brevicornis* interaction as an example of inhibition competition. Inhibition by *T. brevicornis* adults does not appear to alter the absolute performance of *T. castaneum*; however, *T. castaneum* adults impose an extremely large loss upon *T. brevicornis* populations. Rothstein (1979) predicted that selection for inhibitory traits was favored only if the inhibiting species received a disproportionate share of the benefits during interspecific competition. This criterion seems to be satisfied in this *Tribolium* interaction.

8.4 Two Inhibiting Species

8.4.1 *T. madens* versus *T. brevicornis*

A competition experiment has also been conducted for the species *T. madens* and *T. brevicornis*. Single species cultures were initiated by placing 10 newly emerged adult beetles of each sex into a one-half pint bottle containing 20g of standard medium (95% unbleached wheat flour and 5% dried brewer's yeast by weight). The mixed-species cultures were started with 5 male and 5 female adult beetles of each species. Each single and mixed-species combination was replicated four times. The beetles were maintained in incubators in darkness at 33±1°C and 60±5% relative humidity. At regular intervals, the number of adults were counted in each culture. After counting, all living animals were returned to fresh medium.

The observed number of adults in the single-species cultures (Fig. 8.4a) initially increased and under our laboratory conditions each species appeared to be capable of maintaining a reproducing population indefinitely. In the mixed species cultures (Fig. 8.4b), the number of adults of each species initially increased and then maintained

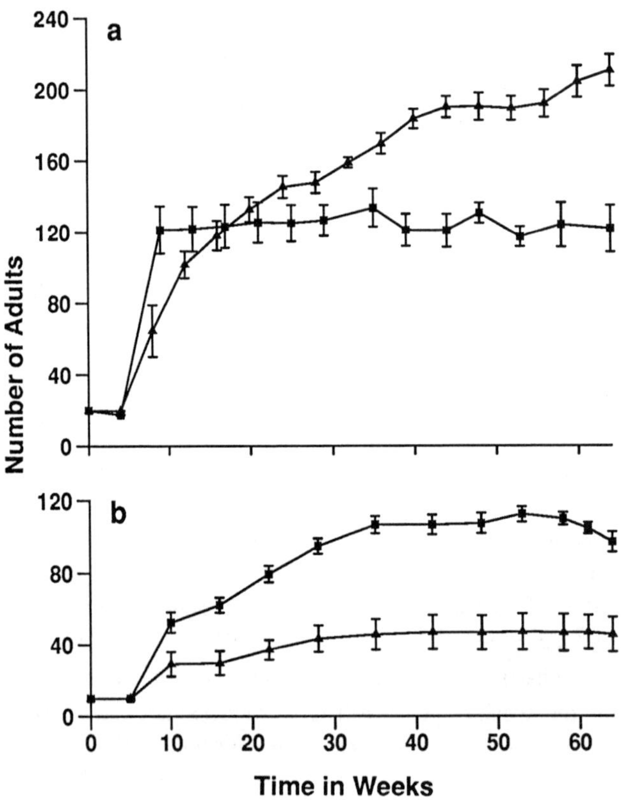

Fig. 8.4. Mean (±s.e.) number of adults of *T. madens* (squares) and *T. brevicornis* (triangles) in the (**a**) single species cultures and (**b**) mixed species cultures.

a relatively constant density until the end of this 64 week experiment. We emphasize that during the period of apparent equilibrium (approximately 7 generations), new adults were recruited by both species which offset losses due to mortality. The equilibrium number of adults for both species was smaller in the mixed cultures than in their single-species counterparts. However, the difference was much larger for *T. brevicornis* than for *T. madens*. Although these data are best viewed as a "pilot study," they do suggest, for the first time, the possibility of stable coexistence for two *Tribolium* species.

8.4.2 A Hypothesis

In the study of *T. madens* and *T. brevicornis*, the mechanism of the interaction is probably not due solely to cannibalism. As discussed in §8.3, there is good evidence that large larvae of *T. brevicornis* delay metamorphosis when adult densities are high (Jillson and Costantino 1980). The ability to delay pupation has also been demonstrated in *T. madens* (Howe 1962), but the effect of adult density has not been examined in this species. Since our data suggest coexistence, we hypothesize that the product of the intraspecific inhibitory effects exceed the interspecific effects, i.e. $c_{11}c_{22} > c_{21}c_{12}$. A comparison of the single and mixed-species cultures does reveal some relative magnitudes for these values. Writing the equilibrium densities for the single-species populations as N_i^o and the mixed-species populations as N_i^*, from (8.3) and (8.20) we have

$$\begin{aligned} \frac{c_{12}}{c_{11}} &= \frac{(N_1^o - N_1^*)}{N_2^*}, \\ \frac{c_{21}}{c_{22}} &= \frac{(N_2^o - N_2^*)}{N_1^*}. \end{aligned} \tag{8.25}$$

Using the mean numbers of adults for the last six census periods as estimates of the equilibrium densities and labelling *T. madens* as species 1 and *T. brevicornis* as species 2 we obtain $c_{12}/c_{11} = 0.32$, $c_{21}/c_{22} = 1.39$, and $(c_{12}c_{21})/(c_{11}c_{22}) = 0.45$. The effect of *T. brevicornis* adults on the larvae and pupae of *T. madens* is approximately one-third that of *T. madens* on itself. Similarly, the effect of *T. madens* adults on the larvae and pupae of *T. brevicornis* is 39% greater than the effect of *T. brevicornis* adults on their own potential recruits. Finally, the overall interspecific effects are roughly one-half the intraspecific effects. These conclusions are tentative and await verification using independent estimates of the c_{ij}'s for these species.

8.4.3 Developmental Inhibition

Is it biologically reasonable to expect that the condition $c_{11}c_{22} > c_{21}c_{12}$ will be satisfied when the species interaction is the inhibition of pupation? Is there an advantage of some sort for a species to delay the pupation of its own larvae? The timing of pupation may be critical to survival in the presence of other cannibalistic life stages. In the view of Tschinkel and Willson (1976, p. 143), "cannibalism may have provided the selective pressure for the delaying of pupation until the larva has found a secluded niche, free of tactile contact with other organisms. A lack of frass in the area might also be a helpful indication of seclusion."

If the mechanism of inhibition is tactile contact, then it is not clear why intraspecific effects should exceed interspecific effects. Since the adults of *T. brevicornis* are larger and more slow moving than *T. madens* adults, *T. brevicornis* larvae may have evolved to be more "sensitive" to the inhibitory mechanism. Likewise, *T. madens* larvae are less effected by *T. brevicornis* adults than adults of their own species. It may simply be the case that the effect of *T. brevicornis* on *T. madens* is so small as to allow coexistence, whereas the reciprocal effect of *T. madens* on *T. brevicornis* is not so large as to prevent coexistence. This is only speculation; further investigations are needed.

We have no data on the origin and evolution of intraspecific density regulating mechanisms such as cannibalism and developmental inhibition. An extensive review of this topic is given by Polis (1981). The flour beetle does provide an interesting and convenient experimental model for these types of inquiries.

8.5 Summary and Conclusions

In this chapter we placed the extensive experimental observations on two species *Tribolium* cultures in the context of an adult-immature interaction model. We looked at data on four species: *T. castaneum*, *T. confusum*, *T. brevicornis*, and *T. madens*. The mechanisms of density regulation are not the same among these species. The critical behavior in *T. castaneum* and *T. confusum* is cannibalism. On the other hand, the inhibition of pupation caused by the mechanical stimulation of adult-larva contacts is central in *T. brevicornis* and *T. madens*. Our thought is that flour beetle species interact primarily via the intraspecific density-regulating mechanism. So species interaction can be classified as being between two cannibalistic species, two inhibiting species, or a cannibalistic and inhibiting species.

We conclude that coexistence between two indiscriminate cannibalistic species is very unlikely. On the other hand, when cannibalism is discriminate, it has always been observed that interspecific predation rates exceed intraspecific predation. A coexistence equilibrium exists, but it is unstable. When a cannibalistic species like *T. castaneum* is placed together with an inhibitory species like *T. brevicornis*, the inhibitory species receives a double threat. Its larvae are "over-inhibited" by the smaller faster-moving cannibalistic adults and its pupae are vulnerable to cannibalism. This makes the necessary conditions for coexistence difficult to satisfy. Stable species coexistence in laboratory cultures of *Tribolium* appears to be most likely to occur between two inhibiting species such as *T. brevicornis* and *T. madens*. If any general conclusion is to be made, it is that one should consider the mechanisms of species interactions to understand the dynamics of competition.

We readily acknowledge that the two species adult recruitment model is a crude caricature of reality. The next logical step is to consider species interactions in the context of an age-structured population. For *Tribolium*, the density-dependent interactions would occur among life stages within and between species. While the model would be quite complex, it may also yield interesting dynamical behavior such as subcritical Hopf bifurcations and multiple attractors (see Chapter 3). The simple model considered in this chapter is but a beginning.

Chapter 9 Dynamics and the *Tribolium* Model

The study of population dynamics has opened the way to new explanations of data, to new procedures for data analyses, and will surely lead to a new phase of research in which experiments are focused directly on phenomena such as stable equilibria, bifurcations, multiple attractors, limit cycles, life not lived, and stationary distributions. In this final chapter we speculate on the emerging discipline of *population dynamics* and the role of *Tribolium* as an experimental animal model.

9.1 Demography

9.1.1 Exploring Parameter Space

Recent research with age-structured population models has revealed a fascinating variety of potential dynamics including stable cycles, sub- and supercritical bifurcations, multiple attractors, and chaos. While the theoretical work accumulates at a dizzying pace, there is little experimental work being done to investigate these phenomena. The results reported in Chapters 2 and 3 barely begin to address these ideas. More experimental research is needed which is focused on the qualitative features of population dynamics.

The *Tribolium* experimental model can play an important role in the study of demographic dynamics. The flour beetle has been used in ecological research for over sixty years. Much basic information on the effects of environmental conditions on vital rates has been accumulated for several species. Many genetic strains have been identified and developed. A number of laboratory studies already exist which demonstrate demographic oscillations (Chapman 1928, 1933, Stanley 1932, Crombie 1946, Strawbridge 1953, Landahl 1955a, Lloyd 1965, 1968, Mertz 1969, Fujii 1975, Desharnais and Liu 1987). Through a careful choice of species, genetic strains, and environmental conditions, one could "sample" parameter space experimentally, obtaining populations which cover the entire spectrum of dynamical behavior. The possibility exists for "nudging" populations across boundaries within parameter space by changing the culture temperature, humidity, and media composition or through the introduction of genetic mutants. This would allow one to document experimentally the transition from, say, a stable equilibrium to a stable limit cycle. If multiple attractors are predicted (Chapter 3), then a variety of initial conditions could be sampled to seek out the attractors. If multiple attractors are discovered, perturbations could be used to shift a population from one domain of attraction to another. The prospects are exciting.

This work will not be easy. As demonstrated in Chapter 4, stochastic forces play an important role in laboratory cultures of *Tribolium*. Furthermore, beetle populations are not genetically homogeneous or static, parameters may "evolve" in unwanted directions. However, these sorts of problems are likely to be important in natural populations as well; they are more easily studied with *Tribolium*. Finally, the largest obstacle will be the great deal of time and effort needed to carefully census many replicate populations. This work can be tedious and mundane. It is our feeling that the potential insights gained from these sorts of studies will make the effort worthwhile.

Of course, the goal of this research is not a thorough understanding of what flour beetles do when cultured in the laboratory. The point of this work is that these demographic phenomena may be an important component of the dynamics of natural populations. That is not to say that species interactions, genetic factors, environmental effects, etc. are unimportant, but they are not the whole story. By analogy, if one wished to understand the behavior of a feather in the wind, it would be foolish to ignore the effects of gravity. But one would not try to test a theory of gravity by dropping a feather in the wind. Instead, one might drop a feather in a vacuum in the laboratory. Similarly, *Tribolium* laboratory populations can be used to test theories of nonlinear demographic dynamics, which may aid our understanding of the dynamics of natural populations.

9.1.2 Strange Attractors and Chaos

Much as geneticists have learned from the study of mutants, perhaps it will be of value to examine exceptional dynamic behavior as well as the more common. For example, nonlinear growth models can undergo a transition from predictable behavior to the unpredictable behavior termed "chaos." The high reproductive rates and the nonlinearities in *Tribolium* make this an ideal organism to study this phenomenon. In the same way that Shaw's (1984) "dripping faucet" is an example of a physical system capable of chaotic transition, the flour beetle could serve as an example in population biology.

A particularly interesting strange attractor occurs for the two age-class *Tribolium* matrix model (1.22). One million iterations of this model are displayed in Fig. 9.1 for parameter values in the range of chaotic behavior. The swirling pattern is quite complex and beautiful. To our knowledge, there is little theory to explain this attractor and, at the same time, there is no experimental work to confirm its existence.

Do *Tribolium* populations have the potential for displaying chaotic dynamics? To address this question we conducted a pilot study using the using the *sooty* genetic strain of *Tribolium castaneum*. Glass vials were filled with 20 g of standard media (95% wheat flour with 5% dried brewer's yeast) and covered with cotton gauze and a plastic cap with four small air holes. A fixed number of adults of 2-4 weeks of age were counted into each of these vials. Forty different densities were used: from 2 to 80 adults at intervals of 2 adults. Each adult density was replicated four times for a total of 160 vials. The vials were placed in a dark incubator at 34°C and 46% relative humidity. After 7 days of egg-laying, the adults were removed and recounted and the number of dead adults was noted. The vials were replaced in the incubator. The second generation was allowed to

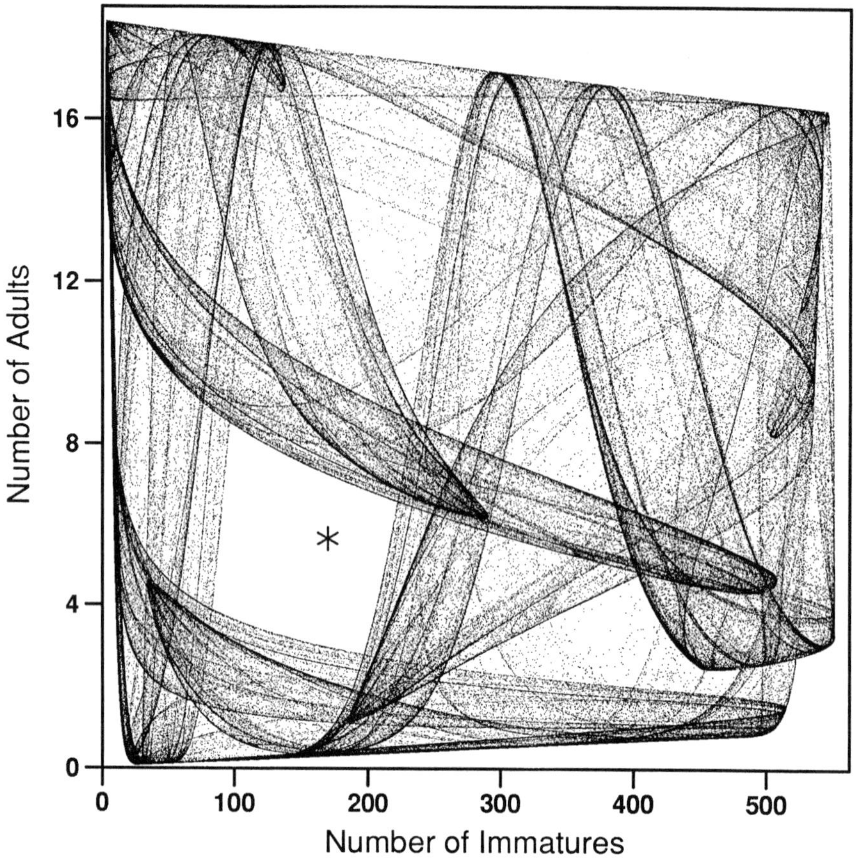

Fig. 9.1. A strange attractor for the two age class model (1.22) with $b=30$, $\mu=0.95$, $c_A=0.01$ and $c_I=0.02$. Each point is the result of one iteration of the model. A total of one million iterations were used to highlight the attractor. The unstable equilibrium point is indicated by the asterisk.

develop undisturbed for four weeks. The number of live and dead adults in the second generation was recorded. These experiments were designed to mimic a seasonally breeding organism with discrete generations.

The mean and standard error of the number of recovered adults are plotted in Fig. 9.2. The results strongly support the general model of a one-humped curve, with the number of recovered adults reaching a peak at about 26 first-generation adults and declining gradually to a low number at about 58 first-generation adults. There was a great deal of variation in the results as indicated by the large standard errors.

The question of chaotic behavior requires a specific mathematical function to describe the relationship between the adult numbers in two consecutive generations. In §1.1.2 we discussed a simple nonlinear model of the form

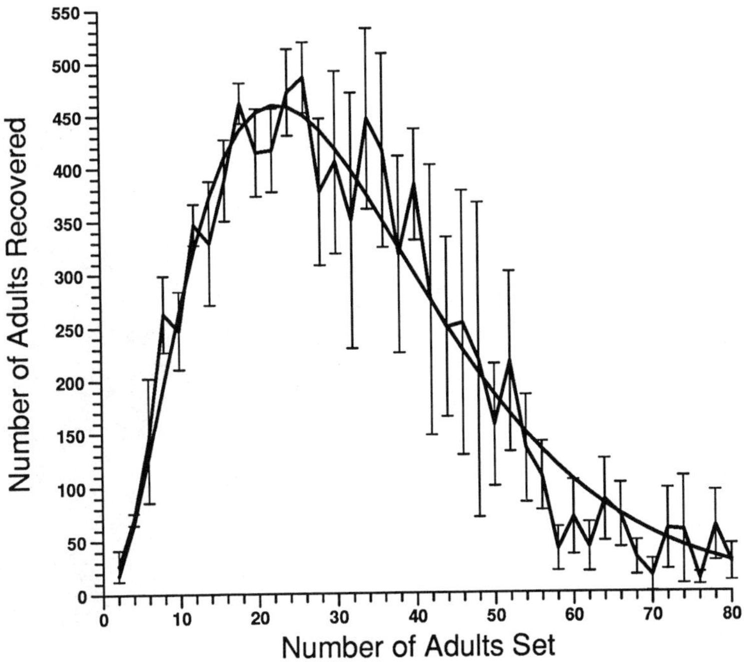

Fig. 9.2. The mean (±s.e.) of the number of *Tribolium* adults recovered *versus* the number of *Tribolium* adults set. The smooth curve represents the function (9.2) with $b = 4.74$, $\alpha = 2.16$, and $c = 0.0957$. These parameter values were obtained using the method of least squares.

$$N' = bN\exp(-cN)\,, \tag{9.1}$$

which has been associated often with *Tribolium*. N is the number of adults in the current generation, N' is the number in the next generation, b is the *per capita* rate of reproduction, and c is a parameter that describes the rate of decrease in reproduction with crowding. We tried fitting (9.1) to the data in Fig. 9.2 using the method of least squares. Unfortunately, the function did not describe the data very well. We then tried the following modification of (9.1):

$$N' = bN^{\alpha}\exp(-cN)\,. \tag{9.2}$$

Using the method of least squares we estimated the parameters b, α, and c. The estimates and their standard errors are $b = 4.74 \pm 2.87$, $\alpha = 2.16 \pm 0.27$, and $c = 0.0957 \pm 0.010$. This fitted function is represented by the smooth curve in Fig. 9.2 which fits the data well.

A bifurcation diagram was generated using the function (9.2) with the parameter b as the bifurcation parameter. This diagram is constructed by choosing a value of b and any initial number of adults. The initial population size N is substituted into the model (9.2) to obtain the population size in the next generation. This new population size is

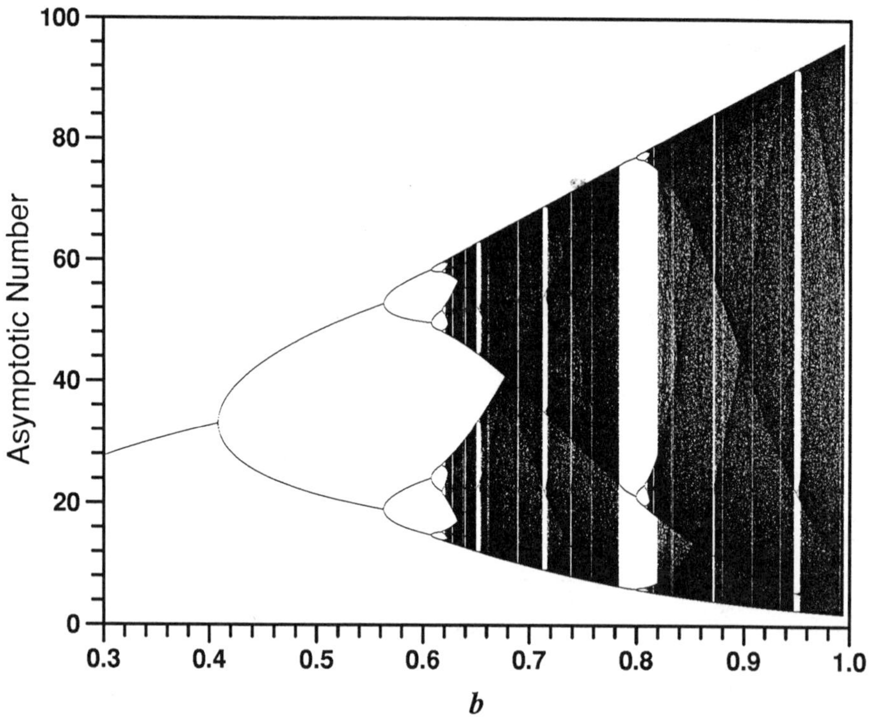

Fig. 9.3. The bifurcation diagram for the *Tribolium* model (9.2) with $\alpha = 2.16$ and $c = 0.0957$. Values of the parameter b were plotted along the abscissa and the asymptotic values for N were plotted along the ordinate.

again substituted into the model. This iteration process is repeated a large number of times. Eventually, the population size approaches some "attractor." For example, N may approach a constant value (stable equilibrium) or a two-point cycle. However, if the dynamics are chaotic, N will never settle down to a repeatable pattern. A bifurcation portrait is a representation of the attractor as it changes with changes in the parameter b. Figure 9.3 is the bifurcation portrait for the model (9.2). The abscissa represents the parameter b and the ordinate is the asymptotic number of adults. We began with $b = 0.3$ and $N = 10$ and iterated the model 2000 times. We then plotted the next 1000 iterations along the ordinate. In this case the attractor was a stable equilibrium point which yields a single point on the graph. We then increased b slightly and repeated the entire process. This yielded another stable equilibrium. Proceeding in this fashion, the series of stable equilibria form a line as b is increased. At the point in Fig. 9.3 where the line "bifurcates," the equilibrium point becomes unstable and gives way to a stable two-point cycle. This two-point cycle persists for a range of b-values until it bifurcates again and yields a stable four point cycle (Fig. 9.3). The bifurcations continue for geometrically smaller increases in b until the model becomes chaotic. Chaos is represented by dark bands in Fig. 9.3

because the sequence of population numbers is aperiodic. Within the chaotic regions are regions of order which appear as light bands. Within these orderly regions the cascade of finite point cycles with period doublings occurs anew.

The range of values for b in Fig. 9.3 is from 0.3 to 1.0. However, the estimated value for the *Tribolium* data is $b = 4.74$. This places the curve in Fig. 9.2 well into the chaotic region. While this observation opens the possibility for experimentally observing chaotic population dynamics in the laboratory, it does not prove that populations will exhibit chaos. This would require a long term experiment in which *Tribolium* populations are cultured over many discrete generations.

There are several possible reasons why chaos may be difficult to obtain under the present experimental conditions. The most obvious problem is extinction. Beetles come in discrete numbers while the models (9.1) and (9.2) treat population size as a continuous variable. In fact the parameter values estimated from Fig. 9.2 lead to dynamics where the population size goes to zero. In a real laboratory population, this would most likely result in no adults to seed the next generation. It is also obvious from Fig. 9.2 that there is a great deal of variation in the results. In a population experiment, this variation would be superimposed upon the chaotic dynamics. Distinguishing between randomness and chaos might be very difficult if the random component is large. Finally, chaotic population models assume that the parameter values themselves remain constant. In fact, a small alteration in the parameters can sometimes move the system from chaos to a fixed point cycle. In living populations the parameter values are controlled by environmental conditions, such as temperature and humidity, and by the genetic composition of the population. Both of these factors may change throughout the course of the experiment, making the chaotic attractor a "moving target." Despite the potential problems the possibility of observing chaotic population dynamics in the laboratory is intriguing.

Perhaps the most significant aspect of the experiment results is the revealed potential for additional experiments. By decreasing temperature, we can decrease the reproductive rate of flour beetles. This gives us the possibility of "sampling" the bifurcation diagram in Fig. 9.3 by culturing populations at different temperatures in different incubators. It might be possible to detect the first few period doublings and the onset of chaos. Again, one must deal with random variation. But the idea of doing experimental studies of population chaos is very exciting. Our pilot study shows that the reproductive potential and crowding effects of *Tribolium* make this a possibility.

9.1.3 Fast *versus* Slow Time Scales

We discussed in Chapter 3 the utility of viewing the demographic dynamics of *Tribolium* as occurring on two different time scales: a "fast" time scale for interactions among the immatures and a "slow" time scale for adult numbers. With this paradigm we considered the egg-larval submodel of Hastings (1987) and Hastings and Costantino (1987) as a fast-time scale phenomenon. In Chapter 4 we introduced a simple nonlinear differential equation for adult numbers that was based on the idea of a slow time scale. This separation of time scales allowed us to study simpler more tractable models and derive results and predictions which would have been impossible with the full nonlinear demographic equations. Can one integrate the results from these time scales in a simple and meaningful way without resorting to the complexity of a full nonlinear age-structured model?

One possible approach might be to modify the recruitment parameter in the equation for adult numbers to reflect the egg-larval oscillations. For example, we could consider the following modification of (4.1):

$$\frac{\mathrm{d}N(t)}{\mathrm{d}t} = N(t)\left(b(t)\exp[-c\,N(t)] - \mu\right), \tag{9.3}$$

where, as before, c is the rate at which adults prevent recruitment and μ is the adult mortality rate. The rate at which potential adult recruits are produced is given by $b(t)$ which is now a function of time. This function could be designed to approximate the egg-larval limit cycle. One simple candidate based on a sinusoidal input is

$$b(t) = b_0\left[1 - \cos\left(\frac{2\pi t}{D}\right)\right]. \tag{9.4}$$

The coefficient b_0 sets the amplitude of the oscillations and the parameter D controls the periodicity. Appropriate choices for b_0 and D can be made using results from the egg-larval submodel.

Models of the form (9.3) might be used to address a number of questions. How do the huge oscillations in immatures lead to "dampened" oscillations in the adult age class? What happens when stochasticity is superimposed upon the oscillations in adult number? Can the oscillations in adult number drive oscillations in allele frequencies? Can concepts like the life not lived (Chapter 7) be worked out for oscillating population densities? Simple approximations like (9.3) are probably the best way to begin to investigate these topics.

9.2 Stochasticity

9.2.1 Parameters of the Gamma Distribution

All of the stochastic models that we have considered lead to a single central idea for the experimentalist: *the probability distribution of population numbers*. These distributions are described by a family of density functions, such as the gamma, whose variation in shape and size depends on a small number of parameters. Through the use of a stochastic model, the parameters of the probability distribution are associated with the parameters of population growth. Might we learn something of the population dynamics from a comparison of the parameters of the stationary distribution?

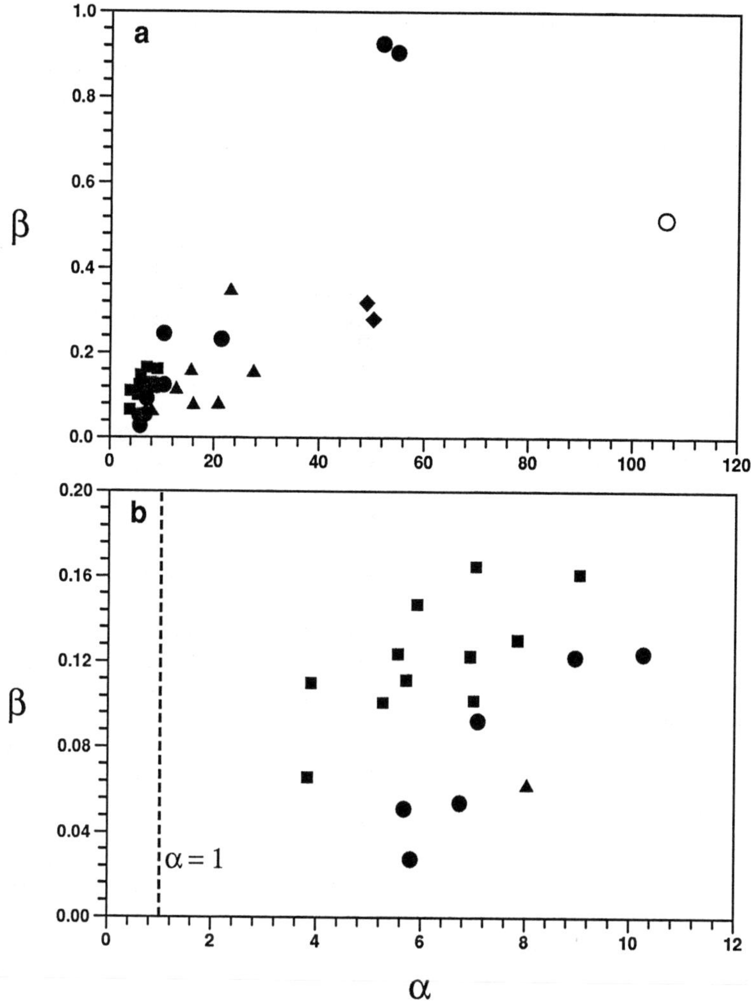

Fig. 9.4. Parameter estimates (α, β) for the gamma density function (4.15) applied to 31 *Tribolium* data sets. See Table 9.1 for the identification of each data set: 1–7 triangles, 8–17 solid circles, 18–28 squares, 29–30 diamonds, 31 open circle.

In this monograph, we apply the gamma distribution to 31 data sets: 7 on *T. confusum*, 23 on *T. castaneum* and, in this chapter, 1 on *T. brevicornis*. The properties of these observed stationary distributions are summarized by the parameters α and β of the gamma (see Chapter 4). While the experiments were not conducted to obtain the overall distribution of α and β for laboratory cultures of *Tribolium*, we might, nevertheless, get some insight from the empirical data. We have collected the estimated values of α and β in Table 9.1 and present a graphical summary in Fig. 9.4.

Table 9.1. Summary of the parameter estimates for the gamma distribution

Identification	Reference in text	α	β
T. confusum[a]			
1. Chicago (24°C)	Fig. 4.5a	12.68091	0.11241
2. Chicago (29°C)	Fig. 4.5b	8.03158	0.06184
3. bI	Fig. 4.7a	20.88335	0.07821
4. bI-a	Fig. 4.12b	27.56627	0.15316
5. bII	Fig. 4.7b	16.04930	0.07665
6. bIII	Fig. 4.7c	15.58278	0.15617
7. bIV	Fig. 4.7d	23.10839	0.34513
T. castaneum[b]			
8. Chicago (24°C, 31% RH)	Fig. 4.10a	54.71710	0.90476
9. Brazil (24°C, 31% RH)	Fig. 4.10b	51.94132	0.92520
10. Chicago (24°C)	Fig. 4.6a	5.67184	0.05133
11. Chicago (29°C)	Fig. 4.6b	6.74342	0.05416
12. cI	Fig. 4.8a	5.79822	0.02752
13. cII	Fig. 4.8b	7.08624	0.09233
14. cIII	Fig. 4.8c	21.40513	0.23227
15. cIV	Fig. 4.8d	10.25296	0.24495
16. cIV-a, 1968	Fig. 4.11b	8.95560	0.12222
17. cIV-a, 1969	Fig. 4.13	10.25861	0.12391
cos genetic polymorphism experiment			
18. $p(0)=0.0$	Fig. 6.11	7.00160	0.10209
19. $p(0)=0.1$	Fig. 6.8	9.03374	0.16136
20. $p(0)=0.2$	Fig. 6.8	3.82764	0.06591
21. $p(0)=0.3$	Fig. 6.8	5.26045	0.10102
22. $p(0)=0.4$	Fig. 6.8	7.83768	0.13026
23. $p(0)=0.5$	Fig. 6.8	6.93378	0.12260
24. $p(0)=0.6$	Fig. 6.8	5.90461	0.14693
25. $p(0)=0.7$	Fig. 6.8	5.54226	0.12377
26. $p(0)=0.8$	Fig. 6.8	5.70670	0.11154
27. $p(0)=0.9$	Fig. 6.8	7.03089	0.16506
28. $p(0)=1.0$	Fig. 6.11	3.88840	0.11015
cos genetic stability experiment			
29. homozygous	Fig. 6.12	49.08821	0.31829
30. polymorphic	Fig. 6.12	50.34501	0.27955
T. brevicornis[c]			
31. standard	Fig. 5.3	109.01700	0.53296

[a] All four b-strains of *T. confusum* were derived from Chicago standard (Park et al. 1961).

[b] The strains cI and cII of *T. castaneum* were derived from the Brazil stock; strains cIII and cIV came from Chicago standard (Park, et al. 1961).

[c] We fit the gamma distribution to these data for comparison. The gamma was accepted at the 0.01 probability level. The large value of α is consistent with the prediction from Chapter 5 of a normal distribution.

In these data sets we find that both parameters span nearly two orders of magnitude: $3.8 < \alpha < 109$ and $0.027 < \beta < 0.93$ (Fig. 9.4a). No data sets were observed with small values of α ($\alpha < 30$) and large values of β ($\beta > 0.4$). Nor did we see distributions with large values of α ($\alpha > 60$) together with small values of β ($\beta < 0.4$). This positive association between α and β also seems to hold on a fine scale (Fig. 9.4b). The shape parameter of the gamma is α. As α gets large, the gamma approaches a normal distribution. Five data sets have large α values: *T. brevicornis* with $\alpha = 109.0$; four *T. castaneum* strains, Chicago with $\alpha = 54.72$, Brazil with $\alpha = 51.94$, and the homozygous and polymorphic strains in the genetic stability experiment with $\alpha = 49.09$ and $\alpha = 50.35$. For $\alpha = 1$, the gamma is the exponential distribution. The smallest α values are associated with the *cos* strain (in the genetic polymorphism experiment) with $\alpha = 3.83$ and $\alpha = 3.89$. Curiously, in the *cos* strain there was a ten-fold increase in the values of α from the experiment of Moffa and Costantino (1977) to the experiment of Desharnais and Costantino (1985). Does this represent a genetic change or small differences in the culturing procedures? In contrast, differences in temperature and genetic strains among the *T. castaneum* data sets do not lead to such dramatic differences in α and β (triangles in Fig. 9.4). As more data are generated, it may be worthwhile to continue this comparison of the parameters of the stationary distribution.

9.2.2 Stochasticity and Age-structure

The adage that research identifies many more new questions than it resolves is certainly true for stochastic models. Can the concept of a stationary distribution, successfully applied to adult number, be extended to include all age classes? Some work is available. Witten (1983) sketched one approach which included stochastically varying birth and death rates in the McKendrick-von Foerster model (Chapter 2). Tuljapurkar (1989) reviews many of the most recent results of stochastic demography. Unfortunately, most of the models are linear. The specific application to the biology of the flour beetle remains to be done.

Many new questions could be addressed with a stochastic nonlinear age-structured model. What can we learn from the statistical correlations among the age classes? How do phenomena like demographic limit cycles and multiple attractors translate to a stochastic framework? The data can be obtained; more theoretical work must be done. *Tribolium* could have an important role in testing and evaluating these general theoretical developments.

9.2.3 Extinction

The discipline of conservation biology has drawn attention to the practical importance of extinction in the management of nature reserves (Soule 1987). Along with other problems associated with the intervention of humans on natural ecosystems, there has been a renewed theoretical interest in extinction (Hallam and Zhien 1987, Zhien et al. 1989). Unfortunately, experimental data on this topic are sparse. *Tribolium* could serve as experimental system for testing predictions related to population extinction.

Many possibilities are open. Extinction in laboratory cultures of *Tribolium* is regularly observed when two species are grown together. Environmental conditions can be manipulated to study extinction in single species cultures. For example, low concentrations of pesticide can be added to the media to vary recruitment and mortality rates near an extinction threshold. Extinction times could be measured. The distribution of extinction times could be obtained. Environmental variability could be added to see if an acceleration in extinction rates is observed. Inbred populations could be compared to outbred populations to examine the genetic components of extinction.

In Chapter 5 we derived estimates of extinction times from a simple nonlinear "birth-death process." None of these predictions have been tested. Indeed, the entire topic of population extinction has yet to be explored with the *Tribolium* experimental system.

9.2.4 Scale of Stochastic Forces

It is extremely difficult to unambiguously assign a biological interpretation to chance phenomena. It is an oxymoron: defined uncertainty. Indeed, while we have some ideas, we do not really know what forces drive the apparently happenstance population fluctuations in flour beetle cultures. Demographic and environmental stochasticity are two possibilities. *Tribolium*, with its replicate cultures and accurate and long census records, is a good candidate for new experiments that are focused explicitly on studying the *scale* of the stochastic forces.

In Chapter 4 we considered the stochastic differential equation (SDE),

$$\mathrm{d}N(t) = N(t)(b\exp[-c\,N(t)] - \mu)\,\mathrm{d}t + \sigma N(t)\mathrm{d}\omega(t)\,. \tag{9.5}$$

Note that population size, $N(t)$, is the scale multiplier of $\mathrm{d}\omega(t)$. Alternate SDE versions exist (Dennis and Costantino 1988), for example,

$$\mathrm{d}N(t) = N(t)(b\exp[-c\,N(t)] - \mu)\,\mathrm{d}t + \sigma\,\mathrm{d}\omega(t)\,, \tag{9.6}$$

$$\mathrm{d}N(t) = N(t)(b\exp[-c\,N(t)] - \mu)\,\mathrm{d}t + \sigma\sqrt{N(t)}\,\mathrm{d}\omega(t)\,. \tag{9.7}$$

Each of these models lead to stationary distributions of population size. How can one decide among them?

One possibility would be to conduct experiments using different population sizes. This could easily be accomplished with *Tribolium* by varying the size of the container and the volume of flour. If the variance of the fluctuations were found to be more or less constant in the region of the steady-state, then the additive model (9.6) is appropriate. Under this model, the expected stationary distribution is normal. On the other hand, the "square root" scaling in (9.7) is typical of SDE models which approximate demographic stochasticity (see §5.5.2). In this case, the variance of the population fluctuations would increase linearly with the mean population size. Lastly, if the noise is "multiplicative" as modeled using (9.5), then the steady state population variance will increase at a faster than linear rate (eg. with N^2) with increases in population size N. One could test between these and other alternative hypotheses using a carefully designed and replicated experimental study.

Another experimental possibility, pointed out to us by Brian Dennis, would be to attempt to detect a noise-induced transition (see §4.2.3 and §4.6) from a gamma to an exponential distribution. For this experiment, one could deliberately vary the environmental conditions such as temperature and humidity in an attempt to increase the parameter σ. At a certain critical value for σ, $\alpha = 1$ in the gamma distribution and the population density function becomes a negative exponential. In effect, one would be shifting the populations in Fig. 9.4b to the left beyond the dashed curve. A noise induced transition is a peculiar emergent property of multiplicative noise and detection of such a transition would rule out additive models such as (9.6) where the noise term in the population growth equation is not multiplied by any function of population size.

9.2.5 Stochastic Disequilibrium

In our discussion of natural selection in Chapter 7, we introduced the idea of comparing an "evolving population" with nonequilibrium initial allele frequencies to a "reference population" which remains at genetic equilibrium. The deviation of the evolving population from the reference population was measured using an "entropy" function $H(t)$. Iwasa (1988) has introduced a similar function which can be used with stochastic distributions.

Let $f(N,t)$ represent the probability density function of population size N at time t. Let $f^*(N)$ represent the stationary distribution; i.e. $f(N,t) \to f^*(N)$ as $t \to \infty$. We will call the deviation of $f(N,t)$ from $f^*(N)$ "stochastic disequilibrium." Iwasa (1988) introduced the following entropy function to measure the deviation between $f(N,t)$ and $f^*(N)$:

$$I(t) = \int_0^\infty f(N,t) \log\left[\frac{f(N,t)}{f^*(N)}\right] dN . \tag{9.8}$$

For any Markovian process, $I(t)$ monotonically decreases with time and, as $t \to \infty$, $I(t) \to 0$. $I(t)$ is also a Lyapunov function which means that the stationary state $f^*(N)$ is globally asymptotically stable (Horsthemke and Lefever 1984, p. 112). Note that in (9.8) we have removed a negative sign from Iwasa's (1988) expression to make it comparable to the genetic disequilibrium function $H(t)$.

The stochastic disequilibrium function $I(t)$ can be computed empirically. In §4.6 we used the gamma probability density function with time-dependent parameters $\alpha(t)$ and $\beta(t)$ as a transient distribution of population numbers. We then fit this model to data from the Chicago strain of *T. castaneum* using the method of maximum likelihood. To illustrate the computation of $I(t)$, we will use the data for the "moving histograms"

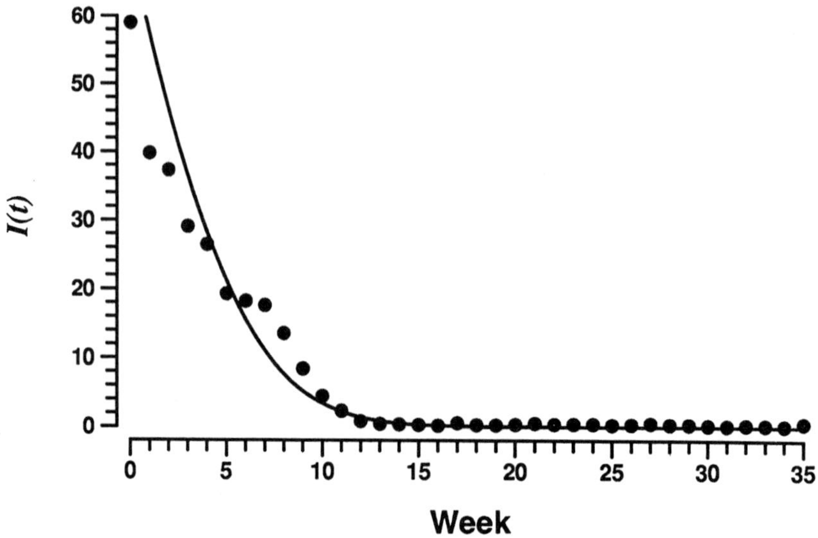

Fig. 9.5. Measure of stochastic disequilibrium for the Chicago strain of *T. castaneum*. The circles are observed values of $I(t)$ computed using (9.9) on the histograms in Fig. 4.3b. The solid curve is the theoretical value for $I(t)$ in (9.8) which was computed using the time-dependent gamma density function (4.46) with $\lambda=0.277410$, $N^*=61.824$, and $\sigma^2=0.009129$. These parameter values were obtained using the method of maximum likelihood as described in §4.6.

presented in Fig. 4.3b for the Chicago strain. At each census, adult numbers for the 12 replicates were grouped as 0-5, 6-10, 11-15, etc. Although each histogram is only based on 12 observations, it does give an observed probability distribution for each value of time. To obtain a discretized analog of $f(N,t)$, we let $p_1(t)$, $p_2(t)$, ... equal the number of observations satisfying $\{N(t)\in[0,5]\}$, $\{N(t)\in(5,10]\}$, ..., respectively, divided by the total number of observations ($n=12$) at time t. We used the asymptotic α^* and β^* values obtained from the time-dependent gamma for $f^*(N)$. We numerically integrated the gamma, for the intervals [0,5], (5,10], ... to obtain $p_1^*, p_2^*, \ldots$. This is a discretized analog of $f^*(N)$. For each time value t, we computed

$$I(t) = \sum_i p_i(t)\log\left[\frac{p_i(t)}{p_i^*}\right]. \tag{9.9}$$

Since $p_i\log[p_i/p^*]\to 0$ as $p_i\to 0$, we omitted from the summation terms where $p_i(t)=0$. Expression (9.9) gives the empirical values for $I(t)$. For comparison, we used the estimated parameters for the time dependent gamma (see §4.6) and numerical integration to compute the continuous form of $I(t)$. The results are presented in Fig. 9.5. $I(t)$ goes from a large positive value to a value that is near zero. Although the data show systematic departures from the solid curve, the overall trend supports the idea of a monotonic decrease in the stochastic disequilibrium. Ideally, larger data sets should be used to evaluate these predictions more rigorously.

Many questions remain. Can the function $I(t)$ be used to test the validity of a stochastic model? How is the change in stochastic disequilibrium related to the idea of population stability? Can the function $I(t)$ be applied to deterministic models whose distribution is the result of a chaotic strange attractor? Can the concepts of genetic disequilibrium and stochastic disequilibrium be combined using a stochastic model of natural selection? The ability to measure $I(t)$ empirically using censuses from replicated populations makes *Tribolium* an excellent research organism for addressing these types of questions.

9.3 Natural Selection

9.3.1 Natural Selection and Age Structure

A general theme of the research presented in Chapters 6 and 7 has been the dynamical interplay of natural selection and population growth. In particular, we have been interested in how allele frequencies, $p_i(t)$, and population size, $N(t)$, interact. However, our simple models do not consider age structure. In this section we present some preliminary results which deal with selection on recruitment and mortality rates in the context of age structure.

The theory for natural selection in age-structured populations can be quite complex. Texts such as those by Charlesworth (1980) and Roughgarden (1979) provide a good introduction.

We consider the simplest model based on several assumptions: one autosomal genetic locus, random union of gametes, and no differences in fertility and survivorship among males and females with the same genotype. The population is described by a set of functions $n_{ij}(x,t)$, where $n_{ij}(x,t)\,dx$ gives the number of individuals between the ages of x and $x+dx$ at time t. Each genotype has an age-specific mortality rate $\mu_{ij}(x)$ and fertility rate $m_{ij}(x)$. Changes in age structure are described using the McKendrick-von Foerster equation (see §2.1.1)

$$\frac{\partial n_{ij}(x,t)}{\partial t}+\frac{\partial n_{ij}(x,t)}{\partial x} = -\mu_{ij}(x)\,n_{ij}(x,t)\,. \qquad (9.10)$$

The probability for genotype ij to survive to age x at time t is

$$l_{ij}(x) = \exp\left[-\int_0^x \mu_{ij}(s)\,\mathrm{d}s\right], \qquad (9.11)$$

Total reproduction is given by the boundary condition

$$B(t) = \sum_{ij} \int_0^\infty n_{ij}(x,t) m_{ij}(x) \mathrm{d}x \,. \tag{9.12}$$

The allele frequencies for newly formed zygotes equal the allele frequencies among the gametes which formed those zygotes. This is given by

$$p_i(t) = \frac{\sum_j \int_0^\infty n_{ij}(x,t) m_{ij}(x) \mathrm{d}x}{B(t)} \tag{9.13}$$

for the ith allele. With random union of gametes, the boundary condition for each genotype is

$$n_{ij}(0,t) = p_i(t) p_j(t) B(t) \,. \tag{9.14}$$

Equations (9.10)–(9.14) provide a complete specification of the model.

Now we consider recruitment and mortality rates at genetic and demographic equilibrium. Let $N(t)$ represent the total number of individuals of all genotypes and all ages. We define the following age specific "weighting functions" for each genotype:

$$q_{ij}(x) = \frac{B(t-x) l_{ij}(x)}{N(t)} \,. \tag{9.15}$$

If all individuals had the same mortality rate, $\mu_{ij}(x)$, then $q_{ij}(x)$ would equal the age distribution of the population. The genotype-specific "birth" and "death" rates are defined as

$$b_{ij} = \int_0^\infty q_{ij}(x) m_{ij}(x) \mathrm{d}x \,, \tag{9.16}$$

$$d_{ij} = \int_0^\infty q_{ij}(x) \mu_{ij}(x) \mathrm{d}x \,, \tag{9.17}$$

and the average population values are obtained using

$$b = \sum_{ij} p_i p_j b_{ij} \,, \tag{9.18}$$

$$d = \sum_{ij} p_i p_j d_{ij} \,. \tag{9.19}$$

At demographic equilibrium, all age classes increase at the same exponential rate, r, which is given by

$$r = b - d\,. \tag{9.20}$$

It can be shown that genetic equilibrium corresponds to a maximization of the intrinsic rate of increase, r (Roughgarden 1979, Charlesworth 1980). Thus natural selection maximizes the *difference* between the average population birth and death rates. If density dependent mechanisms adjust the growth rate so that $r=0$, then it can also be shown that the equilibrium allele frequencies must satisfy

$$\sum_{ij} p_i p_j \left(\frac{b_{ij}}{d_{ij}}\right) - \sum_j p_j \left(\frac{b_{ij}}{d_{ij}}\right) = 0\,. \tag{9.21}$$

This result is equivalent to the expression $W - W_i = 0$ for equilibrium allele frequencies in the classic model of natural selection (see equation 6.1). In this case, the equilibrium allele frequencies will maximize the average of the *ratio* b_{ij}/d_{ij}. These predictions are similar to the results obtained in §6.2.4 for the genetic model without age structure.

More work is needed. Explicit density-dependent recruitment and mortality terms could be included in the fertility and mortality schedules $m_{ij}(x)$ and $\mu_{ij}(x)$. The effects of differential fertility and mortality in males and females should also be examined. It would also be interesting to see what happens to allele frequencies when demographic limit cycles and multiple attractors are among the possible dynamical behaviors.

The *Tribolium* laboratory model could play an important role in this research. Outbred populations could be maintained at different phases of population growth. One treatment might involve the regular and random culling of population numbers to keep $r>0$ while another population could be cultured at demographic equilibrium ($r=0$). Fertility and mortality schedules could be monitored to see the effects of natural selection on birth and death rates. One could also test to see if selection at demographic equilibrium maximizes the equilibrium population size. Other traits such as cannibalism could also be examined. The experimental possibilities are numerous.

9.3.2 Joint Distributions of Population Numbers and Allele Frequencies

In Chapter 6 we discussed the need for a stochastic theory of natural selection and population growth. In this section we present some ideas on how this theory could be developed and demonstrate the type of data that could be collected using *Tribolium*.

There are several approaches one could take in developing a stochastic model with genetic and demographic variables. Genotypic fitness functions could include a random component. For example, we could write

$$W_{ij}(N,t) = w_{ij}(N) + \gamma_{ij}(t)\,, \tag{9.22}$$

where $w_{ij}(N)$ is the density-dependent deterministic component of fitness and $\gamma_{ij}(t)$ is a random noise variable. Substitution of (9.22) into (6.1)–(6.4) would yield a stochastic model where both allele frequencies and population numbers fluctuate. In the formulation of this model, it would be necessary to specify the variance-covariance of the genotypic noise variables. At one extreme, all γ_{ij} could have a perfect positive correlation. In this case, the random components would cancel in the genetic equations (6.1) and allele frequencies would behave as deterministic variables. At the other extreme, all γ_{ij} could be identically and independently distributed. Recall that natural selection and population growth are determined by the average population and allelic fitnesses W and $W_{i\cdot}$. In the computation of these averages, the central limit theorem suggests that the random effects would tend to cancel one another as the number of alleles increased. Increased genetic diversity might result in decreased population size and allele frequency variability. Another possibility would be to combine the SDE approach of Chapter 4 with the traditional theory of genetic drift. The result would be set of equations of the form

$$\frac{\mathrm{d}p_i}{\mathrm{d}t} = p_i\,(W_{i\cdot} - W) + \phi_i\,, \tag{9.23}$$

$$\frac{\mathrm{d}N}{\mathrm{d}t} = N\,(W + \gamma)\,. \tag{9.24}$$

The random "sampling" factor ϕ_i would be determined using a binomial or multinomial distribution based on a population size of $N(t)$. The random factor γ is the usual "white noise" variable (see Chapter 4). In this formulation, fluctuations in population size are coupled with fluctuations in allele frequencies. As a final possibility, one could model natural selection as a "multidimensional birth-death process." The state space would be represented by numbers of the different possible alleles. Transitions would involve single births or deaths which increase or decrease the allele numbers by one (heterozygote) or two (homozygote) units. The transition rates would depend on total population size and genotype-dependent recruitment and mortality values. Genotype frequencies could be based on the assumption of random mating. Stationary or quasi-stationary distributions for number of alleles would yield the distributions for allele frequencies and population size.

Regardless of the type of model that is used, a fundamental prediction is the multivariate probability distribution for population size and allele frequencies. Let $f^*(N,\mathbf{p})$ represent the equilibrium density function for population size N and the vector of allele frequencies $\mathbf{p}$. Let X be the region defined by $a_0 < N < b_0$ and $a_i < p_i < b_i$. The probability that the population is in the region X is given by

$$\mathrm{Prob}\,\{(N,\mathbf{p}) \in X\} = \int_X f^*(N,\mathbf{p})\,\mathrm{d}\mathbf{p}\,\mathrm{d}N\,. \tag{9.25}$$

Associated with this density function are the marginal distributions for allele frequency,

$$f^*(\mathbf{p}) = \int_0^\infty f^*(N,\mathbf{p})\,\mathrm{d}N\,, \tag{9.26}$$

and population size,

$$f^*(N) = \int_\Delta f^*(N,\mathbf{p})\,\mathrm{d}\mathbf{p}\,, \tag{9.27}$$

where Δ is the simplex of all possible allele frequencies. Of particular interest is how these marginal distributions compare with the equivalent distributions based on a purely genetic or demographic model. For example, one can compare the density function (9.27) with the prediction of a gamma distribution as derived in Chapter 4. Can experimental data on population numbers alone be used to distinguish between these two possibilities? A well-developed theory would open the door to further experimentation using the *Tribolium* laboratory model.

Finally, to illustrate the type of data that can be obtained, we present some preliminary data based on an experiment with the corn oil sensitive (*cos*) mutant (Magill 1978). The experimental design was similar to the genetic polymorphic experiment described in §6.3.2. The initial *cos* allele frequencies were $p(0)=0.1, 0.3, 0.5, 0.7$, and 0.9 and the initial demographic array consisted of 28% newly emerged adults, 3% pupae, 7% large larvae, and 62% small larvae. This age structure corresponds to the equilibrium stage class distribution observed by Moffa and Costantino (1977). The populations were censused and the *cos* allele frequencies were obtained at weeks 5, 9, 13, and 17. There were three replicates per treatment. The 60 paired observations for *cos* allele frequency and adult number are presented in Fig. 9.6. The "wire mesh" was constructed using a distance-weighted least-squares algorithm. The marginal distributions (9.26) and (9.27) are represented by the histograms in the side panels. While it is obvious that more data are needed, these results demonstrate the possibility of using flour beetles to study joint distributions of population numbers and allele frequencies.

9.3.3 Estimating the Life Not Lived

In Chapter 7 we derived predictions based on the comparison of an *evolving population* with a *reference population.* The evolving population has nonequilibrium initial allele frequencies; it is initially in *genetic disequilibrium.* The reference population has the same fitness values and initial population number as the evolving population, but its initial allele frequencies equal the equilibrium values. The total difference in population size between the reference and the evolving populations is called the *life not lived.* We showed that the life not lived is directly proportional to an entropy measure of the deviation of the evolving population's initial allele frequencies from their equilibrium values. This prediction is independent of the initial population size and the strength of selection.

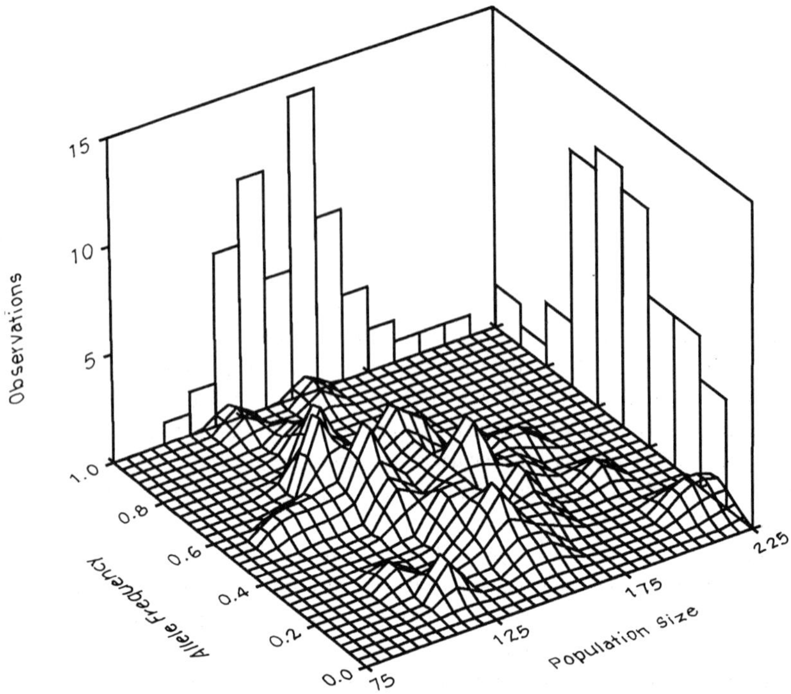

Fig. 9.6. A three dimensional plot of the paired observations on adult population size and *cos* allele frequency. The histograms in the side panels are the marginal distributions.

This section describes briefly some recent results from an experiment designed to test predictions based on the life not lived. A full report on this experiment is being prepared for publication (Desharnais 1990).

The experimental work involves two genetic strains of flour beetles belonging to the species *Tribolium castaneum*. One of the strains is homozygous for an allele, *R*, conferring resistance to the pesticide malathion. The other strain is homozygous for the susceptible wild type allele, *S*. Malathion resistance in these strains is a dominant, autosomal trait which lies in the sixth linkage group. The *S*/*R*, *R*/*S*, and *R*/*R* genotypes require malathion concentrations 500–1000 times greater than *S*/*S* to achieve the same levels of mortality.

The basic prediction is given by (7.32). Let $p(t)$ represent the frequency of the *R* allele and let $N(t)$ represent the adult numbers in an "evolving population" segregating for both the *R* and *S* alleles. Since *R* is selected to fixation, $p^* = 1$ and the "reference population," $\tilde{N}(t)$, is composed entirely of *R*/*R* individuals. When $p^* = 1$ the entropy measure of genetic disequilibrium (7.1) reduces to $H(t) = -\log[p(t)]$. Hence we have the following prediction:

$$H(0) = -\log[p(0)] = \lambda \int_0^{\infty} [\tilde{N}(t) - N(t)]\,dt , \tag{9.28}$$

where λ is a constant. The prediction (9.28) should hold for different initial population numbers and different intensities of selection.

A population experiment was designed to test three predictions from our theoretical work: (i) the relationship between the initial entropy, $H(0)$, and the life not lived is linear with no intercept as predicted in (9.28); (ii) the relationship (9.28) is independent of the intensity of selection, and (iii) the relationship (9.28) is independent of the initial population density. These three hypotheses are embodied in the three-way factorial design which is summarized in Table 9.2. Each combination of treatments was replicated three times. In the experimental design, the initial frequencies of the resistant allele correspond to entropy values of 0, 1, 2, 3, 4, and an *S/S* control. The initial genotype frequencies were in Hardy-Weinberg equilibrium. The initial population densities in the low and high treatments represent an approximate equilibrium age distribution as determined from the work of Desharnais and Liu (1987). The three malathion concentrations represent conditions of strong (3.0 ppm) and moderate (1.5 ppm) selection and a control (0.0 ppm). Each population was contained in a 120 ml glass vial with 20 g of media. The media consisted of 94% wheat flour, 5% brewer's yeast, and 1% sunflower oil. Technical grade malathion was diluted in the sunflower oil to provide the desired malathion concentrations in the media. All populations were cultured in an unlighted incubator at 33°C and 60% relative humidity.

Table 9.2. Factorial design for the population experiment

	Frequency of the resistant allele					
Malathion concentration	0.0000	0.0183	0.0498	0.1353	0.3679	1.0000
0 ppm	Low[a] High[b]	Low High	Low High	Low High	Low High	Low High
1.5 ppm	Low High	Low High	Low High	Low High	Low High	Low High
3.0 ppm	Low High	Low High	Low High	Low High	Low High	Low High

[a] Initial population composed of 50 adults, 2 pupae, 12 large larvae, and 16 small larvae.
[b] Initial population composed of 200 adults, 8 pupae, 48 large larvae, and 64 small larvae.

The population experiment ran for 32 weeks. Adults and pupae were censused every two weeks and at that time all life stages, including eggs, were transferred to fresh media. Every four weeks, 25 pupae were sampled from all the populations and the frequencies of the R and S alleles were estimated.

The control populations are the cultures with 100% susceptible or resistant alleles. In the S/S control populations with no malathion, adult numbers approached a mean equilibrium of approximately 250 adults. At 1.5 ppm of malathion, the S/S control populations approached a mean equilibrium of approximately 125 adults. At 3.0 ppm, adults became extinct in all of the S/S controls. All the R/R populations maintained densities which were comparable to the S/S controls with no malathion. The same was true of the polymorphic populations which contained no malathion.

In the populations segregating for the R and S alleles, the frequency of the R allele increased quickly and reached high levels by the end of the experiment. The change in adult numbers depended strongly on the initial allele frequency. The lower the initial frequency of the R allele, the slower the population growth.

The life not lived was computed for all the treatment combinations involving selection for the resistant allele (malathion concentration = 1.5 ppm, 3.0 ppm; $N(0)=50$, 200 adults; $p(0)=0.0183, 0.498, 0.1353, 0.3679$). For each malathion concentration and initial population density, the trajectory of the reference population, $\tilde{N}(t)$, was defined using the mean number of adults for the 100% resistant replicates ($p(0)=p^*=1.0$). For each of the polymorphic populations, the life not lived was computed using

$$L = \sum_{t=0}^{16} [\tilde{N}(t)-N(t)] , \tag{9.29}$$

where $N(t)$ is the observed number of adults in the evolving population and t is the census period. This life not lived was then compared the initial entropy, $H(0)=-\log[p(0)]$, and the prediction of a linear relationship between $H(0)$ and L was evaluated.

The results are summarized in Fig. 9.7. This shows a plot of the life not lived versus the initial entropy for the treatments with malathion concentrations of 1.5 ppm and 3.0 ppm and initial population sizes of 50 and 200 adults. Note the strong linear relationship between $H(0)$ and L. These data were analyzed statistically using the general linear model procedures of SYSTAT (Wilkinson 1988). An analysis of variance was conducted where malathion concentration and initial population size were categorical variables and $H(0)$ was a covariate. This statistical model provides regressions of L versus $H(0)$ and allows a simultaneous comparison of the slopes of these regressions. The results of the ANOVA are given in Table 9.3. The high significance of fitness entropy as a main effect reflects the strong linear relationship between $H(0)$ and L. The significant interaction between the initial density and fitness entropy indicates that the initial population size does affect the slope of the regressions (**a** versus **c** and **b** versus **d** in Fig. 9.7). The lack of any other significant effects indicates that the intensity of selection (malathion concentration) does not affect the slopes of the regression (**a** versus **b** and **c** versus **d** in Fig. 9.7). Further statistical analyses suggest that none of the

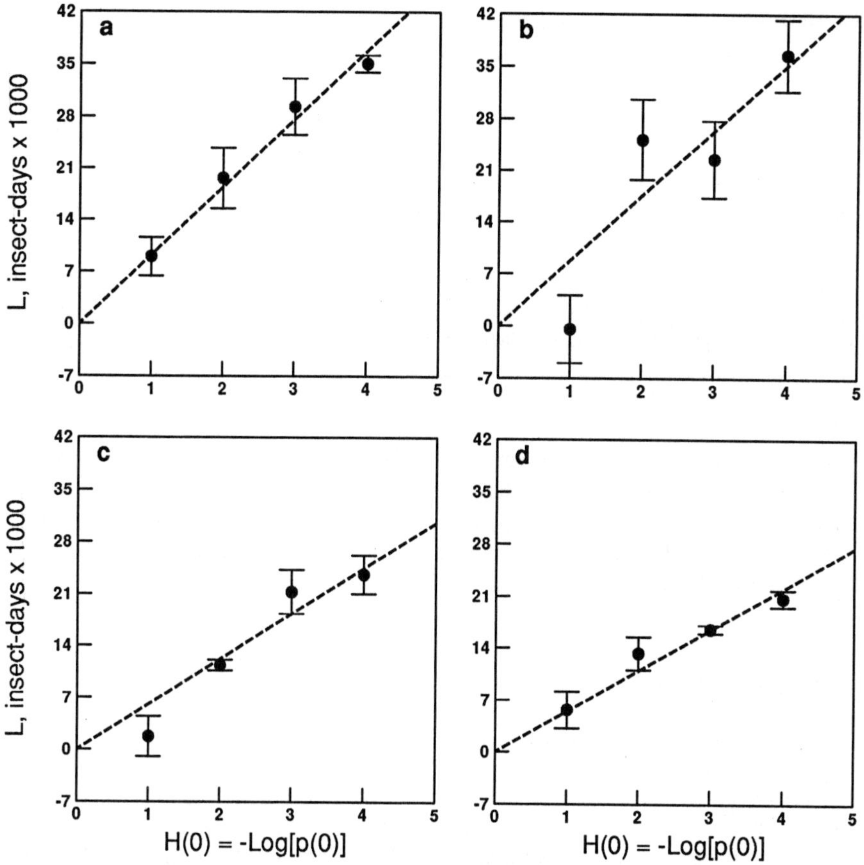

Fig. 9.7. The estimated mean (±s.e.) values of L, life not lived, versus $H(0)$, the initial fitness entropy. In **a** and **b** the initial number of adults is 50; in **c** and **d** the initial number is 200. In **a** and **c** the malathion concentration is 1.5 ppm; in **b** and **d** the concentration is 3.0 ppm. The dashed lines represent least squares regressions through the origins.

regressions deviate significantly from the origin. Except for the relatively small effect due to initial density, the experimental results strongly support the theoretical prediction $H(0)=\lambda L$.

Other hypotheses concerning life not lived can be tested using the *Tribolium* experimental system. For example, the prediction (9.28) holds for any arbitrary number of malathion resistant alleles, provided one dominant allele is selected to fixation. A experiment could be designed using fixed frequencies of the dominant allele and different combinations of the subdominant alleles. In §7.5.2, we discuss an extension of the theory to *communities* of several interacting species. In fact, as long as the surviving species is the one undergoing selection, coexistence is not a prerequisite for the application of the theory (Desharnais 1986). This means that experiments with two or more *Tribolium*

Table 9.3. Analysis of variance for the population experiment

Source of variation	Degrees of freedom	F-ratio	Probability
Malathion concentration	1,40	0.016	0.900
Initial density	1,40	0.081	0.777
Fitness entropy	1,40	99.900***	<0.001
Malathion concentration × initial density	1,40	2.466	0.124
Malathion concentration × fitness entropy	1,40	0.042	0.838
Initial density × fitness entropy	1,40	5.110*	0.029
Initial density × malathion concentration × fitness entropy	1,40	2.213	0.145

species are feasible. In §7.5.3, we present some of the results of Desharnais and Cohen (1986) for life not lived in age-structured populations. Although this theory must be extended to models with sexual reproduction, one can estimate demographic disequilibrium experimentally using marking techniques to monitor the adult age structure. As the theory is developed, more opportunities for experimental work will be revealed.

9.4 Species Interactions

9.4.1 Complex Demographic Models

Much biological detail remains to be incorporated into the two species interaction equations; most obvious is the need to include the immature life stages. In the case of the two cannibalistic species, *T. castaneum* and *T. confusum*, both larvae and adults consume eggs and pupae discriminately. Adults also prefer to eat larger sized eggs (Teleky 1980). In addition, these rates are variable between species and among genetic strains within a species. An extensive theoretical literature on two species interactions is available, with many variations on the theme of "egg-eating predators," such as satiable versus nonsatiable predators and age-structured prey species (Smith and Mead 1974, 1979, Gurtin and Levine 1979, Levine 1981, Thompson, DiBiasio and Mendes 1982, Coleman and Frauenthal 1983, Saleem 1983, 1984, Bosh and Diekmann 1986, Saleem, Siddiqui and Gupta 1987, Saleem and Pandey 1988). The possibilities for experimental research using *Tribolium* have not yet been fully explored.

In *T. brevicornis* and *T. madens* the implications of the adult-larval inhibition hypothesis are largely uncharted. Does the extension of the larval interval mean that the rate of egg cannibalism is increased? Very high rates of larval cannibalism can lead to multiple demographic attractors (Fig. 3.3). On the other hand, extending the duration of the larval stage favors stability (Fig. 3.3). Is the population dynamic role of larval egg-eating altered in a species possessing the inhibition mechanism? As the model is expanded new qualitative features are likely to emerge. Are more than two locally stable attractors possible? Can a species which would lose in competition on the slow time scale survive and coexist on the fast time scale? Can multiple attractors provide a new mechanism to explain indeterminate results in competition experiments? Many interesting questions remain.

9.4.2 Stochastic Dynamics of Coexistence

Thirty years ago, Leslie and Gower (1958) computed a bivariate stationary distribution for two competing species. The species interaction between *T. castaneum* and *T. confusum* did not allow coexistence and so the concept was not applied to flour beetles. But *T. madens* and *T. brevicornis* offer the possibility of studying coexistence, thus, the idea of a univariate stationary probability distribution for adult numbers should be extended to the bivariate case. New questions emerge. How do the intensity of the species interactions affect these bivariate distributions? What happens if the stochastic growth rates of the two species are correlated? What information can be obtained from the means, variances, covariances, and autocorrelations at the steady state? More experiments involving *T. madens* and *T. brevicornis* should be conducted.

9.5 Summary and Conclusions

An important feature of our research effort, which is repeated throughout the monograph, is the integration of biological data with mathematical analysis. Data and theory are together; nevertheless, it is just a beginning. In our opinion, the study of population dynamics is such a complex undertaking that the collaboration between experimentalists and theoreticians must be strengthened. As the general theory continues to be developed, some of which may be motivated by the flour beetle, new uses for the *Tribolium* model are sure to be found. Much exciting work remains to be done.

References

Aagaard-Hansen H, Yeo GF (1984) A stochastic discrete generation birth, continuous death population growth model and its approximate solution. J Math Biol 20:69–90

Abramowitz M, Stegun IA (1965) Handbook of Mathematical Functions. US Dept Commerce, Washington

Ahmed MS (1963) A stochastic model for the tunnelling and retunnelling of the flour beetle. Biometrics 19:341–351

Allee WC (1931) Animal Aggregations. Univ Chicago Press

Anderson W (1971) Genetic equilibrium and population growth under density-regulated selection. Am Nat 105:489–498

Arrigoni M, Steiner A (1985) Logistic growth in a fluctuating environment. J Math Biol 21:237–241

Asmussen MA, Feldman MW (1977) Density dependent selection 1: A stable feasible equilibrium may not be attainable. J Theor Biol 64:603–618

Bailey VA (1931) The interaction between hosts and parasites. Quant J Math 68–77

Barnett VD (1962) The Monte Carlo solution of a competing species problem. Biometrics 18:76–103

Bartlett MS (1946) On the theoretical specification and sampling properties of autocorrelated time-series. J R Stat Soc 8:27–41

Bartlett M S (1957) On theoretical models for competitive and predatory biological systems. Biometrika 44:27–42

Bartlett MS (1960a) Stochastic Population Models. Wiley, NY

Bartlett MS (1960b) Monte Carlo studies in ecology and epidemiology. In: Neyman J (ed) Proc Fourth Berkeley Symp Math Stat and Prob, Univ California Press, pp 39–55

Bartlett MS, Gower JC, Leslie PH (1960) A comparison of theoretical and empirical results for some stochastic population models. Biometrika 47:1–11

Basawa IV, Prakasa-Rao BLS (1980) Statistical Inference for Stochastic Processes. Academic Press, NY

Bell AE (1982) The *Tribolium* model and animal breeding. Second World Cong Genet App Livestock Prod 5:26–42

Bell AE (1983) Laboratory animals for animal breeding research-experimental models versus model experiments. In: Genetics: New Frontiers. Oxford and IBH Publishing Co, pp 159–168

Bell AE, Moore CH (1958) Further comparisons of reciprocal recurrent selection with conventional methods of selection for improvement of quantitative characteristics. Proc Xth Inter Cong Genet 2:20–21

Bernardelli H (1941) Population waves. J Burma Res Soc 31:1–18

Birch LC, Park T, Frank MB (1951) The effect of intra-species competition on the fecundity of two species of flour beetles. Evolution 5:116–132

Bishop YMM, Fienberg SE, Holland PW (1975) Discrete Multivariate Analysis: Theory and Practice. The MIT Press, Cambridge

Bloomfield P (1976) Fourier Analysis of Time Series: An Introduction. John Wiley, NY

Bosch F, Diekman O (1986) Interactions between egg-eating predator and prey: the effect of the functional response and of age structure. IMAJ Math Appl Med Biol 3:53–69

Botella LM, Mensua JL (1986) Larval arrest in development of *Tribolium castaneum* (Coleoptera: Tenebrionidae). Environ Entomol 15:1264–1267

Botsford LW, Wickham DE (1978) Behaviour of age specific, density dependent models and the Northern California Dungeness Crab (*Cancer magister*) fishery. J Fish Res Bd Can 35:833–843

Boyce JM (1946) The influence of fecundity and egg mortality on the population growth of *T. confusum* Duv. Ecology 27:299–302

Brues AM (1964) The cost of evolution versus the cost of not evolving. Evolution 18:379–383
Brues AM (1969) Genetic load and its varieties. Science 164:1130–1136
Capocelli RM, Ricciardi LM (1977) A diffusion model for population growth in a random environment. Theor Pop Biol 5:28–41
Chapman RN (1928) Quantitative analysis of environmental factors. Ecology 9:111–122
Chapman RN (1933) The causes of fluctuations of populations of insects. Proc. Hawaiian Ent Soc 8:279–297
Chapman RN, Baird L (1934) The biotic constants of *Tribolium confusum* Duval. J Exp Zool 68:293–304
Chapman RN, Whang WY (1934) An experimental analysis of the cause of population fluctuations. Science 80:297–298
Charlesworth B (1971) Selection in density-regulated populations. Ecology 52:469–474
Charlesworth B (1980) Evolution in Age-structured Populations. Cambridge Univ Press
Chiang CL (1954) Competition and other interactions between species. In: Kempthorne O (ed) Statistics and Mathematics in Biology. Hafner, NY
Chiang CL (1957) An application of stochastic processes to experimental studied on flour beetles. Biometrics 13:79–97
Clarke B (1972) Density-dependent selection. Am Nat 106:1–13
Clarke B (1973) Mutation and population size. Heredity 31:367–379
Cohen JE (1976) Ergodicity of age structure in populations with Markovian vital rates, I. Countable states. J Am Stat Assoc 71:335–339
Cohen JE (1977a) Ergodicity of age structure in populations with Markovian vital rates, II. General states. Adv Appl Prob 9:18–37
Cohen JE (1977b) Ergodicity of age structure in populations with Markovian vital rates, II. Finite state moments and growth rates; illustration. Adv Appl Prob 9:462–475
Cohen JE (1979a) Long-run growth rates of discrete m ultiplicative processes in Markovian environments. J Math Anal Appl 69:243–251
Cohen JE (1979b) Ergodic theorems in demography. Bull Am Math Soc 1:275–295
Cohen JE, Christensen SW, Goodyear CP (1983) A stochastic age-structured population model of striped bass (*Morone saxatilis*) in the Potomac River. Can J Fish Aquat Sci 40:2170–2183
Cole, LC (1960) Competitive exclusion. Science 132:348–349
Coleman CS, Frauenthal JC (1983) Satiable egg eating predators. Math Biosci 63:99–119
Comins HN, Hassell MP (1976) Predation in multi-prey communities. J Theor Biol 62:93–114
Costantino RF, Bell AE, Rogler JC (1966) Genetic control of lipid metabolism in *Tribolium.* Nature 210:221–222
Costantino RF, Bell AE, Rogler JC (1967) Genetic analysis of a population of *Tribolium.* I. Corn oil sensitivity and selection response. Heredity 22:529–539
Costantino RF, Rogler JC, Bell AE (1968) Genetic analysis of a population of *Tribolium* II. Metabolic pattern of corn oil sensitive anomaly. Heredity 23:477–483
Costantino RF, Mumma RO, Bruszewski TE (1970) Genetic analysis of a population of *Tribolium.* III. Fatty acid composition of unsaturated acid sensitive mutant. Heredity 25:411–418
Costantino RF, Rowe PM (1972) Genetic analysis of a population of *Tribolium.* IV. Gene expression modified by corn oil and relative humidity. Heredity 29:247–252
Costantino RF, Ginzburg LR, Moffa AM (1977) An experimental check of fitness entropy versus selective delay. J Theor Biol 68:317–320
Costantino RF, Desharnais RA (1981) Gamma distributions of adult numbers for *Tribolium* populations in the region of their steady states. J An Ecol 50:667–681
Cox DR, Smith WL (1957) On the distribution of *Tribolium confusum* in a container. Biometrika 44:328–335
Craig DM (1986) Stimuli governing intraspecific egg predation in the flour beetles *Tribolium confusum* and *T. castaneum.* Res Pop Ecol 28:173–183
Crombie AC (1943) The effect of crowding upon the natality of grain-infesting insects. Proc Zool Soc Lond A 113:77–98
Crombie AC (1946) Further experiments on insect competition. Proc Roy Soc Lond B Biol Sci 133:76–109
Crombie AC (1947) Interspecific competition. J An Ecol 16:14–73

Crow JF (1968) The cost of evolution and genetic loads. In: Dronamraju KR (ed) Haldane and Modern Biology. Johns Hopkins Press, Baltimore, pp 165–178

Crow JF, Kimura M (1970) An Introduction to Population Genetics Theory. Harper and Row, NY

Cull P (1981) Global stability of population models. Bull Math Biol 43:47–58

Cullum J, Willoughby RA (1986) A practical procedure for computing eigenvalues of large sparse nonsymmetric matrices. In: Cullum J and Willoughby RA (eds) Large Scale Eigenvalue Problems, Elsevier Science Publishers B V, North-Holland, pp 193–240

Daly PJ, Ryan MF (1983) Density-related mortality of the flour beetle, *Tribolium castaneum* Duval. Res Pop Ecol 25:210–219

Darroch JN, Seneta, E (1967) On quasi-stationary distributions in absorbing continuous-time finite Markov chains. J Appl Prob 4:192–196

Darwin C (1859) The Origin of Species. Mentor Book, NY (reprinted 1958), p 84

Dawson PS (1970) A further assessment of the role of founder effects on the outcome of *Tribolium* competition. PNAS, USA 66:1112–1118

Dawson PS, Lerner IM (1962) Genetic variation and indeterminism in interspecific competition. Am Nat 96:379–380

Dawson PS, Lerner IM (1966) The founder principle and competitive ability in *Tribolium* PNAS, USA 55:1114–1117

DeAngelis DL, Cox D, Coutant C (1979) Cannibalism and size dispersal in young-of-the-year-largemouth bass: Experiment and model. Ecol Med 8:138–148

Delgado MA, Englert DC (1986) Genetic and demographic effects of single wild-type immigrants on a mutant population of *Tribolium castaneum*. Can J Genet Cytol 28:1003–1008

Dennis B (1981) Extinction and waiting times in birth-death processes: applications to endangered species and insect pest control. In: Taillie C, Patil GP, Baldessari BA (eds) Statistical Distributions in Scientific Work. Reidel, Holland, pp 289–301

Dennis B (1988) Stochastic differential equations as insect population models. In: McDonald LL, Manly BJ, Lockwood JA, Logan JA (eds) Estimation and Analysis of Insect Populations. Springer-Verlag, NY

Dennis B, Costantino RF (1988) Analysis of steady-state populations with the gamma abundance model and its application to *Tribolium*. Ecology 69:1200–1213

Dennis B, Patil GP (1984) The gamma distribution and weighted multimodal gamma distributions as models of population abundance. Math Biosc 68:187–212

Dennis B, Patil GP (1988) Applications in ecology. In: Crow EL, Shimizu K, (eds) Lognormal Distributions: Theory and Applications. Marcel Dekker, NY, Chapter 12

Desharnais RA (1986) Natural selection, fitness entropy, and the dynamics of coevolution. Theor Pop Biol 30:309–340

Desharnais RA (1990) Effects of genetic selection on population growth. Fourth Intl Cong Syst Evol Biol, Univ Maryland, College Park (abstract)

Desharnais RA, Dennis B, Costantino RF (1990) Genetic analysis of a population of *Tribolium*. IX. Maximization of population size and the concept of a stochastic equilibrium. Genome (in press)

Desharnais RA, Costantino RF (1980) Genetic analysis of a population of *Tribolium*. VII. Stability: Response to genetic and demographic perturbations. Can J Genet Cytol 22:577–589

Desharnais RA, Costantino RA (1982a) The approach to equilibrium and the steady-state probability distribution of adult numbers in *Tribolium brevicornis*. Am Nat 119:102–111

Desharnais RA, Costantino RF (1982b) Natural selection and fitness entropy in a density regulated population. Genetics 101:317–329

Desharnais RA, Costantino RF (1983) Natural selection and density-dependent population growth. Genetics 105:1029–1040

Desharnais RA, Costantino RF (1985) Genetic analysis of a population of *Tribolium*. VIII. The stationary stochastic dynamics of adult numbers. Can J Genet Cytol 27:341–350

Desharnais RA, Cohen JE (1986) Life not lived due to disequilibrium in heterogeneous age–structured populations. Theor Pop Biol 29:385–406

Desharnais RA, Liu L (1987) Stable demographic limit cycles in laboratory populations of *Tribolium castaneum*. J An Ecol 56:885–906

Devaney RL (1986) An Introduction to Chaotic Dynamical Systems. Benjamin/Cummings, Menlo Park

Diekmann O, van Gils SA (1984) Invariant manifolds for Volterra integral equations of convolution type. J Diff Equ 54:139–180

Ebenman B (1988) Competition between age classes and population dynamics. J Theor Biol 131:389–400

Ellner S (1984) Asymptotic behavior of some stochastic difference equation population models. J Math Biol 19:169–200

Ellner S (1985) The accuracy of Bartlett's small-fluctuation approximation for stochastic-difference-equation population models. Math Biosci 74:233–246

Elmes GW (1973) Observations on the density of queens in natural colonies of *Myrmica rubra* L. (Hymenoptera: Formicidae). J An Ecol 42:761–771

Feller W (1966) On the influence of natural selection on population size. PNAS USA 55:733–737

Feller W (1967) On fitness and the cost of natural selection. Genet Res 9:1–15

Fisher RA (1930) The Genetical Theory of Natural Selection. Dover, NY (reprinted 1958), p 37

Fisher ME, Goh BS, Vincent TL (1979) Some stability conditions for discrete time single species models. Bull Math Biol 41:861–875

Forney J (1976) Year class formation in the walleye (*Stizostedion vitreum*) population of Oneida Lake, New York, U.S.A. 1966–1973. J Fish Res Bd Can 33:783–792

Fox LR (1975) Cannibalism in natural populations. Ann Rev Ecol Sys 6:87–106

Fujii K (1975) A general simulation model for laboratory insect populations. I. From cohort of eggs to adult emergences. Res Pop Ecol 14:85–133

Fujii K (1978) Computer simulation study on the cyclicities of *Tribolium* population dynamics. Res Pop Ecol 19:155–169

Fujita H, Utida S (1953) The effect of population density on the growth of an animal population. Ecology 34:488–498

Gantmacher FR (1959) The Theory of Matrices, Vol. 1. Chelsea, NY

Gause GF (1931) The influence of ecological factors on the size of population. Am Nat 65:70–76

Gause, GF (1934) The Struggle for Existence. Hafner, NY (reprinted 1964)

Gill DE (1974) Intrinsic rate of increase, saturation density, and competitive ability. II. The evolution of competitive ability. Am Nat 108:103–116

Ginzburg LR (1972) Diversity of fitness and generalized fitness. J General Biol 33:77–81 (in Russian)

Ginzburg LR (1974) A macro-equation of the dynamical theory of biological communities. In: Cybernetics Models in Biology, Novosibirsk, pp 51–79 (in Russian)

Ginzburg LR (1977a) A macro-equation of natural selection. J Theor Biol 67:677–686

Ginzburg LR (1977b) The equilibrium and stability for n-alleles under the density-dependent selection. J Theor Biol 68:545–550

Ginzburg LR (1983) Theory of Natural Selection and Population Growth. Benjamin/Cummings, Menlo Park

Ginzburg LR, Costantino RF (1979) On the rate of genetic adaptation under natural selection. J Theor Biol 77:307–316

Gleser LJ, Moore DS (1985) The effect of positive dependence on chi-squared tests for categorical data. J Roy Stat Soc B 47:459–465

Goel NS, Maitra SC, Montroll EW (1971) On the Volterra and Other Nonlinear Models of Interacting Populations. Academic Press, NY

Goel NS, Richter-Dyn N (1974) Stochastic models in biology. Academic Press, NY

Goodman LA (1968) Stochastic models for the population growth of the sexes. Biometrika 55:469–487

Goodman D (1987) The demography of chance extinction. In: Soulé ME (ed) Viable Populations for Conservation. Cambridge University Press, pp 11–34

Gregorius HR (1979) Deterministic single-locus density-dependent selection. J Math Biol 8:375–391

Griffel DH (1976) Age-dependent population growth. J Inst Math App 17:141–152

Guckenheimer J, Oster G, Ipaktchi A (1977) The dynamics of density dependent population models. J Math Biol 4:101–147

Gurney WSC, Nisbet RM (1980) Age dependent population dynamics in static and variable environments. Theor Pop Biol 17:321–344

Gurney WSC, Nisbet RM (1985) Fluctuation periodicity, generation separation, and the expression of larval competition. Theor Pop Biol 28:150–180

Gurtin ME, Levine DS (1979) On predator-prey interactions with predation dependent on age of prey. Math Biosci 47:207–219

Gurtin ME, MacCamy RC (1974) Nonlinear age-dependent population dynamics. Arch Rat'l Mech Anal 54:281–300

Gurtin ME, MacCamy RC (1979) Some simple models for nonlinear age-dependent population dynamics. Math Biosci 43:119–211

Haldane JBS (1957) The cost of natural selection. J Genet 55:511–524

Haldane JBS (1953) Animal population and their regulation. In: Penguin New Biology 15:9–24

Hallam TG, Ma Z (1986) Persistence in population models with demographic fluctuations. J Math Biol 24:327–339

Hallam TG, Zhien M (1987) On density and extinction in continuous population models. J Math Biol 25:191–201

Hamada Y (1981) Dynamics of the noise-induced phase transition of the Verhulst model. Prog Theor Phys 65:850–860

Hanson FB, Tuckwell HC (1981) Logistic growth with random density independent disasters. Theor Pop Biol 19:44–18

Hardin G (1960) The competitive exclusion principle. Science 131:1292–1297

Hassell MP (1978) The Dynamics of Arthropod Predator-Prey Systems. Princeton Univ Press

Hassell MP, Comins HN (1976) Discrete time models for two-species competition. Theor Pop Biol 9:202–221

Hastings A (1978) An evolutionary optimization principle. J Theor Biol 75:519–525

Hastings A (1987) Cycles in cannibalistic egg-larval interactions. J Math Biol 24:651–666

Hastings A, Costantino RF (1987) Cannibalistic egg-larva interactions in *Tribolium*: An explanation for the oscillations in population numbers. Am Nat 130:36–52

Heyde CC, Cohen JE (1985) Confidence intervals for demographic projections based on products of random matrices. Theor Pop Biol 27:120–153

Ho FK, Dawson PS (1966) Egg cannibalism by *Tribolium* larvae. Ecology 47:318–322

Hofbauer J, Hutson V, Jansen W (1987) Coexistence for systems governed by difference equations of Lotka-Volterra type. J Math Biol 25:553–570

Holcik J (1977) Changes in fish community of Kikava Reservoir with particular reference to Eurasian perch. J Fish Res Bd Can 34:1734–1747

Holdaway FG (1932) An experimental study of the growth of populations of the flour beetle *Tribolium confusum* Duv. as affected by atmospheric moisture. Ecol Monogr 2:261–304

Hoppensteadt F (1975) Mathematical theories of populations: demographics, genetics and epidemics. CBMS Vol 20, SIAM, Philadelphia

Horsthemke W, Lefever R (1984) Noise-induced Transitions, Theory and Applications in Physics, Chemistry, and Biology. Springer-Verlag, NY

Howe RW (1962) The effect of temperature and relative humidity on the rate of development and mortality of *Tribolium madens* (Charp.) (Coleoptera, Tenebrionidae). Ann Appl Biol 50:649–660

Hunt, F (1980) On the persistence of spatially homogeneous solutions of a population genetics model with slow selection. Math Biosci 52:185–206

Iosifescu, M (1980) Finite Markov Processes and Their Applications. Wiley, NY

Iwasa Y (1988) Free fitness that always increases in evolution J Theor Biol 135:265–281

Jennrich RI, Moore RH (1975) Maximum likelihood estimation by means of nonlinear least squares. Am Stat Assoc Proc 1975:52–65

Jensen RV (1987) Classical chaos. Am Sci 75:168–181

Jillson D (1977) Population biology of *Tribolium*: distribution, abundance and competition in fine-grained environments. PhD Thesis, Univ Rhode Island

Jillson D, Costantino RF (1980) Growth, distribution, and competition of *Tribolium castaneum* and *Tribolium brevicornis* in fine-grained habitats. Am Nat 116:206–219

Johnson NL, Kotz S (1970) Continuous Univariate Distributions – 1. Houghton Mifflin, Boston

Karlin S, McGregor J (1958) Linear growth, birth and death processes. J Math Mech 7:643–662

Karlin S, Taylor HM (1975) A First Course in Stochastic Processes, 2nd Ed. Academic Press, NY
Karlin S, Taylor HM (1981) A Second Course in Stochastic Processes. Academic Press, NY
Keiding N (1975) Extinction and exponential growth in random environments. Theor Pop Biol 8:49–63
Kendall DG (1949) Stochastic processes and population growth. J Roy Stat Soc B 11(2):230–264
Kendall MG, Stuart A (1968) The Advanced Theory of Statistics, Volume 1, 3rd Ed. Hafner, NY
Keyfitz N (1968) Introduction to the Mathematics of Population. Addison-Wesley, Reading, MA
King CE, Dawson PS (1972) Population biology and the *Tribolium* model. Evol Biol 5:133–227
Kingsland SE (1985) Modeling Nature: Episodes in the History of Population Ecology. Univ Chicago Press
Kollros CL (1944) A study of the gene, pearl, in populations of *Tribolium castaneum* Herbst. PhD Thesis, Univ Chicago
Landahl HD (1955a) A mathematical model for the temporal pattern of a population structure, with particular reference to the flour beetle. Bull Math Biophys 17:63–77
Landahl HD (1955b) A mathematical model for the temporal pattern of a population structure, with particular reference to the flour beetle. II. Competition between species. Bull Math Biophys 17:131–140
Lasota A, Mackey MC (1985) Probabilistic Properties of Deterministic Systems. Cambridge Univ Press
Leon JA, Charlesworth B (1978) Ecological versions of Fisher's fundamental theorem of natural selection. Evolution 59:457–464
Lerner IM, Ho FK (1961) Genotype and competitive ability of *Tribolium* species. Am Nat 95:329–343
Lerner IM, Dempster ER (1962) Indeterminism in interspecific competition. Proc Nat Acad Sci USA 48:821–826
Leslie PH (1945) On the use of matrices in certain population mathematics. Biometrika 33:183–212
Leslie PH (1948) Some further notes on the use of matrices in population mathematics. Biometrika 35:213–245
Leslie PH (1957) An analysis of the data for some experiments carried out by Gause with populations of the protozoa, *Paramecium aurelia* and *Paramecium caudatum*. Biometrika 44:314–327
Leslie PH (1958) A stochastic model for studying the properties of certain biological systems by numerical methods. Biometrika 45:16–31
Leslie PH (1959) The properties of a certain lag type of population growth and the influence of an external random factor on a number of such populations. Physiol Zool 32:151–159
Leslie PH (1960) A note on some approxiations to the variance in discrete-time stochastic models for biological systems. Biometrika 47:196–197
Leslie P H (1962) A stochastic model for two competiting species of *Tribolium* and its application to some experimental data. Biometrika 49:1–25
Leslie PH, Gower JC (1958) The properties of a stochastic model for two competing species. Biometrika 45:316–330
Leslie PH, Gower JC (1960) The properties of a stochastic model for the predator-prey type of interaction between two species. Biometrika 47:219–234
Leslie PH, Park T, Mertz DB (1968) The effect of varying the initial numbers on the outcome of competition between to *Tribolium* species. J An Ecol 37:9–23
Levin SA, Goodyear CP (1980) Analysis of an age-structured fishery model. J Math Biol 9:245–274
Levine DS (1981) On the stability of a predator-prey system with egg-eating predators. Math Biosci 56:27–46
Levins R (1969) The effect of random variations of different types on population growth. Proc Nat Acad Sci USA 62:1061–1065
Lewis EG (1942) On the generation and growth of a population. Sankhya 6:93–96
Lewontin RC, Cohen D (1969) On population growth in a randomly varying environment. Proc Nat Acad Sci USA 62:1056–1060
L'Hertier P, Neefs Y, Teissier G (1937) Apterisme des insects et selection naturelle. Compt Rend Soc Biol 204:907–909
Lloyd M (1965) Laboratory studies with confined cannibalistic populations of flour beetles (*Tribolium castaneum*) in a cold-dry environment. I. Data for 24 unmanipulated populations. Trib Infor Bull 8:89–123

Lloyd M (1968) Self regulation of adult numbers by cannibalism in two laboratory strains of flour beetles (*Tribolium castaneum*). Ecology 49:245–259

Lotka AJ (1925) Elements of Physical Biology. Williams and Wilkins, Baltimore

Lotka AJ (1932) The growth of mixed populations: Two species competing for a common food supply. J Wash Acad Sci 21:461–469

Ludwig D (1974) Stochastic Population Theories. Springer-Verlag, NY

MacArthur RH (1962) Some generalized theorems of natural selection. Proc Nat Acad Sci USA 38:1893–1897

MacLagan SD (1932) The effect of population density upon rate of reproduction with special reference to insects. Proc Roy Soc Lond B Biol Sci 111:437–454

MacLagan DS, Dunn E (1935) The experimental analysis of the growth of an insect population. Proc Roy Soc Edin 55:126–139

Magill, AJ (1978) Selection and population size in *Tribolium castaneum*. MS Thesis, Univ Rhode Island

Mangel M, Ludwig D (1977) Probability of extinction in a stochastic competition. SIAM J Appl Math 33:256–266

Mangel M (1985) Decision and Control in Uncertain Resource Systems. Academic Press, NY

May RM (1973) Stability in randomly fluctuating environments. Am Nat 107:621–650

May RM (1974a) Biological populations with nonoverlapping generations: stable points, stable cycles, and chaos. Science 186 645–647

May RM (1974b) Stability and Complexity in Model Ecosystems. Princeton Univ Press

May RM (1974c) Ecosystem patterns in randomly fluctuating environments. In Rosen R, Snell F (eds) Progress in Theoretical Biology. Academic Press, NY

May RM (1976) Simple mathematical models with very complicated dynamics. Nature 261:459–467

May RM (1980) Mathematical models in whaling and fisheries management. In Oster GF (ed) Some mathematical questions in biology, Vol 13. Am Math Soc, Providence, RI

May RM, Oster GF (1976) Bifuractions and dynamic complexity in simple ecological models. Am Nat 110:573–599

Maynard Smith J (1968a) Mathematical Ideas in Biology. Cambridge Univ Press

Maynard Smith J (1968b) Haldane's dilemma and the rate and evolution. Nature 29:1114–1116

Maynard Smith J (1974) Models in Ecology. Cambridge Univ Press

McDonald DJ, Peer NJ (1960) Natural selection in experimental populations of *Tribolium*. I. Preliminary experiments with population cages. Genetics 45:1317–1333

McDonald DJ, Peer NJ (1961) Natural selection in experimental populations of *Tribolium*. II. Factors affecting the fitness of the mutant "split" of *Tribolium confusum*. Heredity 16:317–330

McKelvey R, Hankin D, Yanosko K, Snygg C (1980) Stable cycles in multi-stage recruitment models: An application to the Northern California Dungeness crab (*Cancer magister*) fishery. Can Jour Fish Aquat Sci 37:2323–2345

McKendrick AG (1926) Applications of mathematics to medical problems. Proc Edin Math Soc 54:98–130

Menshutkin VV (1964) Population dynamics studied by representing the population as a cybernetic system. Voprosy Ikhthiologii:23–33

Mertz DB (1969) Age-distribution and abundance in populations of flour beetles. I. Experimental studies. Ecol Monogr 39:1–31

Mertz DB (1972) The *Tribolium* model and the mathematics of population growth. Ann Rev Ecol Syst 3:51–78

Mertz DB, Park T, Youden WJ (1965) Mortality patterns in eight strains of flour beetles. Biometrics 21:99–114

Mertz DB, Davies RB (1968) Cannibalism of the pupal stage by adult flour beetles: An experiment and a stochastic model. Biometrics 24:247–275

Mertz DB, Robertson JR (1970) Some developmental consequences of handling, egg eating, and population density for flour beetle larvae. Ecology 24:247–276

Mertz DB, Cawthon DA, Park T (1976) An experimental analysis of competitive indeterminacy in *Tribolium*. Proc Nat Acad Sci USA 73:1368–1372

Metz JAJ, Diekmann O (1986) The Dynamics of Physiologically Structured Populations. Lecture Notes in Biomath Vol 68, Springer-Verlag, NY

Miller RS (1967) Pattern and process in competition. In: Cragg JB (ed) Advances in Ecological Research. Academic Press, NY, pp 1–74
Moffa AM (1976) Genetic polymorphism and demographic equilibrium in *Tribolium castaneum*. PhD Thesis, Univ Rhode Island
Moffa AM, Costantino RF (1977) Genetic analysis of a population of *Tribolium*. VI. Polymorphism and demographic equilibrium. Genetics 87:785–805
Moran PAP (1950) Some remarks on animal population dynamics. Biometrics 6:250–258
Moran PAP (1962) The Statistical Processes of Evolutionary Theory. Clarendon, Oxford
Moran PAP (1970) "Haldane's dilemma" and the rate of evolution. Ann Hum Genet Lond 33:245–249
Murphy GI (1968) Pattern in life history and the environment. Am Nat 108:391–403
Namkoong G, Selgrade JF (1986) Frequency-dependent selection in logistic growth models. Theor Pop Biol 29:64–86
Neyman J, Park T, Scott EL (1956) Struggle for existence, the *Tribolium* model: Biological and statistical aspects. In: Neyman J (ed) Proc Third Berkeley Symp Math Stat Prob, Univ of California Press, pp 41–79
Nicholson AJ (1933) The balance of animal populations. J An Ecol 11:131–178
Nicholson AJ, Bailey VA (1935) The balance of animal populations. Part I Proc Zool Soc Lond 1935:551–598
Nisbet RM, Gurney WSC (1982) Modelling Fluctuating Populations. Wiley, NY
Niven BS (1967) The stochastic simulation of *Tribolium*. Physiol Zool 40:67–82
Niven BS (1969) Simulation of two interacting species of *Tribolium*. Physiol Zool 42:248–255
O'Donald, P (1969) "Haldane's dilemma" and the rate of natural selection. Nature 221:815–816
Orzack SH (1985) Population dynamics in variable environments. V. The genetics of homeostasis revisited. Am Nat 125:550–572
Oster GF (1976) Internal variables in population dynamics. In: Levin, SA (ed) Lectures on Mathematics in the Life Sciences, Vol 8. Am Math Soc, Providence, RI, pp 37–68
Oster G (1978) The dynamics of nonlinear models with age structure. Studies in Math Biol 16:411–38
Park T (1932) Studies in population physiology: The relation of numbers to initial population growth in the flour beetle *Tribolium confusum* Duval. Ecology 13:172–181
Park T (1934) Observations on the general biology of the flour beetle, *Tribolium confusum*. Quart Rev Biol 9:36–54
Park T (1948) Experimental studies of interspecies competition. I. Competition between populations of the flour beetles *Tribolium confusum* Duval and *Tribolium castaneum* Herbst. Ecol Monogr 18:265–308
Park T (1954) Experimental studies of interspecies competition. II. Temperature, humidity and competition in two species of *Tribolium*. Physiol Zool 27:177–238
Park T (1957) Experimental studies of interspecies competition. III. Relation of initial species proportion to competitive outcome in populations of *Tribolium*. Physiol Zool 30:22–40
Park T (1962) Beetles, competition, and populations. Science 138:1–7
Park T, Frank MB (1950) The population history of *Tribolium* free of sporozoan infection. J An Ecol 19:7–105
Park T, Mertz DB, Petrusewicz K (1961) Genetic strains of *Tribolium*: Their primary characteristics. Physiol Zool 37:97–162
Park T, Leslie PH, Mertz DB (1964) Genetic strains and competition in populations of *Tribolium*. Physiol Zool 37:97–162
Park T, Mertz DB, Grodzinski W, Prus T (1965) Cannibalistic predation in populations of flour beetles. Physiol Zool 38:289–321
Park T, Mertz DB, Nathanson M (1968) The cannibalism of pupae by adult flour beetles. Physiol Zool 41:228–253
Park T, Nathanson M, Ziegler JR, Mertz DB (1970) Cannibalism of pupae by mixed species of adult *Tribolium*. Physiol Zool 43:166–184
Park T, Ziegler JR, Ziegler DL, Mertz DB (1974) The cannibalism of eggs by *Tribolium* larvae. Physiol Zool 47:37–58
Pielou EC (1977) Mathematical Ecology. Wiley, NY

Polis GA (1981) The evolution and dynamics of intraspecific predation. Ann Rev Ecol Sys 12:225–251
Popova OA, Sytina LA (1977) Food and feeding relations in Eurasian perch (*Perca fluviatilis*) and pike perch (*Stizostedian lucioperca*). J Fish Res Bd Can 34:1559–1570
Prajneshu (1980) Time-dependent solution of the logistic model for population growth in random environment. J Appl Prob 17:1083–1086
Provine WB (1986) Sewall Wright and Evolutionary Biology. Univ Chicago Press
Radovich J (1962) Effects of sardine spawning size and environment on year-class production. California Fish and Game 48:123–140
Rao, CR (1973) Linear Statistical Inference and Its Applications, 2nd Ed. Wiley, NY
Rich EL (1956) Egg cannibalism and fecundity in *Tribolium*. Ecology 37:109–120
Ricker WE (1954) Stock and recruitment. J Fish Res Bd Can 11:559–623
Roozen H (1987) Equilibrium and extinction in stochastic population dynamics. Bull Math Biol 49:671–696
Rorres C (1976) Stability of an age specific population with density dependent fertility. Theor Pop Biol 10:26–46
Rosenblatt, M (1974) Random Processes. Springer-Verlag, NY
Rosenzweig M (1971) Paradox of enrichment: destabilization of explotation ecosystems in ecological time. Science 171 385–387
Ross, SM (1972) Introduction to Probability Models. Academic Press, NY
Rothstein SI (1979) Gene frequencies and selection for inhibitory traits, with special emphasis on the adaptiveness of territoriality. Am Nat 113:317–331
Roughgarden J (1971) Density-dependent natural selection. Ecology 106:683–718
Roughgarden J (1979) Theory of Population Genetics and Evolutionary Ecology: An Introduction. Macmillan, NY
Ryan MF, Park T, Mertz DB (1970) Flour beetles: responses to extract of their own pupae. Science 170:178–180
Ryan MF, Behan M (1973) The sensory receptors of *Tribolium* larvae. Physiol Zool 46:238–244
Saleem M (1983) Predator-prey relationships: egg-eating predators. Math Biosci 65:187–197
Saleem M (1984) Egg-eating age-structured predators in interactions with age-structured prey. Math Biosci 70:91–104
Saleem M, Siddiqui SU, Gupta V (1987) A mathematical model with young predation. J Math Biol 25:89–101
Saleem M, Pandey RK (1988) Egg-eating age-structured predator and prey interactions: some simple cases. Math Biosci 89:209–224
Schaffer WM, Ellner S, Knot M (1986) Effects of noise on some dynamical models in ecology. J Math Biol 24:479–523
Schlager G (1963) The ecological genetics of the mutant sooty in populations of *Tribolium castaneum*. Evolution 17:254–273
Scudo FM, Ziegler JR (1978) The Golden Age of Theoretical Ecology:1923–1940. Lect Notes Biomath, Springer-Verlag, NY
Scully AMM, Costantino RF (1975) Genetic analysis of a population of *Tribolium*. V. Viability of the unsaturated fatty acid sensitive mutant. Can J Gen Cyt 17:585–590
Searle SR (1971) Linear Models. Wiley, NY
Selgrade JF, Namkoong G (1984) Dynamical behavior of differential equation models of frequency-and density-dependent populations. J Math Biol 19:133–146
Sharpe FR, Lotka AJ (1911) A problem in age-distribution. Philos Mag 21:435–438
Shaw R (1984) The Dripping Faucet as a Model Chaotic System. Aerial Press, Santa Cruz, CA
Slade NA, Levenson H (1982) Estimating population growth rates from stochastic Leslie matrices. Theor Pop Biol 22:299–308
Slatkin M, Maynard Smith J (1979) Models of coevolution. Quart Rev Biol 54:133–262
Slobodkin LB (1961) Growth and Regulation of Animal Populations. Holt, Rinehart and Winston, NY
Smith RH, Mead R (1974) Age structure and stability in models of prey-predator systems. Theor Pop Biol 6:308–322

Smith RH, Mead R (1979) On predicting extinction in simple population models I. Stochastic linearization. J Theor Biol 80:189–203

Smith RH, Mead R (1980) The dynamics of discrete-time stochastic models of population growth. J Theor Biol 86:607–627

Snedecor GW, Cochran WG (1967) Statistical Methods. Iowa State Univ Press, Ames

Sokal RR, Sonleitner FJ (1968) The ecology of selection in hybrid populations of *Tribolium castaneum*. Ecol Monogr 38:345–379

Sokal RR, Huber I (1963) Competition among genotypes in *Tribolium castaneum* at varying densities and gene frequencies (the *sooty* locus). Am Nat 97:169–184

Sokal RR, Karten I (1964) Competition among genotypes in *Tribolium castaneum* at varying densities and gene frequencies (the *black* locus). Genetics 49:195–211

Sokoloff A (1972) The Biology of *Tribolium*, Vol. 1. Oxford Univ Press

Sokoloff A (1974) The Biology of *Tribolium*, Vol. 2. Oxford Univ Press

Sokoloff A (1977) The Biology of *Tribolium*, Vol. 3. Oxford Univ Press

Sonleitner EJ (1961) Factors affecting egg cannibalism and fecundity in populations of adult *Tribolium castaneum* Herbst. Physiol Zool 34:133–255

Sonleitner FJ (1977) A stochastic computer model for simulating population growth. Res Pop Ecol 19:10–32

Soong TT (1973) Random Differential Equations in Science and Engineering. Academic Press, NY

Soule ME (1987) Viable Populations for Conservation, Cambridge Univ Press

Stanley J (1932a) A mathematical theory of the growth of populations of the flour beetle, *Tribolium confusum*, Duv. Can J Res 6:632–671

Stanley J (1932b) A mathematical theory of the growth of populations of the flour beetle, *Tribolium confusum*, Duv. II. The distribution by ages in the early stages of populati on growth. Can J Res 7:426–433

Stanley J (1934) A mathematical theory of the growth of populations of the flour beetle *Tribolium confusum*, Duv. III. The effect upon the early stages of population growth of changes in the nutritive value, palatability and density of packing of the flour medium. Can J Res 11:728–732

Stanley J (1941) A mathematical theory of the growth of populations of the flour beetle *Tribolium confusum*, Duv. IV. A modified theory descriptive of the relation between the limiting value of egg-populations in the absence of hatching, and the volume (or weight) of flour used in the culture. Ecology 22:23–37

Stanley J (1942) A mathematical theory of the growth of populations of the flour beetle *Tribolium confusum*, Duv. V. The relation between the limiting value of the egg-population in the absence of hatching and the sex-ratio of the group of adult beetles used in a culture. Ecology 23:24–31

Stanley J (1943) A mathematical theory of the growth of populations of the flour beetle *Tribolium confusum*, VI. Egg populations in which the initial number of eggs is above the limiting value. Ecology 24:323–328

Stanley J (1949) A mathematical theory of the growth of populations of the flour beetle, *Tribolium confusum* Duv. VII. A study of the re-tunnelling problem. Ecology 30:209–222

Stanley J (1964) A mathematical theory of the growth of populations of the flour beetle, *Tribolium confusum* Duv. VIII. A further study of the re-tunneling problem. Can J Zoo 42:201–227

Stanley J (1966) A mathematical theory of the growth of the flour beetle *Tribolium confusum*, Duval. IX. The persistence of tunnels in the flour mass. J Theor Biol 13:379–411

Strawbridge, DW (1953) Population dynamics of the flour beetle *Tribolium castaneum* Herbst. PhD Thesis, Univ Chicago

Strebel DE (1985) Environmental fluctuations and extinction – single species. Theor Pop Biol 27:1–26

Sved JA (1968) Possible rates of gene substitution in evolution. Am Nat 102:283–293

Swick KE (1976) A model of single species population growth. SIAM J Math Anal 7:565–576

Swick KE (1977) A nonlinear age-dependent model of single species population dynamics. SIAM J Appl Math 32:484–498

Swick KE (1980) Periodic solutions of a nonlinear age-dependent model of single species population dynamics. SIAM J Math Anal 11:901–910

Taylor NW (1965) A theoretical study of population regulation in *Tribolium confusum*. Ecology 46:334–340

Taylor NW (1967) A mathematical model for *Tribolium confusum* populations. Ecology 48:290–294
Taylor NW (1968) A mathematical model for two *Tribolium* populations in competition. Ecology 49:843–848
Taylor NW (1971) Simulation of *Tribolium* populations. Ecol Soc Australia 6:105–115
Teleky SB (1980) Egg cannibalism in *Tribolium* as a model of interference competition. Res Pop Ecol 21:217–227
Thieme HR (1979) Density-dependent regulation of spatially distributed populations and their asymptotic speed of spread. J Math Biol 8:173–187
Thompson WR (1929) On the effect of random oviposition on the action of entomophagous parasites as agents of natural control. Parasitology 21:180–188
Thompson RW, DiBiasio D, Mendes C (1982) Predator-prey interactions: egg-eating predators. Math Biosci 60:109–120
Tier C, Hanson FB (1981) Persistence in density dependent stochastic populations. Math Biosci 53:89–117
Tschinkel WR, Willson CD (1976) Inhibition of pupation due to crowding in some Tenebrionid beetles. J Exp Zool 176:137–146
Tuckwell HC (1974) A study of some diffusion models of population growth. Theor Pop Biol 5:345–357
Tuljapurkar SD (1982) Population dynamics in variable environments. III. Evolutionary dynamics of r-selection. Theor Pop Biol 21:141–165
Tuljapurkar SD (1984) Demography in stochastic environments. I. Exact distributions of age structure. J Math Biol 19:335–350
Tuljapurkar SD (1985) Population dynamics in variable environments. VI. Cyclical environments. Theor Pop Biol 27:1–17
Tuljapurkar SD (1986) Demographic applications of random matrix products. In: Cohen JE, Kesten H, Newman CM (eds) Random Matrices and Their Applications. Am Math Soc, Providence, RI, pp 319–326
Tuljapurkar SD (1987) Cycles in nonlinear age-structured models. I. Renewal equations. Theor Pop Biol 32:26–41
Tuljapurkar SD (1989) An uncertain life: Demography in randon environments. Theor Pop Biol 35:227–294
Tuljapurkar SD, Orzack S (1980) Population dynamics in variable environments. I. Long-run growth rates and extinction. Theor Pop Biol 18:314–342
Turelli M (1977) Random environments and stochastic calculus. Theor Pop Biol 12:140–178
Van Valen L (1963) Haldane's dilemma, evolutionary rates and heterosis. Am Nat 97:185–190
Volterra V (1926) Variazioni e fluttuazioni del numero d'individui in specie animali conviventi. Memorie della R Accademia dei Lincei 2:31–113
von Foerster H (1959) Some remarks on changing populations. In: Stohlman F (ed) The Kinetics of Cellular Proliferation. Grune and Stratton, NY, pp 382–407
Watt KEF (1955) Studies on population productivity. I. Three approaches to the optimum yield problem in populations of *Tribolium confusum*. Ecol Monogr 25:369–290
Webb GF (1985) Theory of Nonlinear Age-Dependent Population Dynamics. Marcel Dekker, NY
West BJ (1985) Lecture Notes in Biomathematics, Vol 62. An essay on the importance of being nonlinear. Springer-Verlag, NY
Wilkinson, L (1988) SYSTAT: The System for Statistics. SYSTAT Inc, Evanston, IL
Witten M (1983) On stochasticity in the von Foerster hyperbolic partial differential equation system. Comp Math Appls 9:447–457
Young AM (1970) Predation and abundance in populations of flour beetles. Ecology 51:602–619
Zhien M, Baojun S, Hallam TG (1989) The threshold of survival for systems in a fluctuating environment. Bull Math Biol 51:311–323
Ziegler JR (1976) Evolution of the migration response: emigration by *Tribolium* and the influence of age. Evolution 30:579–592

Author Index

Subject Index

Monographs on Theoretical and Applied Genetics

(continued from p. ii)